Hydrogen Energy

Since the low energy efficiency and rise in emissions by using fossil fuels, the hydrogen economy has been considered as a unique approach to resolve these problems, making hydrogen an attractive fuel. This book titled *Hydrogen Energy: Production, Storage, and Utilization* describes various technologies for hydrogen production from different sources and storage in liquid, gaseous, and compound forms, which have also been covered in detail. It also highlights the various modes of transportation of hydrogen and its utilization in a variety of engineering applications such as automotive engines, fuel cells, electric power generation, and aerospace. The book also explores the safety challenges, handling issues, and future scope.

This book:

- Covers the methods of hydrogen production including the bioconversion method.
- Explains the hydrogen storage technologies with proper illustrations.
- Includes engineering and non-engineering applications of hydrogen utilization covering the most advanced generation of fuel cells, power generation for hybrid vehicles, and space applications.
- Presents the state-of-the-art research carried out in the domains of hydrogen production, storage, transportation, and utilization along with safety aspects.
- Discusses modes and methods of hydrogen transportation such as in gaseous form, liquid form, and solid form.

The book has a wide scope of reaching out to diverse readers including but not limited to industrial engineers, process engineers, researchers, and academicians. The research fraternity will gain the benefit of being aware and selecting a similar recent research domain in the field of hydrogen production.

Hydrogen Energy

Production, Storage, and Utilization

Edited by

Dilshad Ahmad Khan,
Akhilesh Kumar Choudhary,
and Deepak Sharma

CRC Press
Taylor & Francis Group
Boca Raton London New York

CRC Press is an imprint of the
Taylor & Francis Group, an **informa** business

Designed cover image: Dilshad Ahmad Khan

First edition published 2025
by CRC Press
2385 NW Executive Center Drive, Suite 320, Boca Raton FL 33431

and by CRC Press
4 Park Square, Milton Park, Abingdon, Oxon, OX14 4RN

CRC Press is an imprint of Taylor & Francis Group, LLC

ISBN: 978-1-032-39303-2 (hbk)
ISBN: 978-1-032-93213-2 (pbk)
ISBN: 978-1-003-53781-6 (ebk)

DOI: 10.1201/9781003537816

Typeset in Sabon
by Deanta Global Publishing Services, Chennai, India

Contents

Preface

In today's rapidly evolving world, the escalating global population is driving an unprecedented surge in energy demand. Historical data highlights a staggering increase from 43.5 EJ in 1900 to approximately 575 EJ in 2016, with projections suggesting that this demand could soar to 760 EJ by 2040. Against this backdrop, the quest for alternative energy sources has become imperative. Hydrogen emerges as a promising candidate, offering potential applications in power generation, automotive industries, aerospace, and fuel cells, among others. Notably, hydrogen energy is characterized by zero carbon emissions, rendering it a cleaner and more sustainable option compared to traditional fossil fuels.

Hydrogen stands out with its remarkably high specific energy of 142.2 MJ/kg, significantly surpassing conventional energy sources such as natural gas (52.2 MJ/kg), oil (45.5 MJ/kg), coal (23.9 MJ/kg), and wood (20.6 MJ/kg). This makes hydrogen not only an attractive energy source but also a contender for next-generation fuel, thanks to its high conversion efficiency and non-polluting nature. Given the inefficiencies and environmental impacts associated with fossil fuels, the hydrogen economy presents a unique solution to these challenges.

This book, *Hydrogen Energy: Production, Storage, and Utilization*, provides a comprehensive exploration of the various facets of hydrogen energy. It covers diverse methods of hydrogen production, including those from coal, hydrocarbons, water electrolysis, steam reforming, solar water splitting, thermochemical and photochemical processes, and biomass conversion. The book delves into various storage techniques, such as gaseous, liquid, and solid forms, as well as innovative approaches involving porous materials, nanomaterials, compounds, and hydrates.

Additionally, the book addresses the logistics of hydrogen transportation, ensuring efficiency and safety in its movement across different states. It also investigates a wide array of hydrogen applications. Critical safety concerns associated with the handling, storage, and transportation of hydrogen are thoroughly examined, emphasizing best practices and future challenges.

The book aims to serve a broad audience, including industrial engineers, process engineers, researchers, and academicians. By presenting the latest technological advances and research findings in a coherent and accessible manner, it seeks to foster a deeper understanding and appreciation of hydrogen energy. The reader will gain valuable insights into the potential of hydrogen as a pivotal energy source for the future.

The chapters, authored by experts from various educational institutions and research organizations, provide state-of-the-art knowledge and perspectives, ensuring a well-rounded and comprehensive discussion on hydrogen energy. From production techniques and storage solutions to transportation logistics and practical applications, this book covers all critical areas, highlighting both opportunities and challenges in the field.

In light of the pressing need for sustainable energy solutions, hydrogen energy holds promise for meeting the world's growing energy demands. This book serves as a vital resource for anyone interested in the development and deployment of hydrogen technologies, offering a thorough exploration of the current state and future prospects of this dynamic field.

About the Editors

Dilshad Ahmad Khan is Assistant Professor in the Department of Mechanical Engineering, NIT Hamirpur, India. He is a prominent academician and researcher. He is a chartered engineer (CEng) from the Institution of Mechanical Engineers, London. He has received his PhD in Manufacturing from IIT Delhi, India, in 2018, and Master of Technology from Aligarh Muslim University, India, in 2010. He has received his Bachelor of Engineering in Mechanical Engineering from Dr. B.R. Ambedkar University (formerly Agra University), India, in 2006. He has published a good number of research articles in international journals and conferences. He has published several book chapters on various topics and has filed eight Indian patents too. He is the author/editor of three books. He has received various national and international awards for his research and innovations including Gandhiyan Young Technological Innovation Award (GYTI) hosted by President House (Rashtrapati Bhawan), New Delhi. He has funded research projects. His research interests include energy storage systems, advanced finishing/polishing processes, non-conventional machining, additive manufacturing, and industrial automation.

Akhilesh Kumar Choudhary is working as Assistant Professor in the Department of Mechanical Engineering at National Institute of Technology Hamirpur (HP) since 2018. He received his PhD degree in Mechanical Engineering from PDPM Indian Institute of Information Technology, Design and Manufacturing, Jabalpur. He received his MTech degree from Motilal Nehru National Institute of Technology (MNNIT), Allahabad, UP, and his bachelor's degree from Jabalpur Engineering College Jabalpur, MP. Prior to joining NIT Hamirpur, he was working as Assistant Professor (NPIU) in the Department of Mechanical Engineering at Jabalpur Engineering College, Jabalpur. Under the activity of TEQIP-III, he also worked as a faculty member at National Institute of Technology, Raipur, for 2.5 years in Mechanical Engineering Department. His research interests include condition monitoring, biofuel, diesel engine, and hydrogen fueled engine.

Deepak Sharma is Assistant Professor in the Department of Mechanical Engineering, NIT Hamirpur (Himachal Pradesh). Prior to joining NIT Hamirpur, he has served as a Contract faculty in the Department of Mechanical Engineering at NIT Silchar, Assam. He received his doctoral degree (PhD) in Thermal Engineering from National Institute of Technology Silchar in 2017. Also, he received his MTech from the same institute in 2013 and BTech from Gautam Buddha Technical University, Lucknow, in 2010. His research work was published in reputed national and international journals and conferences. His area of research is thermal energy storage, renewable energy, nanofluid heat transfer, single phase and multiphase flow in channels, combustion, and thermal hydraulics in nuclear reactor.

Contributors

Dilshad Ahmad Khan
Department of Mechanical Engineering
National Institute of Technology Hamirpur
Himachal Pradesh, India

Akhilesh Kumar Choudhary
Department of Mechanical Engineering
National Institute of Technology Hamirpur
Himachal Pradesh, India

Deepak Sharma
Department of Mechanical Engineering
National Institute of Technology Hamirpur.
Himachal Pradesh, India.

Shashi Kant Verma
Department of Mechanical Engineering
National Institute of Technology Durgapur
West Bengal, India

Ajay Kumar Shukla
Institute of Pharmacy
Dr Rammanohar Lohia Avadh University
Uttar Pradesh, India

Ratindra Gautam
Department of Applied Science
Institute of Engineering and Technology, Dr. Rammanohar Lohia Avadh University
Uttar Pradesh, India

Shivani Chaudhary
Department of Physics

Deen Dayal Upadhyaya Gorakhpur University
Uttar Pradesh, India

Anoop Kr. Srivastava
Department of Applied Science
Institute of Engineering and Technology, Dr. Rammanohar Lohia Avadh University
Uttar Pradesh, India

Anil Kumar
Department of Physics and Electronics
Dr. Rammanohar Lohia Avadh University
Uttar Pradesh, India

Sindhu Singh
Department of Physics and Electronics
Dr. Rammanohar Lohia Avadh University
Uttar Pradesh, India

Pramod Kumar Maurya
Department of Pharmacy
Guru Ramdas Khalsa Institute of Science and Techanology
Madhya Pradesh, India

Sukanta Roga
Department of Mechanical Engineering
Visvesvaraya National Institute of Technology Nagpur
Maharashtra, India

Jeet Bhakta
Department of Mechanical Engineering
Visvesvaraya National Institute of Technology Nagpur
Maharashtra, India

Adesh Singh
Department of Mechanical Engineering
Visvesvaraya National Institute of Technology Nagpur
Maharashtra, India

Swapan Bhaumik
Department of Mechanical Engineering
National Institute of Technology Agartala
Tripura, India

Roque Aguado
Departamento de Ingeniería Eléctrica, Escuela Politécnica Superior de Linares
Universidad de Jaén, Avda. de la Universidad s/n
Linares, Spain

Simin Anvari
Departamento de Ingeniería Eléctrica, Escuela Politécnica Superior de Linares
Universidad de Jaén, Avda. de la Universidad s/n
Linares, Spain

Antonio Escámez
Departamento de Ingeniería Eléctrica, Escuela Politécnica Superior de Linares
Universidad de Jaén, Avda. de la Universidad s/n
Linares, Spain

Daniel Sánchez
Departamento de Ingeniería Eléctrica, Escuela Politécnica Superior de Linares
Universidad de Jaén, Avda. de la Universidad s/n
Linares, Spain

David Vera
Departamento de Ingeniería Eléctrica, Escuela Politécnica Superior de Linares
Universidad de Jaén, Avda. de la Universidad s/n
Linares, Spain

Prashant Sharma
Department of Mechanical Engineering
Shri Ram Iinstitute of Science and Technology
Madhya Pradesh, India

Arun Kumar Sharma
Department of Mechanical Engineering
Gyan Ganga College of Technology
Madhya Pradesh, India

Bhupendra Gupta
Department of Mechanical Engineering
Jabalpur Engineering College
Madhya Pradesh, India

Prashant Baredar
Energy Department
Maulana Azad National Institute of Technology (MANIT) Bhopal
Madhya Pradesh, India

Tushar Choudhary
Sustainable Energy Technology Laboratory, Department of Mechanical Engineering
PDPM-Indian Institute of Information Technology Design and Manufacturing Jabalpur
Madhya Pradesh, India

Fatima Jalid
Catalysis Laboratory, Department of Chemical Engineering
National Institute of Technology Srinagar
Jammu and Kashmir, India

Aravi Muzaffar
Metallurgical and Materials Engineering Department
National Institute of Technology Srinagar
Jammu and Kashmir, India

Sangharatna M. Ramteke
Department of Mechanical and Metallurgical Engineering
School of Engineering, Pontificia Universidad Catolica de Chile
Macul, Chile

Sachin Akoji Meshram
Department of Mechanical Engineering
G H Raisoni University
Maharashtra, India

H. Chelladurai
Department of Mechanical Engineering
Indian Institute of Information Technology Design and Manufacturing Jabalpur
Madhya Pradesh, India

Sukanta Roga
Department of Mechanical Engineering
Visvesvaraya National Institute of Technology Nagpur
Maharashtra, India

Swarali Kulkarni

Department of Mechanical Engineering
Visvesvaraya National Institute of Technology Nagpur
Maharashtra, India

Hrushikesh Jadhav
Department of Mechanical Engineering
Visvesvaraya National Institute of Technology Nagpur
Maharashtra, India

Mrunali Ukey
Department of Mechanical Engineering
Visvesvaraya National Institute of Technology Nagpur
Maharashtra, India

Sanket Bahadure
Department of Mechanical Engineering
Visvesvaraya National Institute of Technology Nagpur
Maharashtra, India

Swapan Bhaumik
Department of Mechanical Engineering
National Institute of Technology Agartala
Tripura, India

Harsh Raval
School of Engineering
University of Lincoln
Lincoln, United Kingdom

Aliyu M. Aliyu
School of Engineering
University of Lincoln
Lincoln, United Kingdom

Faiz Iqbal
School of Engineering
University of Lincoln
Lincoln, United Kingdom

Yogesh Kumar Yadav
Department of Physics
Institute of Science, Banaras Hindu University
Uttar Pradesh, India

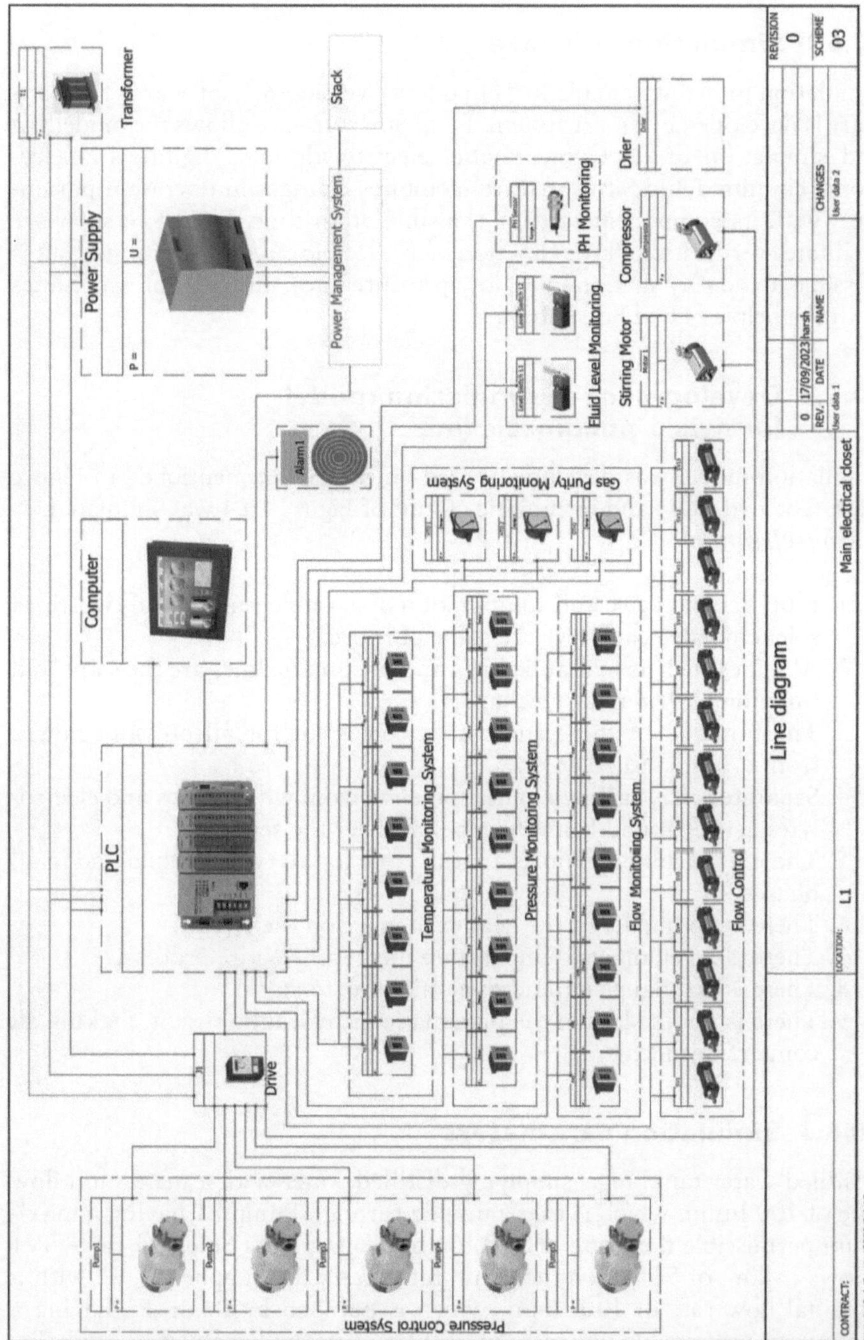

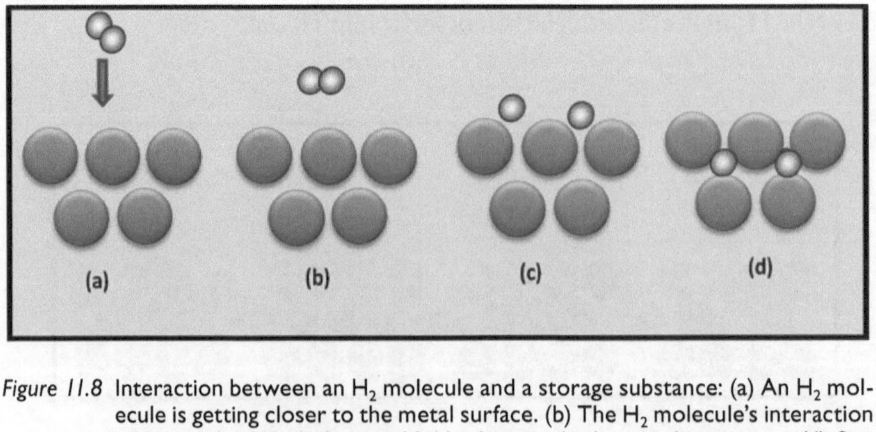

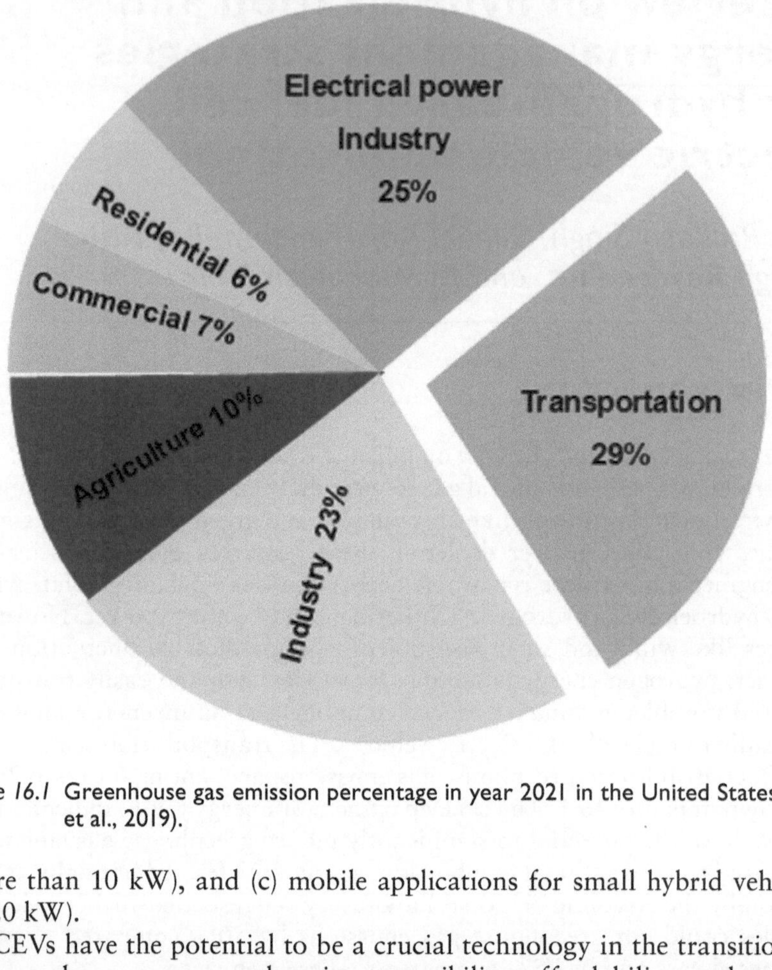

Mohammad Abu Shaz
Department of Physics
Institute of Science, Banaras Hindu University
Uttar Pradesh, India

Thakur Prasad Yadav
Department of Physics
Faculty of Science, University of Allahabad
Prayagraj, UttarPradesh India

Jashveer Singh
Department of Mechanical Engineering
National Institute of Technology, Hamirpur
Himachal Pradesh, India

Rajesh Kumar
Department of Mechanical Engineering
National Institute of Technology, Kurukshetra
Haryana, India

Adegoke C. Ishola
School of Engineering
University of Lincoln
Lincoln, United Kingdom

Ibrahim Albayati
School of Engineering
University of Lincoln
Lincoln, United Kingdom

Behnaz Sohani
School of Engineering
University of Lincoln
Lincoln, United Kingdom

Aliyu M. Aliyu
School of Engineering
University of Lincoln
Lincoln, United Kingdom

Durwesh Jhodkar
Wainganga College of Engineering and Management
Maharashtra, India

Somadatta Karanjekar
Wainganga College of Engineering and Management
Maharashtra, India

Bharat Chede
Wainganga College of Engineering and Management
Maharashtra, India

Tapus Bajpai
Malaviya National Institute of Technology Jaipur
Rajasthan, India

Asish Kumar
University Institute of Engineering and Technology
Babasaheb Bhimrao Ambedkar University
Uttar Pradesh, India

Shivani Chaudhary
Department of Physics
Deen Dayal Upadhyaya Gorakhpur University
Uttar Pradesh, India

C. K. Kaithwas
Department of Mechanical Engineering
Institute of Engineering and Technology, Dr. Rammanohar Lohia Avadh University
Uttar Pradesh, India

Ajay Kumar Shukla
Institute of Pharmacy
Dr Rammanohar Lohia Avadh University Ayodhya
Uttar Pradesh, India

U. B. Singh
Department of Physics
Deen Dayal Upadhyaya Gorakhpur University
Uttar Pradesh, India

Shiv Prakash Singh
Department of Electrical Engineering
Motilal Nehru National Institute of Technology Allahabad
Uttar Pradesh, India

Suneel Raju Pendem
Department of Electrical Engineering
Motilal Nehru National Institute of Technology Allahabad
Uttar Pradesh, India

Brij Kishor
Department of Applied Mechanics
Motilal Nehru National Institute of Technology Allahabad
Uttar Pradesh, India

Dogga Raveendhra
Department of Electrical Engineering
Motilal Nehru National Institute of Technology Allahabad
Uttar Pradesh, India

M. Venkatesh Naik
Department of Electrical Engineering
Motilal Nehru National Institute of Technology Allahabad
Uttar Pradesh, India

Amit Kumar
National Institute of Technology Delhi
Delhi, India

Ramna Tripathi
THDC Institute of Hydropower Engineering and Technology
Uttarakhand, India

Chapter 1

Introduction

Dilshad Ahmad Khan, Akhilesh Kumar Choudhary, and Deepak Sharma

1.1 OVERVIEW OF HYDROGEN PRODUCTION PROCESSES

Hydrogen exists in bound forms in nature, so it needs energy to be liberated from its many compounds. Hydrogen creation is possible using all basic sources of energy. In order to produce hydrogen, individual hydrogen molecules must be extracted and isolated, and the purity of hydrogen is needed as per the requirement of the particular application. The various relevant processes for hydrogen production are demonstrated in Figure 1.1, along with a breakdown of primary energy sources utilized in each step, categorized as renewable and non-renewable (Busby, 2005).

The secondary and primary sources of energy can be utilized for hydrogen production.

Hydrogen generation can be achieved from various feed stocks and processes, which include gasification, electrolysis, photosynthetic algae, fermentation, photoelectrochemical and thermochemical processes, pyrolysis, and high-temperature fuel cells. At the moment, the most popular method for producing hydrogen is electrolysis. Based on fossil raw materials, two commercial processes are now in use, which include natural gas reforming and partial oxidation of lower quality feeds, such as petroleum coke or other refinery byproducts (Busby, 2005; Chong et al., 2022; Navarro et al., 2007).

Over the recent past years, there have been other improvements in the process of gasification of coal to produce hydrogen, which is now a commercially accessible method. Other methods, like bioconversion of biomass and biological treatments for hydrogen production, are still in the underdevelopment phase. However, biomass gasification is anticipated to see commercialization soon. Major hydrogen production from secondary sources is commonly produced by the electrolysis of water using electrical energy (Yoon et al., 2010).

The nuclear energy can be used for hydrogen production in one of the three routes: (1) steam reforming methane using process heat from advance high-temperature reactors; (2) electrolysis using electricity generated by

DOI: 10.1201/9781003537816-1

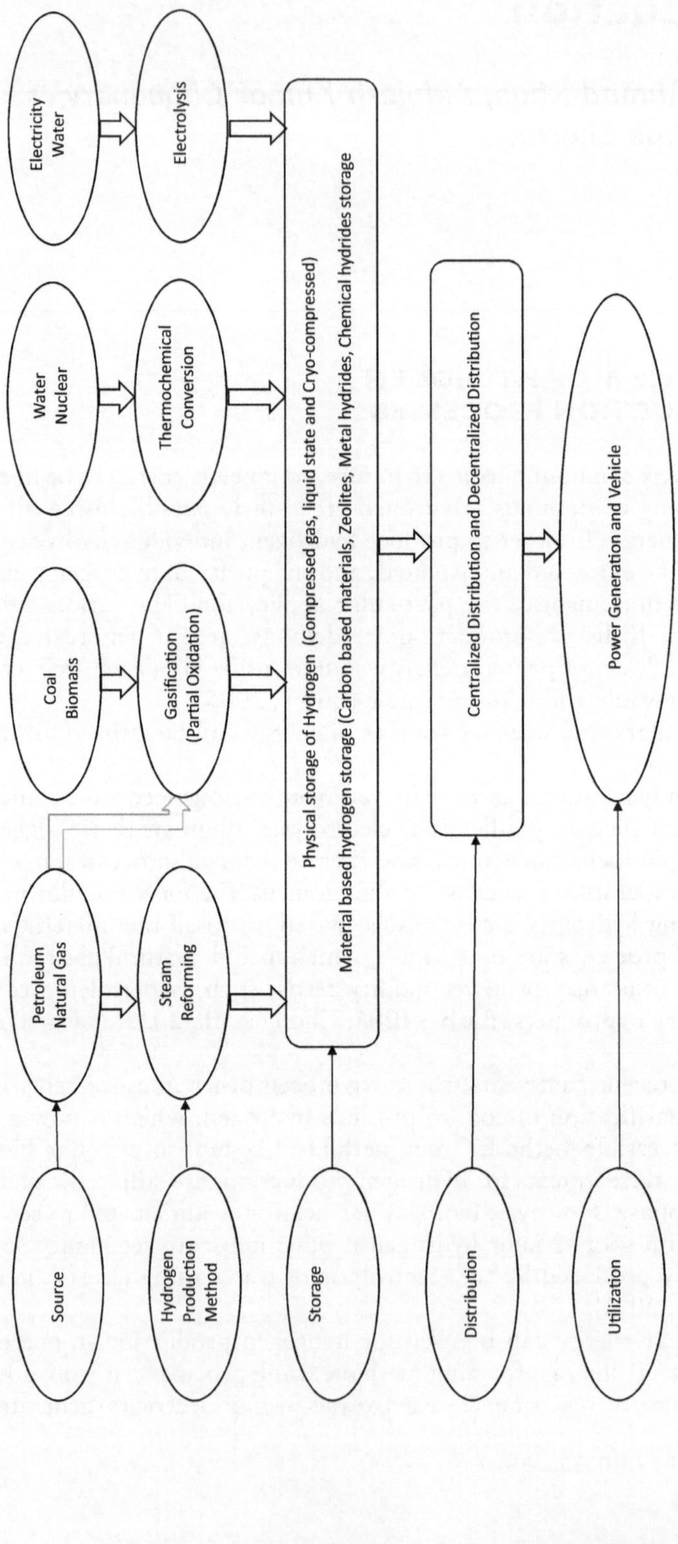

Figure 1.1 Various processes for hydrogen production.

specialized nuclear power plants; or (3) thermochemical cycles including the sulfur–iodine process. At higher temperatures, photo-induced or direct thermal breakdown of water may attract attention; at lower temperatures, more intricate, multistep approaches are needed, such as those attainable from nuclear reactor steam. In the long run, hydrogen can be produced from renewable sources (Busby, 2005; Sorensen & Spazzafumo, 2018). According to the green hydrogen economy, electricity from renewable sources (mainly from solar) is expected to be the main energy source for hydrogen production. In addition, the green hydrogen economy will help reduce CO_2 emission and offer a reliable replacement for the limited availability of gas and oil. Additionally, hydrogen can also be produced from biogas or other secondary energy sources that come from fossil fuels. In this instance, reforming methanol specifically may be advantageous for fuel cells intended for mobile applications (Kraussler et al., 2018; Sorensen & Spazzafumo, 2018).

The real hydrogen production volume data and market surveys frequently only account for captive production or hydrogen produced directly (for example, in steam reformers) in refineries or fertilizer plants. This only partially represents the real production volumes of hydrogen. Furthermore, significant quantities of hydrogen are generated as a byproduct of chemical compound manufacturing, such as ethylene and chlorine, and refinery operations. Hydrogen is either used energetically as fuel to produce heat, electricity, and power or it is vented or flared when it can no longer be used internally—for example, when hydrogenation happens in refineries. The majority of hydrogen utilized in industrial facilities is produced on-site and is referred to as "captive hydrogen" (Busby, 2005).

1.1.1 Steam reformation of fossil fuels and liquids derived from biomass

Reforming technology is mainly used in the petrochemical and fertilizer industries for the production of "on-purpose" hydrogen. Steam reforming of natural gas is a technically and commercially well-established technology on the industrial scale and currently the most economical route (Sorensen & Spazzafumo, 2018). The steam reforming process involves breaking down light hydrocarbons, for example, methane, natural gas, and naphtha that undergo breaking down during the steam reforming process by a catalytic reaction with superheated steam. This finally results in a hydrogen-rich gas mixture. Temperatures above 500°C are required for these reforming procedures. The steam–methane reforming (SMR) typically occurs at 850°C with the help of catalysts such as iron, nickel, or ruthenium, and required pressures are between 2.5 and 5 MPa. This process enhances the number of moles during conversion at higher temperatures and lower pressures that are often preferable. Because of the significant volumes of synthesis gas generated, pressures as high as 5 MPa are selected to minimize the energy needed for compression (Sorensen & Spazzafumo, 2018; Yoon et al., 2010). The carbon monoxide is catalyzed to produce more hydrogen using a slightly exothermic water gas shift reaction, and during the process, carbon deposition is avoided by using steam from the reaction.

This shift reaction normally happens in two stages, at two distinct temperatures: the first step, which uses an iron-oxide catalyst and occurs at 350–450°C, and the second step, which utilizes a copper–zinc or aluminum catalyst and starts at 150–250°C. As a result, the concentration of CO is lowered to 0.5–2% of the dry gas and the concentration of hydrogen is enhanced. To safeguard the catalyst inside the reformer tubes, feed gas is desulfurized. Part of the natural gas is used in the furnace to heat the reformer tubes externally. With the exception of steam, the equilibrium composition of the dry reformer gas is usually 75% H_2, 13% CO, 10% CO_2, and 2% CH_4 (Busby, 2005).

All fuel properties like steam-to-carbon ratio, outlet temperature, and pressure are determined according to the desired output products, and all have an enormous effect on this composition. Methane conversion is made possible and promoted by high steam-to-methane ratios, moderate pressures, and high reforming temperatures.

1.1.2 Coal gasification for hydrogen production

Coal gasification is the process of converting coal into synthesis gas, or syngas, which is composed of many different gases and particles including carbon monoxide, hydrogen, and carbon dioxide. There are four primary processes in the coal-to-hydrogen process: gasification, cooling and cleaning, shifting, and purification as shown in Figure 1.2. This process, which usually includes combining coal with an oxidant such as steam, air, or oxygen, takes place at high temperatures without combustion (Cormos et al., 2008; Seyitoglu et al., 2017). In order to eliminate contaminants like sulfur, particulate matter, and mercury, the syngas is cooled and purified. The syngas's carbon monoxide is subsequently converted into additional hydrogen and carbon dioxide in a shift reactor through the combination of it with steam. The syngas is split into streams of carbon dioxide and hydrogen after purification (Seyitoglu et al., 2017; Stiegel & Ramezan, 2006). The refined hydrogen can be used to make chemicals, electricity generation, fuel for engines, and power, and also many other applications (Huang & Dincer, 2014).

1.1.3 Biomass gasification

The two types of biomasses are organically grown biomass (liquid manure, sludge, process residues, energy crops, dedicated bioenergy crops), where

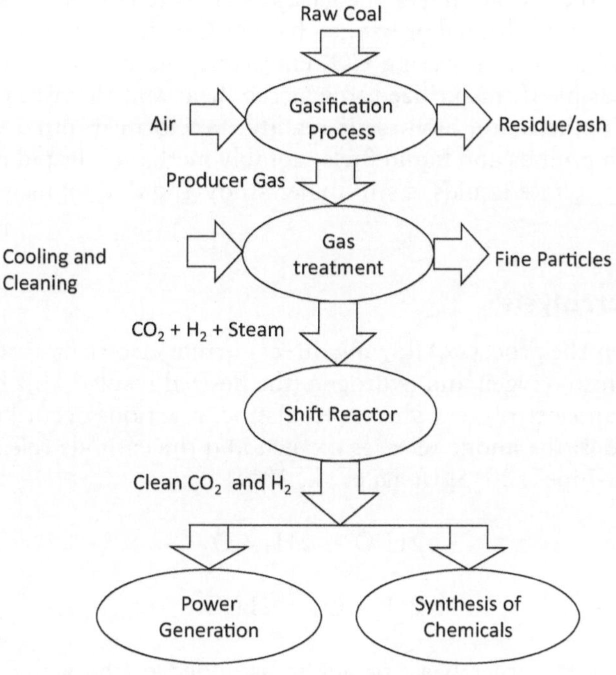

Figure 1.2 Process of coal gasification for hydrogen production.

the hydrogen is indirectly produced by anaerobic fermentation, and massive biomass (harvest residues, energy crops, process residues, dedicated bioenergy crops), which can be directly converted thermochemically to hydrogen (Chang et al., 2011). The hydrogen concentration in biomass falls between 6 and 7 weight%. Biomass usually gets gasified using steam in two different stages. The biomass is first thermochemically broken down, producing tar, char, and gasified char, as well as volatiles breaking and reforming. More light gases are produced in the second stage where carbon in the biomass and hydrocarbon gases react with CO, CO_2, H_2, and H_2O (Cao et al., 2020; Chang et al., 2011).

As the temperature increases from 600 to 900°C, the dry gas yield, carbon conversion efficiency, and hydrogen fraction in the resulting gas also increase. In particular, the hydrogen fraction increases from around 25% to 50%. A gas mixture containing hydrogen is generated whenever biomass, such as peat, wood, and agricultural leftovers, undergoes transformation thermally. The fuel or feedstock, the abundance of oxygen and steam, and the process temperatures significantly affect the amount of H_2 in it. Gasification or pyrolysis are methods for breaking down organic materials, followed by autothermal steam reforming. To prevent nitrogen from appearing in the produced gas, the gasifiers are often operated at low pressures and indirectly heated or oxygen blown (Cao et al., 2020).

In order to assist in lowering CO_2 emissions, biomass conversion seems to be less feasible if it is utilized to generate heat and electricity instead of hydrogen. Furthermore, biomass is readily transformed into a wide range of useful compounds and liquid fuels, notably methanol. Rapid pyrolysis of biomass to generate liquids is still under study and development (Cohce et al., 2010).

1.1.4 Electrolysis

Electrolysis is the process of flowing direct current electricity through water to divide it into oxygen and hydrogen (the desired result). This breakdown happens in an electrolyzer, where two distinct reactions occur between the two electrodes: the anode releases oxygen and the cathode releases hydrogen (Godula-Jopek, 2015; Ursua et al., 2011).

$$2H_2O \rightarrow 2H_2 + O_2$$

$$2H_2 + O_2 \rightarrow 2H_2O$$

An electrolyte, either basic or acidic, is added to the water to make it conductive at the cathode, which is negatively charged and repels electrons. After passing through the water, the current emerges at the anode, which is positively charged and repels protons while attracting electrons. As a result, oxygen concentrates at the anode while hydrogen accumulates at the cathode (Godula-Jopek, 2015).

The process of charge equalization is facilitated by ionic conduction via the electrolyte. The generated hydrogen and oxygen gases are maintained separately by a separator situated in between the electrodes. Direct current is the only operating voltage for electrolyzers.

In the reverse method of electrolysis, hydrogen reacts with pure oxygen and generates only water and heat. According to the fundamental concepts of thermodynamics, this energy release is equivalent to the energy input needed for water splitting (Wang et al., 2014).

Due to the considerable electricity usage, electrolysis is still a costly method of producing hydrogen today. Catalysts, which are usually made of costly materials, are frequently used in commercial water-splitting operations to speed up chemical reactions. Because electrolyzers are modular in nature, scaling them up is not too difficult. Based on the electrolyte used, electrolyzers can be divided into three basic types: high-temperature electrolyzers, membrane electrolyzers (such as PEM electrolyzers), and alkaline water electrolyzers.

The oldest and consequently the most commonly utilized technology is alkaline water electrolysis. Electrodes used in alkaline water electrolysis need to be highly electrically conductive and corrosion resistant and have efficient catalytic characteristics. To satisfy these requirements, electrodes are usually made of nickel or chromium–nickel steel. Individual cells are distinguished by a separator designated as a diaphragm in order to make sure that the byproducts of the reaction, such as hydrogen and oxygen gases, do not combine. Due to its low electrical resistance, this diaphragm has been chosen.

Additionally, electrolysis doesn't require large amounts of water. A car can travel the same amount of distance on hydrogen that is extracted from a gallon of water as gasoline-powered automobiles can travel (Ursua et al., 2011).

1.1.5 Other hydrogen production methods

(i) **Photobiological production**: In this process, certain microorganisms, such as algae and cyanobacteria, use biological processes to produce hydrogen when they are brought into contact with sunshine. Low hydrogen production rates and the need for extensive research and development are the related barriers to commercialization (Asada & Miyake, 1999; Touloupakis & Torzillo, 2019).

(ii) **Thermochemical water splitting:** Through a series of chemical reactions facilitated by high temperatures (usually 500–2000°C), which are often generated by concentrated solar power or nuclear heat, water is divided into hydrogen and oxygen in this process (Safari & Dincer, 2020). This method has several benefits, including the potential for high productivity and the ability to be integrated with current industrial processes. However, there are also some significant drawbacks to this method. For example, it demands high temperatures, which adds complexity and expense to the process, and it requires more research and development to find suitable materials and reactors (Ghorbani et al., 2023; Lee et al., 2022).

(iii) **Photoelectrochemical water splitting:** This approach uses photoelectrochemical cells, which combine photovoltaic and electrolysis technologies, to directly split water into hydrogen and oxygen using sunshine. With higher efficiency, this method directly transforms solar energy into hydrogen (Chen et al., 2013; Joy et al., 2018).

(iv) **Biophotolysis:** It is an additional technique for producing hydrogen that resembles photosynthesis. This process, which is supported by enzymes like hydrogenases, enables some microbes to use sunlight to split water into hydrogen and oxygen. However, this method produces low-cost hydrogen synthesis at the expense of natural processes and renewable energy sources, but it is not very efficient (Ghiasian, 2019; Ghirardi et al., 2014).

(v) **Ammonia cracking:** The process of ammonia cracking, sometimes referred to as thermal breakdown or catalytic cracking, converts ammonia (NH_3) into hydrogen (H_3) and nitrogen (N_3). It is a newly developed technique for producing hydrogen. Ammonia can carry hydrogen, allowing storage and transportation to be quicker, which is why this procedure is becoming more and more popular (Aziz et al., 2020; Chai et al., 2021).

1.2 HYDROGEN STORAGE TECHNOLOGIES

In the context of renewable energy, the ability to store energy at the time of generation or to ensure continuous use without any fluctuations relies on energy storage systems. Such systems enhance the reliability of energy use. At times, energy generation exceeds consumption, leading to a production–consumption mismatch. This imbalance can be addressed by employing a suitable energy storage system, which stores the excess energy for later use when demand surpasses production.

The technologies for hydrogen storage can be categorized into two main types: physical-based hydrogen storage and material-based hydrogen storage. Physical storage techniques encompass compressed gas, liquid hydrogen, and cryo-compressed hydrogen, relying on altering conditions like pressure in the case of compressed gas, temperature for liquid storage, or a combination of both in cryo-compressed gas storage. Conversely, material-based storage, also known as solid-state storage, utilizes processes such as absorption or adsorption.

The following are the major technologies for hydrogen storage.

1.2.1 Physical storage technologies

1.2.1.1 Hydrogen storage in the form of compressed gas

This is the most popular way of storing hydrogen worldwide in hydrogen generation as well as in hydrogen transportation. In this system, the hydrogen is stored at extremely high pressure in high-pressure vessels. The pressure varies from 700 bar to 1000 bar depending on the nature of the application. Hydrogen compression to 700 bar consumes about 10% of the gas energy stored (Burheim, 2017). At such high pressure, the safety issues are of prime concern.

1.2.1.2 Hydrogen storage in liquid state

Cryogenic tanks are used to store liquid hydrogen at –252.8°C (21.2 K) under atmospheric pressure. The primary challenges associated with storing liquid hydrogen include the need for an energy-efficient process to liquefy it and the necessity of adequately insulating the cryogenic storage tanks to minimize hydrogen boil-off (Flynn, 1992). The Joule–Thomson cycle represents the most basic approach to gas liquefaction. Initially, the gas is compressed and subsequently cooled using a heat exchanger. Following this, it is directed through a throttle valve, where it experiences an isenthalpic Joule–Thomson expansion, resulting in the production of some liquid. The now-cooled gas is then separated from the liquid and recycled back to the compressor through the heat exchanger.

The energy required for storing hydrogen in liquid form accounts for approximately 35% of the stored hydrogen's energy content, making it significantly less efficient compared to other hydrogen storage technologies. Consequently, the use of liquid hydrogen is primarily confined to space and aviation applications, where the demand for high gravimetric and volumetric energy storage densities outweighs concerns over its substantial exploitation of energy (Yanxing et al., 2019).

1.2.1.3 Cryo-compressed hydrogen storage

When comparing densities, at a high pressure of about 70 MPa and at normal room temperature, hydrogen exhibits a density of only 39.1 kg/m^3, whereas at a low pressure of 0.4 MPa and at 20 K temperature, hydrogen boasts a density of 71.0 kg/m^3. Interestingly, at the liquefaction point, if the gaseous hydrogen is stored, it surpasses the density exhibited by the hydrogen in liquid state. Thus, to achieve superior performance of hydrogen storage, both temperature and pressure must be optimized.

Cryo-compressed hydrogen storage is a method that combines the benefits of both cryogenic and storing hydrogen in a compressed condition. In this method, hydrogen gas is first compressed to high pressures, typically above 25 MPa. Then, the compressed hydrogen is cooled to cryogenic temperatures, usually below 20 K (–253°C), which is close to the boiling point of liquid hydrogen.

By employing cryogenic temperatures, the density of the compressed hydrogen can be significantly increased, thus approaching or even exceeding the density of liquid hydrogen. This results in higher volumetric energy storage densities compared to traditional compressed hydrogen storage methods.

1.2.2 Material-based hydrogen storage technologies

As previously mentioned, physical hydrogen storage methods exhibit certain drawbacks, including limitations in storage capacity, energy consumption, etc., and, as seen in liquid or cryo-compressed storage, the necessity for vessels capable of preventing heat leakage or withstanding high pressures. Additionally, safety concerns arise due to the risks associated with very high pressure, such as bursting, fire-related hazards, and potential leakage.

Consequently, depending on the chosen storage approach, material-based storage of hydrogen can rely on either adsorption or absorption techniques. These methods offer the potential for achieving substantial storage capacities for hydrogen, all the while functioning at mild pressures and temperatures (Schlapbach & Züttel, 2001).

Material-based methods are expected to have greater room temperature hydrogen storage capacities and demonstrate better safety profiles compared to physical storage since they have lower hydrogen storage pressures and slower rate of hydrogen release.

1.2.2.1 Adsorption technologies

Adsorption is the process of moving molecules from a gaseous or liquid state to a solid surface. It is mainly distinguished by its reversible nature (Artioli, 2008). Covalent-organic frameworks (COFs), metal-organic frameworks (MOFs), carbon nanostructures, zeolites, and polymers of intrinsic microporosity (PIMs) are among the materials that store hydrogen and are based on this phenomenon. These materials have attracted considerable interest due to their swift kinetics, remarkable reversibility, and substantial capacity for storing hydrogen.

(i) *Adsorption in carbon-based materials*

Hydrogen storage properties are exhibited by a variety of carbon-based materials, including porous carbon, carbon nanotubes, and carbon nanofibers with varying topologies (Dou et al., 2023). Activated carbon is a well-known substance that is useful for hydrogen adsorption; it works especially well at cryogenic temperatures and high hydrogen pressures. Carbon nanostructures show significant potential for storing hydrogen, especially those with large specific surface areas (Rödl et al., 2018).

Carbon nanotubes, characterized by their cylindrical shape formed from rolled graphite sheets, exhibit adsorption capacities influenced by factors such as pore size and tube curvature (Geitmann & van der Laak, 2018). Due to their nanoscale diameter, these materials enable the condensation of high-density hydrogen gas, achieving approximately 5–10% by weight (Cormos et al., 2008). At 77 K and pressures between 1 and 20 bar, activated carbon usually exhibits adsorption capabilities of 1–7 weight%, respectively.

(ii) *Adsorption in metal-organic frameworks*

MOFs, or metal-organic frameworks, are a special kind of porous material. They are made up of organic molecules forming a framework with pores and empty spaces connecting metal ions or clusters. MOFs are helpful in a variety of applications, including gas storage, separation, and catalysis, since these pores may trap gases or molecules (Pan et al., 2004). These materials are attractive prospects for hydrogen storage applications because of their variable pore diameters and high surface areas (up to 4800 m^2/g).

In the context of hydrogen adsorption, MOFs offer several advantages. The MOFs' large surface area and distinct pore structures enable efficient adsorption of hydrogen molecules. Additionally, the tunability of MOF properties allows for optimization of adsorption capacities and kinetics.

From a pool of around 4000 different metal-organic framework (MOF) species, those with specific surface areas ranging from 3100 to 4800 m^2/g have been identified as having the highest theoretical capacity for storing substances. Among these, MOF-5, which is composed of zinc oxide clusters connected by 1,4-benzenedicarboxylate molecules, stands out as a particularly well-studied candidate for hydrogen storage (Rosi et al., 2003).

(iii) *Adsorption in zeolites*

Zeolites are structured materials made of crystalline aluminosilicates, possessing well-defined arrangements and consistent pore sizes. These characteristics render them effective adsorbents for a variety of gases, including hydrogen, carbon dioxide, and moisture.

In the context of hydrogen uptake through physisorption, the available surface area for adsorption determines the maximal storage capacity of zeolites (Azzouz, 2012). By a process called encapsulation, hydrogen molecules can be forced into the porous structure of zeolites. This may take place at very high temperatures (up to 350°C) and extremely high pressures (up to 900 bar). Once encapsulated, hydrogen can remain confined within the zeolite structure even at room temperature (Anderson, 2008).

1.2.2.2 Absorption technologies

Hydrogen absorption technologies rely on various types of compounds called hydrides to store hydrogen. These hydrides come in different forms like ionic, covalent, and complex hydrides.

When we talk about reversible hydrogen storage, it means the ability to both absorb hydrogen when needed and release it when required. To achieve this, certain techniques are used for the release, called desorption. However, these desorption techniques often require additional components to be added to the storage system.

The absorption process can occur through various mechanisms, including physisorption (physical adsorption) and chemisorption (chemical bonding).

Once hydrogen is absorbed into the material, it can be stored until needed for use. When the stored hydrogen is required, the absorption material undergoes a process known as desorption, where the hydrogen is liberated from the material and made available for utilization in fuel cells, combustion engines, or other applications. Hydrogen absorption technologies encompass a wide range of materials and methods, including metal hydrides, complex hydrides, carbon-based materials, and more. Each type of absorption material has its own distinct properties and benefits, making it appropriate for different applications and operating conditions.

Researchers and engineers continue to explore new materials, optimize existing processes, and improve system designs to overcome challenges such as storage capacity, efficiency, and safety. Hydrogen absorption technologies are discussed in the following subsections:

(i) *Metal hydrides*

There are several different metallic materials that can absorb hydrogen at low temperatures and with modest pressure (Karkamkar et al., 2007). Among these, elements such as lithium (Li), boron (B), nitrogen (N), magnesium (Mg), and aluminum (Al) are examples of metal hydrides that show great promise for hydrogen storage. These substances generate a great deal of interest for use in hydrogen storage applications.

One notable advantage of metal hydrides is their perceived safety, facilitated by their operation at relatively low temperatures and their tendency to release hydrogen through an endothermic process (He et al., 2016).

Despite these merits, metal hydrides come with drawbacks. One major concern is their weight, which affects the onboard storage of hydrogen. Additionally, metal hydrides exhibit slow reaction kinetics, show limited reversibility, and require high temperatures for hydrogen release, posing challenges for their practical application (Lutz et al., 2019).

Metal hydrides can be grouped into three main categories: elemental, intermetallic, and complex hydrides.

(a) Elemental hydrides: These hydrides consist of a single metal combined with hydrogen, forming a simple compound. Examples include lithium hydride (LiH), sodium hydride (NaH), and magnesium hydride (MgH_2).
(b) Intermetallic hydrides: Intermetallic hydrides are compounds formed by a combination of two or more metals with hydrogen. These

compounds often have complex structures and can exhibit unique hydrogen storage properties. Examples include $LaNi_5$ and Mg_2Ni.

(c) Complex hydrides: Complex hydrides are compounds composed of metal cations bonded to complex hydride anions, often containing boron and/or aluminum. These hydrides typically have high hydrogen storage capacities and can release hydrogen under mild conditions. Examples include sodium borohydride ($NaBH_4$) and ammonia borane (NH_3BH_3).

1.2.3 Chemical hydrogen storage

In contrast to traditional metal hydrides, chemical hydrides, such as ammonia, ammonia borane, metal B–N–H compounds, and liquid organic hydrogen carriers (LOHCs) like methyl cyclohexane-toluene, utilize lighter elements (Smythe & Gordon, 2010), resulting in higher gravimetric densities and easier release of hydrogen. However, they tend to be irreversible and are typically employed as one-time-use fuels, which can pose challenges related to the removal of byproducts (Hwang et al., 2010). Despite this limitation, chemical hydrides offer advantages over liquid hydrogen, including lower costs for storage and transportation, making them increasingly appealing. Furthermore, they present a viable option for transporting energy from regions abundant in renewable resources to areas with high demand.

1.3 OVERVIEW OF HYDROGEN TRANSPORTATION

The transportation of hydrogen from the point of production to its destination of consumption or storage is a vital component of the hydrogen economy (Faye et al., 2022). Certain places in the world have extremely high demand for hydrogen energy, yet these areas are not always able to support the large-scale production that is needed. So, the transportation of hydrogen is a vital and important link in the whole industrial chain of the hydrogen energy sector. Figure 1.3 shows the flowchart of hydrogen production, storage, and transportation. The growth of affordable and safe hydrogen transportation technology is a necessity for the successful and broad deployment of hydrogen energy, as well as the high-quality growth of the hydrogen energy sector. High levels of technology are needed for the delivery of hydrogen due to its combustible and explosive qualities. Therefore, to prevent explosions and leaks, it is necessary to have some understanding of the characteristics of hydrogen when transporting it. Hydrogen can be delivered to the final destination by using a variety of methods from the production site, depending on local conditions, transportation distances, and hydrogen amounts (Ball & Weeda, 2015).

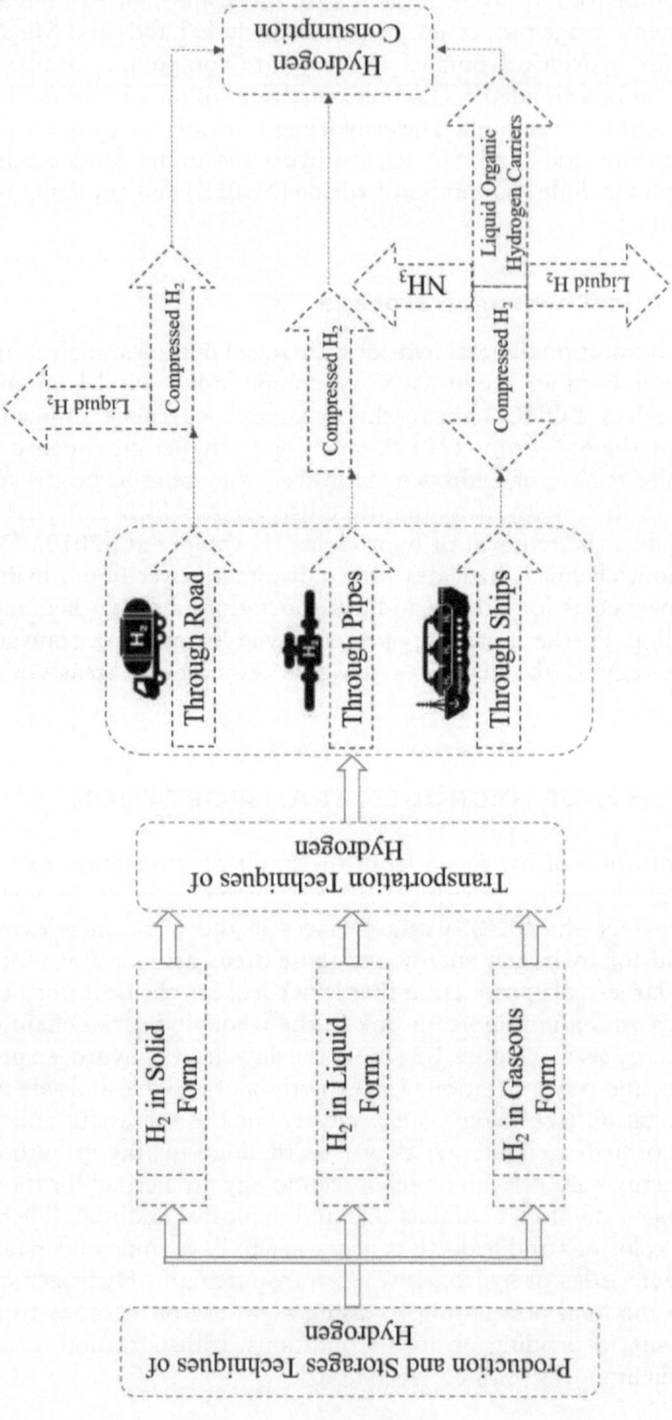

Figure 1.3 Flowchart of hydrogen production, storage, and transportation methods.

The most effective ways of transportation of hydrogen over the shortest distances are hydrogen trucks and pipelines. However, various other techniques are required when distances increase. Nowadays, mainly three primary modes use hydrogen gas in the form of solid, liquid, and gaseous hydrogen. Based on present and developed gas transportation methods, there are three primary categories for hydrogen transportation: shipping, pipelines, and roads (Ma et al., 2008). All three of these approaches have advantages and disadvantages, and they are all the subjects of substantial research at the moment. When hydrogen is transported by road, it is carried in liquid or compressed gaseous form by trailers and trucks. Pressurized gaseous hydrogen may also be transported via pipes, similar to natural gas (J. Li et al., 2021). Hydrogen can be cheaply transported in substantial amounts using pipeline transportation, despite its high initial cost. For this reason, pipeline transportation is a good option for getting hydrogen to large industrial customers. Large-scale hydrogen transportation may also be accomplished using shipping, which is particularly practical for cross-continental travel. As a result, hydrogen may be supplied in both elemental and compound forms.

The following are the main techniques for hydrogen transportation:

(i) Transportation of hydrogen in gaseous form
(ii) Transportation of hydrogen in liquid form
(iii) Transportation of hydrogen through carrier materials

1.3.1 Transportation of hydrogen in gaseous form

The transportation of hydrogen gas in its gaseous state is known as "gaseous hydrogen transportation." In order to ensure that safety criteria are met, this procedure involves compressing hydrogen at very high pressures and transferring it in specific containers (Noh et al., 2023). Because of its affordability and pre-existing infrastructure, it is a widely used strategy. There are two main techniques of compressing gases: centrifugal compression and mechanical methods (such as compressors and diaphragms) (Zhang et al., 2024). Metal hydride-operated hydrogen compressors have grown in popularity over the past few years due to their many advantages, including improved safety, environmental compatibility, low vibration and noise, no friction, efficient sealing, efficient hydrogen purification, and lower maintenance costs (Q. Li et al., 2021).

1.3.1.1 Transportation through pressurized cylinders

For smaller-scale applications, pressurized gas cylinders provide a versatile and transportable method for delivering hydrogen to different places. Using this technique, hydrogen gas is pressurized and kept in high-pressure cylinders made of robust materials like composites or steel that can endure pressures of up to 700 bar (Massarweh et al., 2023). These pressurized hydrogen cylinders or tanks are then securely transported at high pressures using specifically made high-pressure tube trailers. Ships and railways transport are other possible options, depending on the demands and the availability of the necessary infrastructure. The greatest amount of transportable hydrogen is determined by the weight of containers, which are attached with pressurized hydrogen gas tube trailers wrapped in shielding frames for higher amounts of hydrogen. Therefore, in order to facilitate the transportation of greater volumes of hydrogen, the production of lightweight tank compositions becomes essential. The benefit of compressed gas cylinders is that they make it simple to carry hydrogen to those areas where infrastructure for pipelines would not be practical or accessible. Additionally, it offers a practical solution for transient or mobile uses, including distant power generation or hydrogen-powered car refueling stations (Hassan et al., 2021). The weight and size of the high-pressure cylinders might also be an issue, especially when shipping over great distances. To reduce the dangers involved with managing high-pressure cylinders and the possibility of hydrogen leakage, proper handling techniques must be used.

1.3.1.2 Transportation through pipelines

The use of pipelines constitutes one of the main ways that hydrogen is transported across great distances. It entails using specialized pipes made to transport hydrogen gas from production sites to distribution centers or end consumers in a safe and effective manner. It eliminates the prerequisite for regular storage and transportation logistics by offering a steady and dependable source of hydrogen (Makaryan et al., 2022). These approaches are particularly noteworthy because of their affordability for low carbon footprint and the possibility of decades-long pipeline functioning. Furthermore, the pipeline network may be combined with already-existing natural gas pipes to allow for the reuse of such networks for the conveyance of hydrogen (Pellegrini et al., 2020).

There are still several disadvantages to the transportation of hydrogen through pipelines. For instance, in order to minimize hydrogen brittleness and leakage, certain pipeline materials, coatings, and engineering factors are needed since hydrogen exhibits distinct characteristics compared to natural gas (Erdener et al., 2023). The reliability of the pipeline must be guaranteed in order to avoid incidents, and this requires the implementation of precautionary measures like inspection mechanisms, leak identification technologies, and maintenance procedures (Wei et al., 2022). To increase the resistance of a substance to hydrogen breakdown, surface treatments such as shot peening, carburizing, and nitriding can be used (Maleki et al., 2021). By acting as inhibitors to hydrogen penetration, protective coatings like metal or ceramic coatings have a lower danger of breakdown in the underlying material (Laadel et al., 2022).

1.3.2 Transportation of hydrogen in liquid form

With its tremendous efficiency, transportation of liquid hydrogen is an effective option for long distances. For lengths as great as 2000 km, it is more adaptable and cost-effective than the use of pipelines. Transporting hydrogen in the form of a liquid requires cooling down the gas to a very low temperature, as opposed to other methods. Cryogenic carriers, which are specifically made for transporting hydrogen in the form of a liquid, are the main means of carrying liquid hydrogen (Ratnakar et al., 2021). To preserve lower temperatures as well as pressure throughout transit, these tankers are outfitted with insulation and pressure-relief devices. The liquid hydrogen is transported by cryogenic carriers, usually by roadway, to preservation facilities or the final user destination. However, they might also be shipped or carried by rail, contingent upon the particular needs and facilities that are in place (Ahluwalia et al., 2023). Hydrogen, when it is in liquid form, is moved to the suitable storage containers once the cryogenic truck is linked to the final user destination. In the majority of instances, the hydrogen in liquid form is first vaporized and heated to room temperature. Security precautions must be taken at every stage of the shipping process. They involve maintaining an eye on liquid hydrogen's pressure and temperature, utilizing leak detecting devices and abiding by all applicable laws and guidelines (Klebanoff et al., 2017).

1.3.3 Transportation of hydrogen through carrier materials

The transportation of hydrogen depends heavily on hydrogen carrier materials, commonly referred to as materials for storing and transporting the hydrogen in compound form. Ammonia, liquid organic matter, and metal hydrides are typical instances of hydrogen transporters. These elements promote the development of hydrogen-established modes of transport by enabling additional security and adaptable hydrogen distribution options.

1.3.3.1 Blend with ammonia (NH_3)

Ammonia provides an effective means for the preservation and transportation of hydrogen, making it a flexible hydrogen carrier material. Because ammonia has a high proportion of hydrogen by mass and is capable of creating an unchanging liquid at medium temperature and pressure, it can preserve and move hydrogen efficiently (Humphreys et al., 2021). A catalytic mechanism at the point of delivery transforms ammonia back into hydrogen, producing a dependable hydrogen supply for a range of uses, especially in the sector of chemical production. Future progress in creating efficient catalysts will be crucial to the feasibility of employing ammonia (NH_3) as a hydrogen transporter (Smith et al., 2020).

1.3.3.2 Blend with natural gas and liquid organics

The transportation of hydrogen in its pure form to the marketplace is thought to be possible by mixing hydrogen with a natural gas transmission network. Near the point of final use, hydrogen is extracted from natural gas using specific separation and purification techniques. Hydrogen might thus be extracted from hydrogen–natural gas mixes using three different techniques: pressure swing adsorption (PSA), membrane separation, and electrochemical hydrogen separation. Furthermore, according to reports, blends containing below 5–15% hydrogen by weight only cause small problems, depending on the natural gas composition and the pipeline characteristics peculiar to the location. Yet, more significant changes are required for 15–50% hydrogen mixing, such as the addition of large domestic device transformation facilities or increased compressing capability throughout the supply route for customers in industry (Reitenbach et al., 2015).

Chemical substances called liquid organic hydrogen carriers of liquid organic are employed in the preservation and movement of hydrogen. Through chemical procedures, LOHCs can establish reversible bonds among molecules of hydrogen, allowing them to preserve hydrogen in a controlled environment. They consist of a pair of elements: a compound of organic material that is hydrogen-rich (LOHC+) and a molecule of organic material that is hydrogen-lean (LOHC–) (Modisha et al., 2019). The main LOHC substances that are being studied in-depth for the purpose of preservation and shipment are perhydro-dibenzyltoluene, cyclohexane and benzene, decalin and naphthalene, perhydro-*N*-ethyl carbazole, and *N*-ethyl carbazole (Sotoodeh & Smith, 2010). Through a bidirectional dehydrogenation–hydrogenation process, it makes storing hydrogen easier as well as efficient discharge and absorption. Delivering hydrogen to isolated or off-grid areas, where a direct pipeline or alternative means of conveyance would not be practical, is made easier by hydrogen carriers.

1.3.3.3 Blend with metal hydrates

Hydrogen gas may be transported and stored using a family of materials called metal hydrides. At the fabrication site, hydrogen first reacts exothermically with alloys that generate metal hydrides (Bellosta von Colbe et al., 2019). The metal hydride substance absorbs and stores hydrogen as a result of this reaction. The saturated hydrogen metal hydride substance is then delivered to its location by means of standard transportation vehicles, such as carriers, ships, or trains. Thereafter, the metal hydride substance is heated to liberate the hydrogen that has been stored (Tong et al., 2021). Due to their exceptionally high gravimetric and volumetric abilities, light metal borohydrides including sodium, lithium, magnesium, and calcium have been extensively researched as hydrogen carriers. As a result, another interesting hydrogen carrier substance is magnesium borohydride, which is created by combining borane and magnesium and then hydrogenating the mixture under pressure (Zheng et al., 2019).

1.4 CONCLUDING REMARKS OF HYDROGEN TRANSPORTATION

To sum up, the techniques and procedures utilized to distribute and carry hydrogen from manufacturing facilities to final consumers or places of preservation are referred to as hydrogen transporting technologies. In the view of hydrogen supply, several technologies are essential. Transportation of hydrogen through liquid, gaseous, and carrier are the main hydrogen methods of transport.

Transportation of hydrogen in the gaseous form is an unshakable and comparatively inexpensive method for the distribution of hydrogen. However, it is facing various difficulties owing to the higher demand for energy and lower energy density. Its prospects for the upcoming years depend on the development of facilities and better storage substances in order to facilitate broader acceptance and efficient distribution.

The transportation of liquid hydrogen provides a higher energy density and it is a proven method of distributing hydrogen. However, it needs to be stored in a cryogenic environment and uses a lot of energy. The development of improved insulating substances along with more energy-efficient procedures is required to increase the overall effectiveness and viability of transportation of liquid hydrogen in the supply chain of hydrogen energy.

When contrasted with liquid and gaseous hydrogen, the conveyance of hydrogen carriers offers convenience in handling, transit, and preservation. Future prospects for the conveyance of hydrogen through carriers rely on developments in carrier substances and improved techniques for releasing hydrogen, which may make this a more practical choice for broad hydrogen distribution.

REFERENCES

Ahluwalia, R. K., Roh, H.-S., Peng, J.-K., Papadias, D., Baird, A. R., Hecht, E. S., Ehrhart, B. D., Muna, A., Ronevich, J. A., Houchins, C., Killingsworth, N. J., & Aceves, S. M. (2023). Liquid hydrogen storage system for heavy duty trucks: Configuration, performance, cost, and safety. *International Journal of Hydrogen Energy*, *48*(35), 13308–13323. https://doi.org/https://doi.org/10.1016/j.ijhydene.2022.12.152

Anderson, P. A. (2008). 9 - Storage of hydrogen in zeolites. In G. B. T.-S.-S. H. S. Walker (Ed.), *Woodhead publishing series in electronic and optical materials* (pp. 223–260). Woodhead Publishing. https://doi.org/https://doi.org/10.1533/9781845694944.3.223

Artioli, Y. (2008). *Adsorption. Encyclopedia of ecology fath, SEJD*. Academic Press.

Asada, Y., & Miyake, J. (1999). Photobiological hydrogen production. *Journal of Bioscience and Bioengineering*, *88*(1), 1–6.

Aziz, M., Wijayanta, A. T., & Nandiyanto, A. B. D. (2020). Ammonia as effective hydrogen storage: A review on production, storage and utilization. *Energies*, *13*(12), 3062.

Azzouz, A. (2012). Achievement in hydrogen storage on adsorbents with high surface-to-bulk ratio – Prospects for Si-containing matrices. *International Journal of Hydrogen Energy*, *37*(6), 5032–5049. https://doi.org/https://doi.org/10.1016/j.ijhydene.2011.12.024

Ball, M., & Weeda, M. (2015). The hydrogen economy–vision or reality? *International Journal of Hydrogen Energy*, *40*(25), 7903–7919.

Bellosta von Colbe, J., Ares, J.-R., Barale, J., Baricco, M., Buckley, C., Capurso, G., Gallandat, N., Grant, D. M., Guzik, M. N., Jacob, I., Jensen, E. H., Jensen, T., Jepsen, J., Klassen, T., Lototskyy, M. V, Manickam, K., Montone, A., Puszkiel, J., Sartori, S., … Dornheim, M. (2019). Application of hydrides in hydrogen storage and compression: Achievements, outlook and perspectives. *International Journal of Hydrogen Energy*, *44*(15), 7780–7808. https://doi.org/https://doi.org/10.1016/j.ijhydene.2019.01.104

Burheim, O. S. (2017). *Engineering energy storage*. Academic Press.

Busby, R. L. (2005). *Hydrogen and fuel cells*. Pennwell Corporation.

Cao, L., Iris, K. M., Xiong, X., Tsang, D. C. W., Zhang, S., Clark, J. H., Hu, C., Ng, Y. H., Shang, J., & Ok, Y. S. (2020). Biorenewable hydrogen production through biomass gasification: A review and future prospects. *Environmental Research*, *186*, 109547.

Chai, W. S., Bao, Y., Jin, P., Tang, G., & Zhou, L. (2021). A review on ammonia, ammonia-hydrogen and ammonia-methane fuels. *Renewable and Sustainable Energy Reviews*, *147*, 111254.

Chang, A. C. C., Chang, H.-F., Lin, F.-J., Lin, K.-H., & Chen, C.-H. (2011). Biomass gasification for hydrogen production. *International Journal of Hydrogen Energy*, *36*(21), 14252–14260.

Chen, Z., Dinh, H. N., & Miller, E. (2013). *Photoelectrochemical water splitting* (Vol. 344). Springer.

Chong, C. C., Cheng, Y. W., Ng, K. H., Vo, D.-V. N., Lam, M. K., & Lim, J. W. (2022). Bio-hydrogen production from steam reforming of liquid biomass wastes and biomass-derived oxygenates: A review. *Fuel*, *311*, 122623.

Cohce, M. K., Dincer, I., & Rosen, M. A. (2010). Thermodynamic analysis of hydrogen production from biomass gasification. *International Journal of Hydrogen Energy*, *35*(10), 4970–4980.

Cormos, C.-C., Starr, F., Tzimas, E., & Peteves, S. (2008). Innovative concepts for hydrogen production processes based on coal gasification with CO_2 capture. *International Journal of Hydrogen Energy*, *33*(4), 1286–1294.

Dou, Y., Sun, L., Ren, J., & Dong, L. (2023). Opportunities and future challenges in hydrogen economy for sustainable development. In Antonio Scipioni, Alessandro Manzardo, and Jingzheng Ren (Ed.), *Hydrogen economy* (pp. 537–569). Elsevier.

Erdener, B. C., Sergi, B., Guerra, O. J., Lazaro Chueca, A., Pambour, K., Brancucci, C., & Hodge, B.-M. (2023). A review of technical and regulatory limits for hydrogen blending in natural gas pipelines. *International Journal of Hydrogen Energy*, *48*(14), 5595–5617. https://doi.org/https://doi.org/10.1016/j.ijhydene.2022.10.254

Faye, O., Szpunar, J., & Eduok, U. (2022). A critical review on the current technologies for the generation, storage, and transportation of hydrogen. *International Journal of Hydrogen Energy*, *47*(29), 13771–13802. https://doi.org/https://doi.org/10.1016/j.ijhydene.2022.02.112

Flynn, T. M. (1992). A liquification of gases. *McGraw-Hill Encyclopedia of Science & Technology*, *10*, 106.

Geitmann, S., & van der Laak, W. (2018). H2 as option for heavy-duty transport: H 2-Share-Cooperation in the logistic sector. *H2 International*. ISSN 2367-3931, 34.

Ghiasian, M. (2019). Biophotolysis-based hydrogen production by cyanobacteria. In Rastegari, A., Yadav, A., and Gupta, A. (Eds.), *Prospects of renewable bioprocessing in future energy systems: Biofuel and biorefinery technologies* (pp. 161–184), Vol. 10, Springer, Cham.

Ghirardi, M. L., King, P. W., Mulder, D. W., Eckert, C., Dubini, A., Maness, P.-C., & Yu, J. (2014). Hydrogen production by water biophotolysis. In Zannoni, D. and De Philippis, R. (Eds.), *Microbial bioEnergy: Hydrogen production, advances in photosynthesis and respiration* (pp. 101–135), vol. 38.

Ghorbani, B., Zendehboudi, S., Zhang, Y., Zarrin, H., & Chatzis, I. (2023). Thermochemical water-splitting structures for hydrogen production: Thermodynamic, economic, and environmental impacts. *Energy Conversion and Management*, *297*, 117599.

Godula-Jopek, A. (2015). *Hydrogen production: By electrolysis*. John Wiley & Sons.

Hassan, I. A., Ramadan, H. S., Saleh, M. A., & Hissel, D. (2021). Hydrogen storage technologies for stationary and mobile applications: Review, analysis and perspectives. *Renewable and Sustainable Energy Reviews*, *149*, 111311. https://doi.org/https://doi.org/10.1016/j.rser.2021.111311

He, C., Yu, R., Sun, H., & Chen, Z. (2016). Lightweight multilayer composite structure for hydrogen storage tank. *International Journal of Hydrogen Energy*, *41*(35), 15812–15816. https://doi.org/https://doi.org/10.1016/j.ijhydene.2016.04.184

Huang, J., & Dincer, I. (2014). Parametric analysis and assessment of a coal gasification plant for hydrogen production. *International Journal of Hydrogen Energy*, *39*(7), 3294–3303.

Humphreys, J., Lan, R., & Tao, S. (2021). Development and recent progress on ammonia synthesis catalysts for haber–bosch process. *Advanced Energy and Sustainability Research*, 2(1), 2000043. https://doi.org/https://doi.org/10.1002/aesr.202000043

Hwang, H. T., Al-Kukhun, A., & Varma, A. (2010). Hydrogen for vehicle applications from hydrothermolysis of ammonia borane: Hydrogen yield, thermal characteristics, and ammonia formation. *Industrial & Engineering Chemistry Research*, *49*(21), 10994–11000. https://doi.org/10.1021/ie100520r

Joy, J., Mathew, J., & George, S. C. (2018). Nanomaterials for photoelectrochemical water splitting–review. *International Journal of Hydrogen Energy*, *43*(10), 4804–4817.

Karkamkar, A., Aardahl, C., & Autrey, T. (2007). Recent developments on hydrogen release from ammonia-borane. *Material Matters*, 2(2), 6–9. https://doi.org/10.1016/B978-0-12-385142-0.00015-5

Klebanoff, L. E., Pratt, J. W., & LaFleur, C. B. (2017). Comparison of the safety-related physical and combustion properties of liquid hydrogen and liquid natural gas in the context of the SF-BREEZE high-speed fuel-cell ferry. *International Journal of Hydrogen Energy*, *42*(1), 757–774. https://doi.org/https://doi.org/10.1016/j.ijhydene.2016.11.024

Kraussler, M., Binder, M., Schindler, P., & Hofbauer, H. (2018). Hydrogen production within a polygeneration concept based on dual fluidized bed biomass steam gasification. *Biomass and Bioenergy*, *111*, 320–329.

Laadel, N.-E., El Mansori, M., Kang, N., Marlin, S., & Boussant-Roux, Y. (2022). Permeation barriers for hydrogen embrittlement prevention in metals – A review on mechanisms, materials suitability and efficiency. *International Journal of Hydrogen Energy*, *47*(76), 32707–32731. https://doi.org/10.1016/j.ijhydene.2022.07.164

Lee, J. E., Shafiq, I., Hussain, M., Lam, S. S., Rhee, G. H., & Park, Y.-K. (2022). A review on integrated thermochemical hydrogen production from water. *International Journal of Hydrogen Energy*, *47*(7), 4346–4356.

Li, J., Li, G., Ma, S., & Song, J. (2021). An overview on hydrogen energy storage and transportation technology and its typical application in power system. *Modern Electric Power*, *38*(5), 535–545. https://doi.org/10.19725/j.cnki.1007-2322.2021.0023

Li, Q., Peng, Z., Jiang, W., Ouyang, L., Wang, H., Liu, J., & Zhu, M. (2021). Optimization of Ti-Zr-Cr-Fe alloys for 45 MPa metal hydride hydrogen compressors using orthogonal analysis. *Journal of Alloys and Compounds*, *889*, 161629. https://doi.org/https://doi.org/10.1016/j.jallcom.2021.161629

Lutz, M., Bhouri, M., Linder, M., & Bürger, I. (2019). Adiabatic magnesium hydride system for hydrogen storage based on thermochemical heat storage: Numerical analysis of the dehydrogenation. *Applied Energy*, *236*, 1034–1048. https://doi.org/https://doi.org/10.1016/j.apenergy.2018.12.038

Ma, J., Liu, S., Zhou, W., & Pan, X. (2008). Comparison of hydrogen transportation methods for hydrogen refueling station. *Journal of Tongji University Natural Science*, *36*, 615–619.

Makaryan, I. A., Sedov, I. V., Salgansky, E. A., Arutyunov, A. V., & Arutyunov, V. S. (2022). A comprehensive review on the prospects of using hydrogen–methane blends: Challenges and opportunities. *Energies*, *15*(6), 1–27. https://doi.org/10.3390/en15062265

Maleki, E., Unal, O., Guagliano, M., & Bagherifard, S. (2021). The effects of shot peening, laser shock peening and ultrasonic nanocrystal surface modification on the fatigue strength of Inconel 718. *Materials Science and Engineering: A*, *810*, 141029. https://doi.org/https://doi.org/10.1016/j.msea.2021.141029

Massarweh, O., Al-khuzaei, M., Al-Shafi, M., Bicer, Y., & Abushaikha, A. S. (2023). Blue hydrogen production from natural gas reservoirs: A review of application and feasibility. *Journal of CO_2 Utilization*, *70*, 102438. https://doi.org/https://doi.org/10.1016/j.jcou.2023.102438

Modisha, P. M., Ouma, C. N. M., Garidzirai, R., Wasserscheid, P., & Bessarabov, D. (2019). The prospect of hydrogen storage using liquid organic hydrogen carriers. *Energy & Fuels*, *33*(4), 2778–2796. https://doi.org/10.1021/acs.energyfuels.9b00296

Navarro, R. M., Pena, M. A., & Fierro, J. L. G. (2007). Hydrogen production reactions from carbon feedstocks: fossil fuels and biomass. *Chemical Reviews*, *107*(10), 3952–3991.

Noh, H., Kang, K., & Seo, Y. (2023). Environmental and energy efficiency assessments of offshore hydrogen supply chains utilizing compressed gaseous hydrogen, liquefied hydrogen, liquid organic hydrogen carriers and ammonia. *International Journal of Hydrogen Energy*, *48*(20), 7515–7532. https://doi.org/https://doi.org/10.1016/j.ijhydene.2022.11.085

Pan, L., Sander, M. B., Huang, X., Li, J., Smith, M., Bittner, E., Bockrath, B., & Johnson, J. K. (2004). Microporous metal organic materials: Promising candidates as sorbents for hydrogen storage. *Journal of the American Chemical Society*, *126*(5), 1308–1309. https://doi.org/10.1021/ja0392871

Pellegrini, M., Guzzini, A., & Saccani, C. (2020). A preliminary assessment of the potential of low percentage green hydrogen blending in the Italian natural gas network. *Energies*, *13*(21). https://doi.org/10.3390/en13215570

Ratnakar, R. R., Gupta, N., Zhang, K., van Doorne, C., Fesmire, J., Dindoruk, B., & Balakotaiah, V. (2021). Hydrogen supply chain and challenges in large-scale LH_2 storage and transportation. *International Journal of Hydrogen Energy*, *46*(47), 24149–24168. https://doi.org/https://doi.org/10.1016/j.ijhydene.2021.05.025

Reitenbach, V., Ganzer, L., Albrecht, D., & Hagemann, B. (2015). Influence of added hydrogen on underground gas storage: A review of key issues. *Environmental Earth Sciences*, *73*(11), 6927–6937. https://doi.org/10.1007/s12665-015-4176-2

Rödl, A., Wulf, C., & Kaltschmitt, M. (2018). Assessment of selected hydrogen supply chains—Factors determining the overall GHG emissions. In C. Azzaro-Pantel (Ed.), *Hydrogen supply chains* (pp. 81–109). Academic Press.

Rosi, N. L., Eckert, J., Eddaoudi, M., Vodak, D. T., Kim, J., O'Keeffe, M., & Yaghi, O. M. (2003). Hydrogen storage in microporous metal-organic frameworks. *Science*, *300*(5622), 1127–1129. https://doi.org/10.1126/science.1083440

Safari, F., & Dincer, I. (2020). A review and comparative evaluation of thermochemical water splitting cycles for hydrogen production. *Energy Conversion and Management*, *205*, 112182.

Schlapbach, L., & Züttel, A. (2001). Hydrogen-storage materials for mobile atherm pplications. *Nature*, *414*(6861), 353–358.

Seyitoglu, S. S., Dincer, I., & Kilicarslan, A. (2017). Energy and exergy analyses of hydrogen production by coal gasification. *International Journal of Hydrogen Energy*, *42*(4), 2592–2600.

Smith, C., Hill, A. K., & Torrente-Murciano, L. (2020). Current and future role of Haber-Bosch ammonia in a carbon-free energy landscape. *Energy and Environmental Science*, *13*(2), 331–344. https://doi.org/10.1039/c9ee02873k

Smythe, N. C., & Gordon, J. C. (2010). Ammonia borane as a hydrogen carrier: Dehydrogenation and regeneration. *European Journal of Inorganic Chemistry*, *2010*(4), 509–521. https://doi.org/https://doi.org/10.1002/ejic.200900932

Sorensen, B., & Spazzafumo, G. (2018). *Hydrogen and fuel cells: Emerging technologies and applications*. Academic Press.

Sotoodeh, F., & Smith, K. J. (2010). Kinetics of hydrogen uptake and release from heteroaromatic compounds for hydrogen storage. *Industrial & Engineering Chemistry Research*, *49*(3), 1018–1026. https://doi.org/10.1021/ie9007002

Stiegel, G. J., & Ramezan, M. (2006). Hydrogen from coal gasification: An economical pathway to a sustainable energy future. *International Journal of Coal Geology*, *65*(3–4), 173–190.

Tong, L., Yuan, Y., Yang, T., Bénard, P., Yuan, C., & Xiao, J. (2021). Hydrogen release from a metal hydride tank with phase change material jacket and coiled-tube heat exchanger. *International Journal of Hydrogen Energy*, *46*(63), 32135–32148. https://doi.org/https://doi.org/10.1016/j.ijhydene.2021.06.230

Touloupakis, E., & Torzillo, G. (2019). Photobiological hydrogen production. In Francesco Calise, Massimo Dentice D'Accadia, Massimo Santarelli, Andrea Lanzini, and Domenico Ferrero (Eds.) *Solar hydrogen production* (pp. 511–525). Elsevier.

Ursua, A., Gandia, L. M., & Sanchis, P. (2011). Hydrogen production from water electrolysis: Current status and future trends. *Proceedings of the IEEE*, *100*(2), 410–426.

Wang, M., Wang, Z., Gong, X., & Guo, Z. (2014). The intensification technologies to water electrolysis for hydrogen production–A review. *Renewable and Sustainable Energy Reviews*, *29*, 573–588.

Wei, B., Xu, J., Sun, C., & Cheng, Y. F. (2022). Internal microbiologically influenced corrosion of natural gas pipelines: A critical review. *Journal of Natural Gas Science and Engineering*, *102*, 104581. https://doi.org/https://doi.org/10.1016/j.jngse.2022.104581

Yanxing, Z., Maoqiong, G., Yuan, Z., Xueqiang, D., & Jun, S. (2019). Thermodynamics analysis of hydrogen storage based on compressed gaseous hydrogen, liquid hydrogen and cryo-compressed hydrogen. *International Journal of Hydrogen Energy*, *44*(31), 16833–16840.

Yoon, S. J., Choi, Y.-C., & Lee, J.-G. (2010). Hydrogen production from biomass tar by catalytic steam reforming. *Energy Conversion and Management*, *51*(1), 42–47.

Zhang, L., Jia, C., Bai, F., Wang, W., An, S., Zhao, K., Li, Z., Li, J., & Sun, H. (2024). A comprehensive review of the promising clean energy carrier: Hydrogen production, transportation, storage, and utilization (HPTSU) technologies. *Fuel*, *355*, 129455. https://doi.org/https://doi.org/10.1016/j.fuel.2023.129455

Zheng, J., Cheng, H., Xiao, X., Chen, M., & Chen, L. (2019). Enhanced low temperature hydrogen desorption properties and mechanism of $Mg(BH_4)_2$ composited with 2D MXene. *International Journal of Hydrogen Energy*, *44*(44), 24292–24300. https://doi.org/https://doi.org/10.1016/j.ijhydene.2019.07.177

Chapter 2

Universal clean sustainable energy for future—hydrogen

Shashi Kant Verma

2.1 INTRODUCTION

Energy is a vital component of the infrastructure segment, and its accessibility has to be confirmed on a sustainable basis. Hydrogen is one of the renewable energy resources and therefore an environmentally friendly fuel and the greatest possible candidate to meet energy demands. Hydrogen is considered a renewable and clean energy source due to its ability to decarbonize various energy sectors. Hydrogen is widely considered a universal potential solution for future energy security as well as clean sustainable energy. Figure 2.1 shows hydrogen and the sustainable energy future. Figure 2.2 depicts the application of hydrogen in different energy sectors. Hydrogen and its byproducts play a vital role in the mitigation of greenhouse gases. Figure 2.3 represents the hydrogen transformation. Sebastian et al. (2023) provided a broad overview of advances in hydrogen storage and its production. They concluded that there are huge opportunities for hydrogen-related research and development, which is not limited to transportation services but also to the power sector and energy-efficient constructions. Guilbert and Vitale (2021) discussed the present and future applications of hydrogen addressing all the serious challenges encountered in energy sectors. Shenglin Su et al. (2022) presented a review based on different AI techniques in the field of green hydrogen. They compared the types of AI techniques and finally used artificial neural networks to predict the output power of the system. Jamal et al. (2023) discussed the recent trends, opportunities, and challenges in the field of hydrogen fuel cells. They concluded by identifying different research areas that paved the path toward highly energy-efficient and green hydrogen. The purpose of this chapter is to survey the state-of-the-art methodologies used in hydrogen research for the prediction of different parameters. This study comprises five main stages. The first stage is an introduction to hydrogen and sustainable energy future and the application of hydrogen in different energy sectors. Secondly, the section provides diverse properties of solid hydrogen as well as different states of solid hydrogen at cryogenic temperatures. The third stage introduces recent advancements in the field of hydrogen research. Some of the

 DOI: 10.1201/9781003537816-2

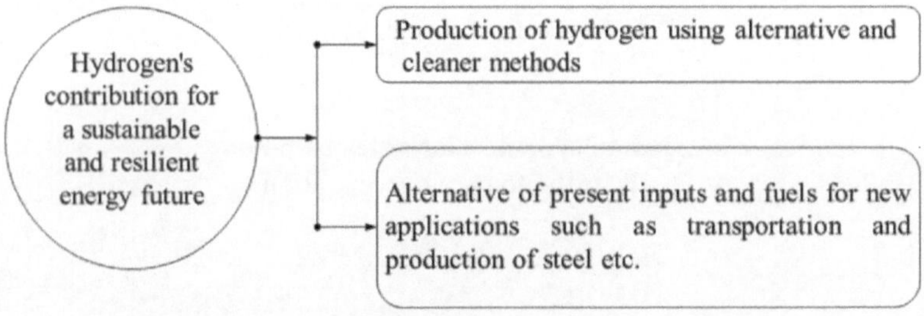

Figure 2.1 Hydrogen and the sustainable energy future.

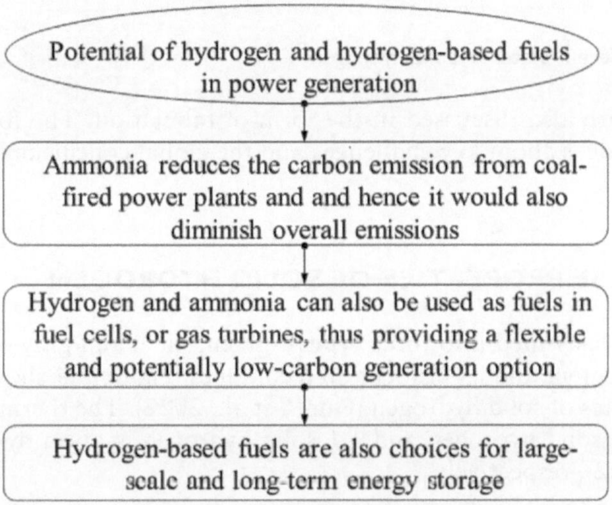

Figure 2.2 Application of hydrogen in different energy sectors.

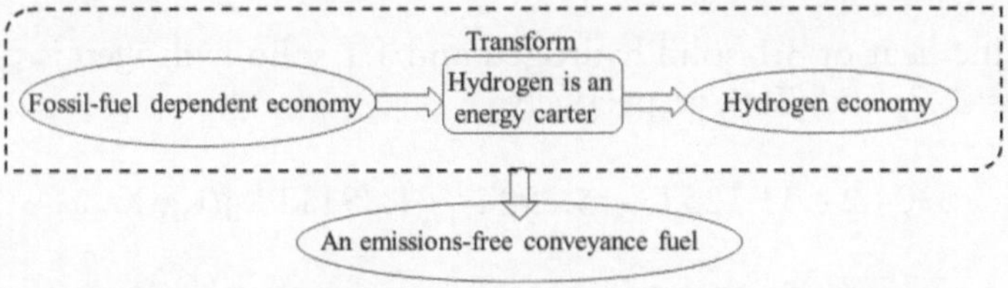

Figure 2.3 Hydrogen transformation.

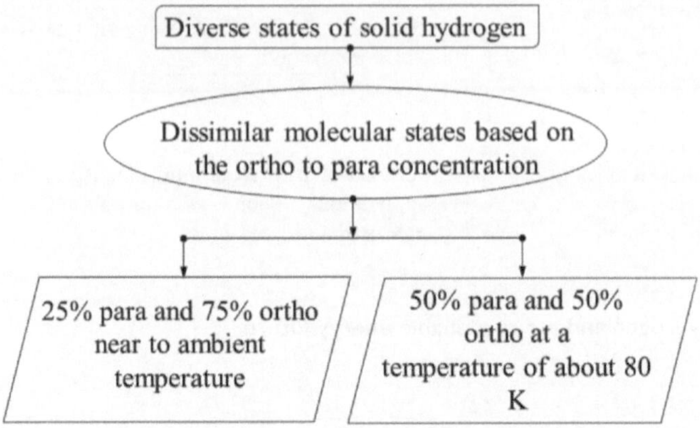

Figure 2.4 Different states of solid hydrogen.

gray areas are also discussed in the form of tabulation. The fourth stage gives some ideas about key challenges and the global conclusion grasped in the chapter.

2.2 DIVERSE PROPERTIES OF SOLID HYDROGEN

Hydrogen is classified into three types—green, blue, and gray—according to carbon content and its production techniques. Figure 2.4 shows the two different states of solid hydrogen (Gupta et al., 2022). The thermal conductivity of 3:1 solid hydrogen and 1:1 solid hydrogen is given by Equations (2.1) and (2.2), respectively.

$$K_{3:1}^{O} = -1.216 + 0.4237T - 0.0204T^2 - 0.0002T^3 \text{ W/m-K} \tag{2.1}$$

$$K_{1:1}^{n} = -8.7753 + 4.0558T - 0.4677T^2 + 0.0163T^3 \text{ W/m-K} \tag{2.2}$$

The specific heat of 3:1 solid hydrogen and 1:1 solid hydrogen is given by Equations (2.3) and (2.4), respectively.

$$C_{P.3:1}^{O} = 1304.2 - 33.125T - 15.293T^2 + 1.8911T^3 \text{ J/Kg-K} \tag{2.3}$$

$$C_{P,1:1}^{n} = 1420.8 - 86.89T - 7.7596T^2 + 1.5641T^3 \text{ J/Kg-K} \tag{2.4}$$

Gupta et al. (2022) defined onset shear stress and dynamic viscosity by using data-fitting techniques, which are dependent on cryogenic operating temperature. Equations (2.5) and (2.6) depict the model of onset shear stress and dynamic viscosity, respectively. Leachman's (2010) experimental data has been used for data fitting.

$$\tau_{O}^{L} = 0.03915\exp\left[-0.191\left(T_{O} - 10\right)\right] \text{ MPa} \tag{2.5}$$

$$\mu_{D}^{L} = 0.0163\exp\left[-0.212\left(T_{o} - 10\right)\right] \text{ MPa.s} \tag{2.6}$$

2.3 RECENT ADVANCEMENTS IN THE FIELD OF HYDROGEN RESEARCH

In this section, various predictive methodologies such as CFD, AI, ML, and CI have been discussed as shown in Table 2.1. Such methods are successfully implemented in different fields like refueling processes in hydrogen stations, fuel cells, small modular reactors, etc.

Table 2.1 Recent applied methodology in the field of hydrogen research

References	*Applied methodology*	*Remarks*
Ardabili et al. (2018)	Computational intelligence (CI)	Review of hydrogen production based on the computational intelligence approach, and comparison of different CI models based on the correlation coefficient.
Cheng and Rosentrater (2019)	Techno-economic analysis (TEA) simulation	Effect of substantial factor on net profit has been identified by sensitivity analysis followed by the TEA model.
Kim et al. (2022)	Machine learning (ML)	Prediction of important parameters of the small modular reactor for the production of hydrogen using an ML-based technique.
Park and Chae (2022)	ML	ML technique has been used for prediction of hydrogen properties.
Fayyazi et al. (2023)	ML	This paper reviews the use of the ML technique on vehicles powered by fuel cell hydrogen-powered.
Ebne-Abbasi et al. (2024)	Computational fluid dynamics (CFD)	They developed a CFD model for the refueling process of the hydrogen station. The model is validated through experimental data related to the refueling process.
Irham et al. (2024)	Bibliometric investigation of hydrogen energy	They have discussed the comprehensive review of hydrogen storage, including key challenges and future research direction.

Sazali (2020) has done a comprehensive review of developing technologies by hydrogen and discussed present technology improvements related to hydrogen research.

2.4 KEY CHALLENGES

The author has identified key challenges or barriers associated with hydrogen research as depicted in Figure 2.5. The existing protocols, morals, and acceptance criteria limit the understanding as well as the use of hydrogen. Salimi et al. (2022) discussed the significance of the value chain in the hydrogen economy in Japan. They suggested implementing a well-organized carbon tax that controls carbon release. Figure 2.6 shows the specific challenges associated with the transformation of hydrogen energy into viscous

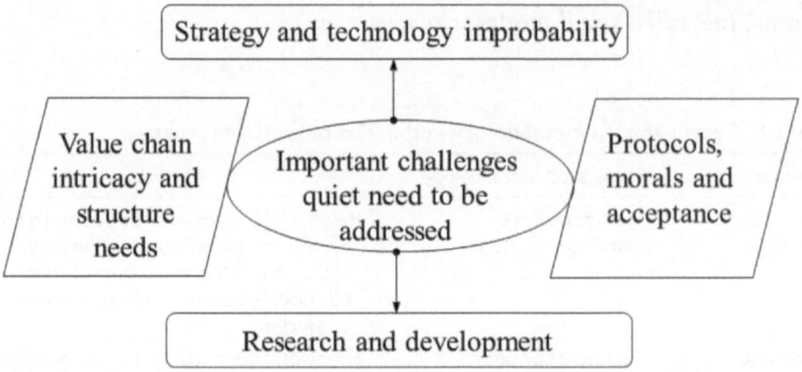

Figure 2.5 Challenges associated with hydrogen research.

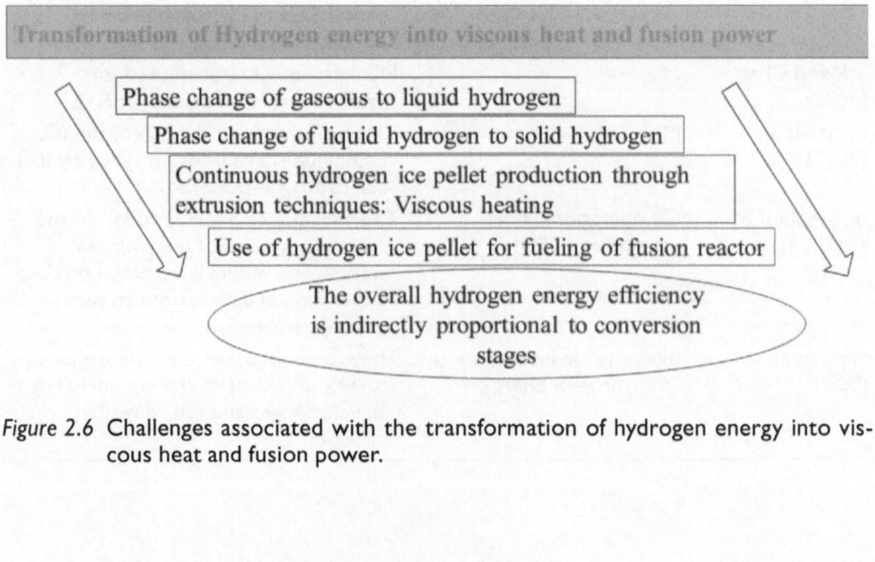

Figure 2.6 Challenges associated with the transformation of hydrogen energy into viscous heat and fusion power.

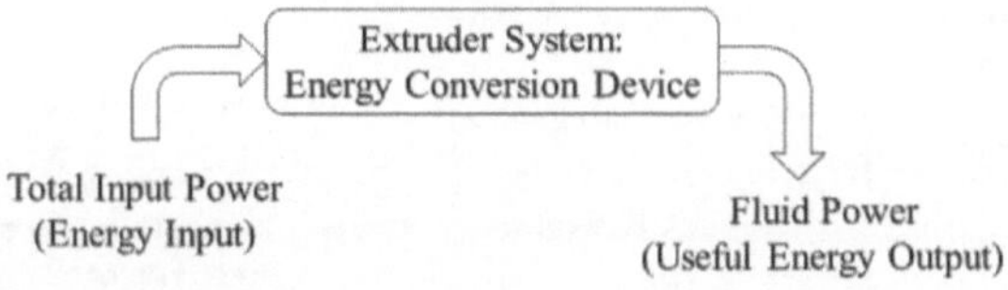

Figure 2.7 Energy conversion in an extruder.

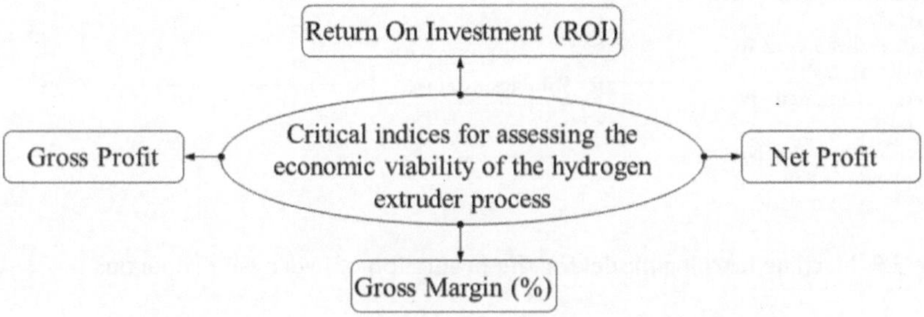

Figure 2.8 Various indices for assessing the economic viability of the hydrogen extruder process.

heat and fusion power in fusion research. This indicates that the overall hydrogen energy efficiency (HEE) can be enhanced through the optimization of conversion stages. This study is vital from a thermal management point of view.

An extruder-type energy conversion device is denoted schematically in Figure 2.7. This device converts mechanical to fluid power. The efficiency of conversion plays an important role for any energy conversion device. The efficiency of energy conversion is an important aspect for the benefits of the nation for energy security, environment, and economics (SEE). The development of an energy-efficient extruder system for hydrogen ice extrusion is one of the key challenges at cryogenic temperatures. From a thermal management point of view, this investigation is vital. Shear-assisted techno-economic analysis of hydrogen extrusion is one of the promising research areas in the future. Figure 2.8 represents the indices for assessing the economic viability of the hydrogen extruder process (Cheng & Rosentrater, 2019). Environmental sustainability and techno-economic viability analysis of the hydrogen extruder system is one of the gray areas in future challenges.

Figure 2.9 represents the machine learning model for the prediction of hydrogen properties. Various thermophysical properties of solid hydrogen such as density, viscosity, and thermal conductivity are required for numerical studies. The reliability of results depends on accuracy of the thermophysical properties of hydrogen at a particular temperature.

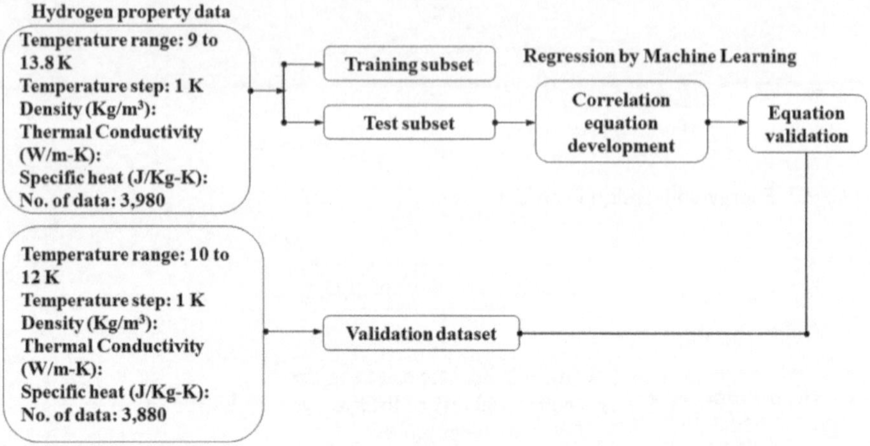

Figure 2.9 Machine learning model for the prediction of hydrogen properties.

2.5 CONCLUSION

In this chapter, the hydrogen properties along with its different states were discussed. Challenges associated with hydrogen research were also discussed. The AI-assisted optimization techniques combined with increased research and development and skilled manpower will be important to green hydrogen. It plays a vital role in the world decarburization effort. Recently, computational intelligence and optimization on hybrid techniques based on predictive methodology are some of the choices for hydrogen research, along with CFD and experimental techniques. The optimization of hydrogen conversion stages is important from a thermal management point of view. Sensible and intelligent use of hydrogen resources for energy efficiency is one of the steps toward a sustainable energy system.

REFERENCES

Ardabili, S., Najafi, B., Shamshirband, S., Minaei-Bidgoli, B., Deo, R. C., & Chau, K. W. (2018, January 1). Computational intelligence approach for modeling hydrogen production: A review. *Engineering Applications of Computational Fluid Mechanics*. https://doi.org/10.1080/19942060.2018.1452296.

Cheng, M., & Rosentrater, K. A. (2019, April 26). Techno-economic analysis of extruding-expelling of soybeans to produce oil and meal. *Agriculture (Basel)*. https://doi.org/10.3390/agriculture9050087.

Ebne-Abbasi, H., Makarov, D., & Molkov, V. (2024, January 1). CFD model of refuelling through the entire equipment of a hydrogen refuelling station. *International Journal of Hydrogen Energy*. https://doi.org/10.1016/j.ijhydene.2023.12.056.

Fayyazi, M., Sardar, P., Thomas, S. I., Daghigh, R., Jamali, A., Esch, T., Kemper, H., Langari, R., & Khayyam, H. (2023, March 15). Artificial intelligence/machine learning in energy management systems, control, and optimization of hydrogen fuel cell vehicles. *Sustainability*. https://doi.org/10.3390/su15065249.

Guilbert, D., & Vitale, G. (2021, December 20). Hydrogen as a clean and sustainable energy vector for global transition from fossil-based to zero-carbon. *Clean Technologies*. https://doi.org/10.3390/cleantechnol3040051.

Gupta, V., Gangradey, R., Mukherjee, S., Verma, S., Dewasi, A., Mishra, J., Panchal, P., & Nayak, P. (2022, June 1). Validation of 3-D non-isothermal CFD simulation with experimental results for single-screw hydrogen extruder. *Cryogenics*. https://doi.org/10.1016/j.cryogenics.2022.103471.

Irham, A., Roslan, M., Ker, P. J., Hannan, M. A., & Mahlia, T. (2024, April 1). Hydrogen energy storage integrated grid: A bibliometric analysis for sustainable energy production. *International Journal of Hydrogen Energy*. https://doi.org/10.1016/j.ijhydene.2024.03.235.

Jamal, T., Shafiullah, G., Dawood, F., Kaur, A., Arif, M. T., Pugazhendhi, R., Elavarasan, R. M., & Ahmed, S. F. (2023, November 1). Fuelling the future: An in-depth review of recent trends, challenges and opportunities of hydrogen fuel cell for a sustainable hydrogen economy. *Energy Reports*. https://doi.org/10.1016/j.egyr.2023.09.011.

Kim, J., Rweyemamu, M., & Purevsuren, B. (2022, September 1). Machine learning-based approach for hydrogen economic evaluation of small modular reactors. *Science and Technology of Nuclear Installations*. https://doi.org/10.1155/2022/9297122.

Leachman, J., Pfotenhauer, J. M., & Nellis, G. F. (2010). Thermophysical property modeling of a hydrogenic pellet production system. In *Transactions of the cryogenic engineering conference*. Advances in Cryogenic Engineering. AIP Conf. Proc. 1218, 1570–1577 (2010) https://doi.org/10.1063/1.342233 .

Park, B. H., & Chae, C. K. (2022, January 1). Development of correlation equations on hydrogen properties for hydrogen refueling process by machine learning approach. *International Journal of Hydrogen Energy*. https://doi.org/10.1016/j.ijhydene.2021.11.053.

Salimi, M., Hosseinpour, M., & Borhani, T. N. (2022, August 21). The role of clean hydrogen value chain in a successful energy transition of Japan. *Energies*. https://doi.org/10.3390/en15166064.

Sazali, N. (2020, July 1). Emerging technologies by hydrogen: A review. *International Journal of Hydrogen Energy*. https://doi.org/10.1016/j.ijhydene.2020.05.021.

Sebastian, S. P., Wijewardane, S., & Srinivasan, S. S. (2023, December 1). Recent advances in hydrogen production, storage, and fuel cell technologies with an emphasis on inventions, innovations, and commercialization. *Solar Compass*. https://doi.org/10.1016/j.solcom.2023.100065

Su, S., Yan, X., Agbossou, K., Chahine, R., & Zong, Y. (2022, March). Artificial intelligence for hydrogen-based hybrid renewable energy systems: A review with case study. *Journal of Physics: Conference Series*, Volume 2208, The 4th International Symposium on Hydrogen Energy and Energy Technologies 2021 (HEET 2021) 19/11/2021–20/11/2021 Online, Shenglin Su et al 2022 J. Phys.: Conf. Ser. 2208 012013DOI 10.1088/1742-6596/2208/1/012013.

Chapter 3

Hydrogen energy

Addressing challenges and exploring future prospects

Ajay Kumar Shukla, Ratindra Gautam, Shivani Chaudhary, Anoop Kr. Srivastava, Anil Kumar, Sindhu Singh, and Pramod Kumar Maurya

3.1 INTRODUCTION

Hydrogen has surfaced as a viable substitute for conventional fossil fuels, possessing the ability to completely transform the energy sector. Hydrogen, an effective and environmentally friendly energy source that is also called the "fuel that will power the future," possesses the capacity to mitigate the rate of global warming and reduce dependence on limited resources. To reach its full potential, however, the widespread use of hydrogen energy must overcome a number of obstacles despite its great promise (Ahad, Bhuiyan et al., 2023). The most common element in the universe, hydrogen, is extremely adaptable and can be created in a number of ways, such as through the electrolysis of water, natural gas, biomass, and renewable energy sources like solar and wind power. Fuel cells convert it into electricity or permit direct combustion, rendering it exceedingly versatile for heating, commercial processes, conveyance, and electricity generation, among other applications (Bhandari & Shah, 2021).

In contemporary times, the reliance on petroleum fuels in energy and transportation systems is unsustainable. Forecasts by the International Energy Agency (IEA) indicate a projected increase in global energy demand exceeding 50% by 2030 (Leaver, 2019). To mitigate anthropogenic climate impact, there is a pressing need to curb the escalating greenhouse gas (GHG) emissions. Hydrogen emerges as a promising solution, as it can be industrially produced in substantial quantities and is deemed crucial in the energy sector (Pivovar et al., 2018). Hydrogen is considered zero-emission since it transfers chemical energy into electrical energy in fuel cells while only producing water as a waste product. In accordance with the goal of the Paris Agreement to restrict the increase in global temperature to no more than 2°C, efforts are currently being made to attain net-zero emissions by 2050 (Agreement, 2015). Goldman Sachs' analysis highlights the increasing importance of clean hydrogen and its crucial role in reducing up to 45% of global human-caused emissions (Vu et al., 2021).

DOI: 10.1201/9781003537816-3

The chemical, transportation, and power-generating sectors might all benefit from using harmless hydrogen as an energy carrier since it produces no carbon dioxide. Launched on June 7, 2021, the Hydrogen Shot program seeks to slash the price of pure hydrogen to USD 1 per kilogram within a decade, an 80% reduction (McNaul et al., 2023). The concept of a "hydrogen economy" encompasses various facets including production, storage, infrastructure, emissions, and supply chain management, delineating critical areas for hydrogen research. Problems with green hydrogen generation, transportation, safety, combustion, performance, life cycle evaluation, and power-to-gas technologies have dominated recent research.

The establishment of hydrogen fuel on a commercial level remains fraught with significant obstacles, notwithstanding scientific advancements. Material compatibility stands out as a significant hurdle due to the unique properties of hydrogen gas. Fuel cells, sensors, combustion engines, transportation, and storage of hydrogen need distinctive engineering concerns in contrast to natural gas. Efforts are underway to develop or identify materials suitable for hydrogen applications. Hydrogen embrittlement is a growing problem that is causing material deterioration in many different businesses, including those dealing with metals processing, mechanics, chemicals, and refineries. The manufacturing, transportation, storage, and distribution systems are all affected by this phenomenon, which may be classified into three types: hydrogen reaction breakdown, internal reversible hydrogen breakdown, and hydrogen ambient breakdown (Gray et al., 1978). Addressing these material challenges and advancing research in hydrogen compatibility are imperative for realizing the full potential of hydrogen as a clean energy solution (Ahad, Yazdan et al., 2023).

3.2 HYDROGEN AND ITS SOURCES

Green hydrogen, produced via electrolysis driven by renewable sources like wind or solar, splits water into oxygen and hydrogen using an electric current. Stored hydrogen is directed to a fuel cell, combining with oxygen to generate electricity, with water as the sole byproduct. This process emits zero CO_2, offering a clean, sustainable energy solution. Two main electrolysis techniques were found by Marcelo and Dell'Era (2008): alkaline electrolyzers and polymer electrolyte membranes (PEMs). The stated efficiency of both technologies spanned 52% to 85% (Marcelo & Dell'Era, 2008). The rapid expansion of power-to-gas plants around the globe is indicative of the encouraging progress in the field of electrolysis, which converts electrical energy to chemical energy. Electrochemical applications are well known for proton exchange membranes (PEMs), alkaline electrolyzers, solid oxide electrolysis cells (SOECs), and molten carbonate electrolysis cells (MCECs).

For the production of hydrogen, several methods exist that may be used to create hydrogen from various sources, such as water, fossil fuels, and

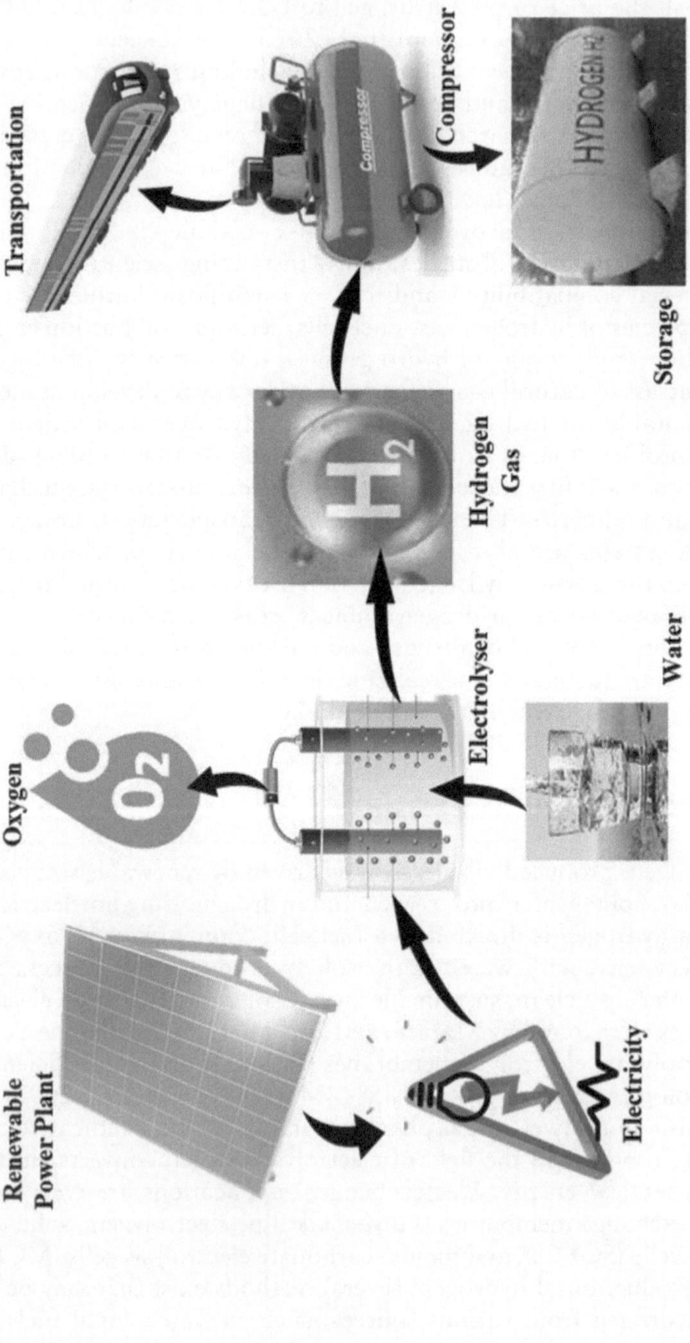

Figure 3.1 Illustration depicting the schematic diagram of the process for producing green hydrogen.

Table 3.1 Sources and techniques for obtaining various forms of hydrogen

Source of H_2	*H_2 type*	*Emission type*	*Method*
Renewable/nuclear and biomass	Green H_2	O_2	Electrolysis
Nuclear energy	Pink H_2	Nuclear waste	Electrolysis
Electricity from grid	Pure H_2		Electrolysis
Natural gas	Gray H_2	CO_2	SMR
Coal	Blue H_2	CO_2	Gasification (with CCUS)
Coal	Brown H_2	CO_2	Gasification (without CCUS)
Coal	Black H_2	CO_2	Gasification (without CCUS)

biomass. But up until recently, fossil fuels were the mainstay in the generation of hydrogen. As the cost of renewable electricity continues to decline, the water electrolysis approach is becoming increasingly attractive for producing green hydrogen. Natural gas accounts for the production of over 76% of the world's hydrogen gas. Coal accounts for 23% of the remainder, while electrolysis accounts for 1%. There has been ongoing work to reduce the quantity of CO_2 emissions during hydrogen synthesis, which has been a concern for some time. Projects that collect, use, and store carbon dioxide (CCUS) are useful in this respect. Numerous types of routes for hydrogen production are outlined in Table 3.1.

Hydrogen has zero-carbon emissions and can be used as a fuel in the chemical, power, and transportation industries. The Energy Earth Shot, also known as the Hydrogen Shot, was the first goal that was introduced on June 7, 2021, with a purpose to reduce the cost of clean hydrogen by 80% to just $1 per kilogram over a ten-year period (Gray et al., 1978). By 2022, the hydrogen gas market is projected to reach a value of approximately USD 154.74 billion worldwide and continue to grow (Binder et al., 2018; Kar et al., 2023).

3.3 HYDROGEN PRODUCTION

The current industry feedstock has the biggest demand potential going forward; in the United States, this potential will rise, while in Europe, it will fall. The transportation sector, where demand has increased dramatically in both the United States and Europe, is where the 2030 and 2050 potentials diverge the most (Qazi, 2022). Green hydrogen is created by electrolyzing water using renewable energy sources like solar or wind power. The technique of electrolysis involves utilizing electrodes and an electrical current to divide a water molecule into hydrogen and oxygen. Fuel cells receive hydrogen that has been stored in designated tanks. Then, it re-bonds with atmospheric oxygen to produce energy. As a result, the process only produces water as a byproduct, creating a clean, sustainable system that produces electricity without emitting any CO_2. Marcelo and Dell'Era (2008) distinguished between two main types of electrolysis processes in their investigation:

Alkaline electrolyzers and polymer electrolyte membrane electrolyzers (PEMEs) are the first two types. Binder et al. (2018) showed that the operational efficiency of these electrolyzers ranges from 52% to 85%. One promising technical development is the process of electrolysis, which turns electricity into chemical energy. This explains why power-to-gas facilities are proliferating throughout the world. Alkaline electrolyzers, proton exchange membrane electrolyzers (PEMEs), and other electrolyzer types including molten carbonate electrolysis cells (MCECs) and solid-oxide electrolysis cells (SOECs) have drawn a lot of attention for their electrochemical applications.

3.3.1 Production of hydrogen from natural gas: Obstacles and prospects

Steam methane reforming is the most widely used technique for producing hydrogen from natural gas resources (SMR). Currently, there are three techniques for producing hydrogen from natural gas supplies. These are autothermal reforming (ATR), partial oxidation of methane (POM), and selective metal redox (SMR). SMR is the most widely used method of producing H_2 on a large scale (Dawood et al., 2020; Kalamaras & Efstathiou, 2013). Currently, the SMR process releases CO_2 into the atmosphere. It can be used, nevertheless, as a byproduct in the packaging and processing of food. In POM, oxygen in the air serves as an oxidant, but in SMR, water serves as both a source of hydrogen and an oxidant. The ATR approach combines both SMR and POM components. With no need for oxygen, SMR is the most advanced industrial process. On the other hand, CO_2 emissions are very high. The CO_2 generated needs to be captured and stored in order to decarbonize these methane-based activities. In this sense, carbon capture, utilization, and storage (CCUS) initiatives are beneficial. It is a method that keeps carbon dioxide emissions out of the atmosphere by capturing them from various sources. Pumping CO_2 into geological reservoirs, such as exhausted oil and gas fields, is a component of CCUS projects (Ahad, Bhuiyan et al., 2023; Bartlett & Krupnick, 2020). Since SMR has a low current cost and is sustainable, it has gained popularity in the generation of H_2. To make SMR a more effective method of manufacturing H_2, however, issues with the automated control system and feedstock management system must be resolved.

3.3.2 Opportunities and challenges in electrolysis

Less than 0.1% of the world's devoted hydrogen production is produced globally by the electrochemical process of electrolysis (Collodi et al., 2017;

Lee et al., 2017). Primarily, the three main kinds of electrolysis include solid-oxide electrolysis cells (SOECs), alkaline electrolysis, and proton exchange membrane (PEM) electrolysis. Since 1920, the fertilizer and chlorine industries have been producing hydrogen by alkaline electrolysis, an established and commercial method. In order to get around some of the operational issues with alkaline electrolysis, General Electric developed proton exchange membrane (PEM) electrolysis in 1960. Solid-oxide electrolysis cells (SOECs) are the least developed electrolysis technology; they have not yet been brought to market.

The integration of heat from nuclear power plants, geothermal or solar heat, and industrial processes makes the integration of heat into hydrogen production an intriguing prospect. Heat produced by nuclear power stations at 300°C can be utilized to produce steam and electricity for solid-oxide electrolysis cells (SOECs).

Electricity and water are needed for electrolysis. One kilogram of H_2 requires about 9 L of water, and 8 kg of oxygen are created as a byproduct. Actually, since the SMR process uses about twice as much water as other processes, freshwater supplies may be a problem in water-stressed locations. In coastal regions, seawater may be a possible supply for electrolysis. However, using seawater directly could result in corrosive damage and chlorine generation. Furthermore, costly electrode catalysts such as iridium, platinum, and membrane materials are required for PEM electrolysis. Once again, the electrode lifespan is likewise lower than that of alkaline electrolyzers (Marcelo & Dell'Era, 2008). A fuel cell is comparatively wet in comparison to the electrolysis process. Polymer electrodes absorb water, causing it to swell. Without the use of thick membranes ranging from 175 to 250 μm, polymer strands are wider apart and mechanically weaker at excessive swelling and operating conditions at 30 bar (430 psi) or more.

Despite the aforementioned constraints, fresh chances to close the gaps arise every day. In order to increase the efficiency of electrolysis cells, thinner membrane materials (50–60 μm) can be used. Additionally, they help to improve mechanical strength (Palucka & Ingram, 2019).

Scholars are examining various polymer compositions in an effort to maximize ionic conductivity and minimize water absorption. Hydrocarbons have been considered as a possible substitute for Nafion's fluorine-based backbone. Enhancements are also required for electrolysis catalysts, which are platinum on the hydrogen side and iridium on the oxygen side (Palucka & Ingram, 2019). Research has to be done on the selection of materials that would work with the temperature ranges of nuclear energy heat sources. In the near future, small modular reactors will also be investigated as a potential addition to SOEC electrolysis. Long-term, sophisticated nuclear reactors will also be a desirable alternative (Palucka & Ingram, 2019).

3.3.3 Producing hydrogen from coal: Difficulties and prospects

The chemical and fertilizer industries use a well-established method called gasification to manufacture hydrogen from coal during the ammonia production process (Fasihi et al., 2019; Muradov, 2017). However, the creation of hydrogen from coal is most likely the cause of the increase in CO_2 emissions. Additionally, using CCUS can result in hydrogen with a low hydrogen-to-carbon ratio and significant amounts of contaminants (sulfur, nitrogen, and minerals) in the feedstock (Fasihi et al., 2019; Muradov, 2017). The gasifier unit's released heat needs to be used more effectively for heating and electricity generation. However, due to material limitations, cooling hot gas above 1400°C with heat exchangers is equally difficult. The absence of recognized standard codes presents another difficulty. To mitigate greenhouse gas emissions, it is necessary to reduce the carbon content carried in syngas.

Sulfur is present in coal, and during the gasification process, sulfur oxide is produced. Acid rain and environmental damage are caused by sulfur oxide. For a sustainable environment, a sulfur reduction system must be included in the coal gasification process. In order to devolatilize coal, various fuel types might be utilized. Moreover, coal segregation can incorporate the devolatilization temperatures. The plasma co-gasification method can be utilized to further use waste heat and hydrogen for power production, storage, and cooling applications.

3.3.4 Hydrogen generation from biomass: Difficulties and prospects

Hydrogen production from biomass has been accomplished in a variety of ways. The biochemical route is one of the routes where biogas is produced by the interaction of an organic substance and a microbe. The process, which involves a mixture of acids, alcohols, and gases, is sometimes referred to as anaerobic digestion. Hydrogen is produced through the breakdown of organic materials by microorganisms like bacteria. Organic matters include refined sugars, unprocessed biomass sources such as maize stover, and even wastewater. These techniques are frequently known as "dark fermentation" approaches because of their lack of need for light.

The microorganisms directly generate hydrogen in direct hydrogen fermentation. These bacteria are capable of breaking down complex compounds in a variety of ways. Enzymes can combine the byproducts of some of these routes to create hydrogen.

Efforts are underway to explore renewable energy as a replacement for fossil fuels. Biohydrogen, with its clean and high energy content, holds promise as a future fuel. Sustainability is a key challenge in establishing a hydrogen economy. Biohydrogen offers renewable biofuel potential and can substitute current hydrogen production methods reliant on fossil fuels. Algal biomass conversion is a promising source of biohydrogen due to its abundance, cleanliness, and renewability. While still in early development compared to bioethanol and biodiesel, various technologies for algal hydrogen production exist, showing potential at laboratory and pilot scales for full-scale implementation. This chapter discusses advancements in biohydrogen development, covering biological pathways, bioreactor designs and operations, and technoeconomic evaluations, while also outlining challenges and prospects for biohydrogen production. Despite the fact that there are numerous biomass gasification plants worldwide, the technology is still in its infancy. The issue of catalyst poisoning brought on by tar formation has not been fully resolved. Whatever the method of generation, additional processing is required to separate hydrogen from the generated gas. The inability to get inexpensive biomass further hinders the large-scale synthesis of hydrogen derived from biomass (Ahad, Bhuiyan et al., 2023). Hydrogen can be more easily stored, transported, and used by converting it into hydrogen-based fuels and feedstocks. Ammonia, synthetic hydrocarbons, and synthetic methanol are a few examples (Fasihi et al., 2019; Philibert, 2017).

Efforts are underway to explore renewable energy as a replacement for fossil fuels. Biohydrogen, with its clean and high energy content, holds promise as a future fuel. Sustainability is a key challenge in establishing a hydrogen economy. Biohydrogen offers renewable biofuel potential and can substitute current hydrogen production methods reliant on fossil fuels. Algal biomass conversion is a promising source of biohydrogen due to its abundance, cleanliness, and renewability. While still in early development compared to bioethanol and biodiesel, various technologies for algal hydrogen production exist, showing potential at laboratory and pilot scales for full-scale implementation (Show et al., 2018).

3.3.5 Methane splitting: Difficulties and possibilities

The current three-phase plasma is alternated in the methane-splitting process, which was first introduced around 1990. It produces hydrogen and solid carbon without releasing CO_2 from natural gas by using electricity as the energy source and methane as the feed (Fasihi et al., 2019; Philibert, 2017). When producing the same amount of hydrogen, methane splitting requires three to five times less electricity compared to electrolysis. The method has drawbacks, though, as it lowers overall efficiency and necessitates high-temperature plasma and a large temperature loss (Fasihi et al., 2019; Philibert, 2017). The industrial plant being built by Monolith Materials in Nebraska has a lower overall efficiency than using natural gas directly in the power plant. Monolith Materials currently runs a prototype methane-splitting plant in California. Even though it uses more natural gas than electrolysis, this method can help lower the pollutants from gas burning, and it may also provide additional income from the sale of carbon black, which is used in rubber, tires, printers, and plastics (Bazzanella & Ausfelder, 2017; Fulcheri, 2018).

3.4 UTILIZING CO_2 CAPTURING TECHNOLOGY TO PRODUCE HYDROGEN

Advanced technological processes for producing gray hydrogen are raising concerns about its carbon dioxide emissions. However, the system's use of carbon capture, utilization, and storage (CCUS) might reduce CO_2 emissions by up to 90%. Many obstacles must be removed in order to maximize demand for an efficient, ecologically friendly pathway for the generation of hydrogen worldwide. Significant CO_2 emissions have been noted when generating H_2 from various sources. Further study is needed to implement CO_2 capture technology for producing H_2 from coal and natural gas sources, even though a number of CO_2 capturing systems have been investigated for the SMR and ATR processes in an effort to increase efficiency at a reasonable cost. According to research, the cost of producing H_2 from natural gas in the region, including OPEX and CAPEX, was less than USD 1/kg of H_2 (Bartlett & Krupnick, 2020).

Hydrogen functions as an energy carrier, facilitating the transportation and storage of energy in a usable form. As previously stated, hydrogen may be produced by several methods such as natural gas, biomass, renewable energy sources, hydroelectric power, and nuclear energy. Hydrogen is a feasible energy source due to its diverse supply sources. Furthermore, it can be manufactured in medium-sized, small-sized, or decentralized facilities located in close proximity to the user, such as permanent power sites or refueling stations. As a result, a lot of research is being done to create hydrogen that is economically viable from various sources. In addition, methods for absorbing CO_2 emissions should be developed in order to boost the effectiveness of producing H_2 from coal and natural gas sources at a cheap cost. Both the partial oxidation and steam methane reforming processes have seen a noticeable advancement in technology, and both have been applied commercially to produce hydrogen. In order to guarantee that greenhouse gas emissions from the generation of hydrogen based on coal are as low as possible, it is also necessary to establish delivery strategies and upgrade infrastructure. To further lower the cost of biomass resources utilized in hydrogen generation, improving biomass growth, harvesting, and handling is also required.

It has long been discussed how to use carbon sequestration technology to slow down global warming while producing hydrogen. Monoethanolamine (MEA) is used in chemical absorption to capture CO_2

during post-combustion. With this technology, the exhaust of boilers that burn coal or gas is converted into high-purity CO_2. After the regenerating absorbent is recycled back to the absorber, the CO_2 is compressed and dried for transportation (usually at 100–150 bar). Moreover, CO_2 from flue gas can be extracted using membranes, adsorption, and low-temperature distillation. At standard pressures and temperatures, the physical adsorption of CO_2 at a solid surface (zeolite) requires a significant amount of energy. When compared to chemical absorption, using commercially available polymeric membranes requires comparatively high energy needs and CO_2 avoidance costs because of the low driving force caused by the low CO_2 partial pressure in flue gas (Ishibashi et al., 1999; Pradier & Pradier, 2014).

3.5 EXPLORING FUTURE PROSPECTS

Hydrogen production holds immense promise as a pivotal component in our journey toward a sustainable energy future. The outlined future prospects highlight key avenues for its development: Firstly, the shift toward green hydrogen production, powered by renewable sources, underscores a vital pathway to minimize reliance on fossil fuels and mitigate carbon emissions. Secondly, advancements in technology and economies of scale are poised to drive down production costs, enhancing competitiveness with conventional fuels. Because of its abundant supply and flexible manufacturing, hydrogen is a major green energy source that is attracting a lot of interest from the general public, businesses, and governments due to its potential as a fuel in the future. Substantial investment flows into hydrogen research, focusing on transmission, storage, fuel cells, and safety measures. Material compatibility and safety in hydrogen facilities remain critical concerns. Challenges, roadmaps, and scenarios serve as guiding factors for long-term planning and technological outlooks. Strategic visions ensure that activities and policies are coordinated to effectively address both societal and ecological aspects. The fusion of hydrogen's potential with advancements in materials is transforming the sustainable energy environment. Collective perspectives on potential benefits are crucial. This chapter explores challenges in hydrogen technology comprehensively, including production with CO_2 capture, material interactions, and embrittlement in storage, transmission, and advancements in fuel cells. Industrial safety considerations are also addressed. This chapter provides a panoramic view of the evolving hydrogen landscape. Moreover, the expansion of hydrogen applications across industries and transportation, coupled with its role in energy storage and grid balancing, signifies its multifaceted potential. International collaboration and infrastructure development are imperative to support widespread adoption, with policy support crucial for creating an enabling environment. Overall, the promising outlook for hydrogen production reflects its pivotal role in decarbonization, energy security, and sustainable economic growth. Continued innovation, investment, and collaboration are essential to fully harness its benefits as a clean energy solution.

REFERENCES

Agreement, P. (2015). *Paris agreement.* Paper presented at the report of the conference of the parties to the United Nations framework convention on climate change (21st session, 2015: Paris). Retrieved December (Vol. 4, No. 2017, p. 2). HeinOnline, Getzville, NY.

Ahad, M. T., Bhuiyan, M. M. H., Sakib, A. N., Becerril Corral, A., & Siddique, Z. J. M. (2023). An overview of challenges for the future of hydrogen. *Materials, 16*(20), 6680.

Ahad, M. T., Yazdan, M. M. S., Selvaratnam, T., Siddique, Z., & Rahman, A. (2023). Biohydrogen from biomass fermentation pathway and economic aspects. In Abu Yousuf and Elia Tomás-Pejó (Eds.), *Microbiology of green fuels* (pp. 97–114). CRC Press.

Bartlett, J., & Krupnick, A. J. R. f. t. F. (2020). Decarbonized hydrogen in the US power and industrial sectors: Identifying and incentivizing opportunities to lower emissions. *Resources for the Future*, 20–25.

Bazzanella, A., & Ausfelder, F. (2017). *Low carbon energy and feedstock for the European chemical industry: technology study.* DECHEMA, Gesellschaft für Chemische Technik und Biotechnologie eV.

Bhandari, R., & Shah, R. R. J. R. E. (2021). Hydrogen as energy carrier: Techno-economic assessment of decentralized hydrogen production in Germany. *Renewable Energy, 177*, 915–931.

Binder, M., Kraussler, M., Kuba, M., Luisser, M., & Rauch, R. J. I. P., France. (2018). *Hydrogen from biomass gasification.* IEA Bioenergy.

Collodi, G., Azzaro, G., Ferrari, N., & Santos, S. J. E. P. (2017). Techno-economic evaluation of deploying CCS in SMR based merchant H2 production with NG as feedstock and fuel. *Energy Procedia, 114*, 2690–2712.

Dawood, F., Anda, M., & Shafiullah, G. J. I. J. o. H. E. (2020). Hydrogen production for energy: An overview. *International Journal of Hydrogen Energy, 45*(7), 3847–3869.

Fasihi, M., Efimova, O., & Breyer, C. J. J. o. c. p. (2019). Techno-economic assessment of CO2 direct air capture plants. *Journal of Cleaner Production, 224*, 957–980.

Fulcheri, L. (2018). *Direct decarbonization of methane by thermal plasma for the co production of hydrogen and carbon nanostructure.* Paper presented at the HTPP 15, 15th International High-Tech Plasma Processes Conference.

Gray, H., Nelson, H., Johnson, R., McPherson, W., Howard, F., & Swisher, J. J. I. J. o. H. E. (1978). Potential structural material problems in a hydrogen energy system. *International Journal of Hydrogen Energy, 3*(1), 105–118.

Ishibashi, M., Otake, K., Kanamori, S. J. G. G. C. T., & Elsevier Science Ltd.: Amsterdam, T. N. (1999). *Study on CO2, removal technology from flue gas of thermal power plant* (p. 95). Elsevier Science Ltd.

Kalamaras, C. M., & Efstathiou, A. M. (2013). Hydrogen production technologies: Current state and future developments. In Y. Al-Assaf and A. Poullikkas (Eds.), *Conference papers in science* (Vol. 2013, No. 1, p. 690627). Hindawi Publishing Corporation.

Kar, S. K., Sinha, A. S. K., Bansal, R., Shabani, B., Harichandan, S. J. W. I. R. E., & Environment. (2023). Overview of hydrogen economy in Australia. *Wiley Interdisciplinary Reviews: Energy and Environment*, *12*(1), e457.

Leaver, J. (2019). *International Energy Agency (IEA) strategic initiatives and activities for hydrogen*. https://hdl.handle.net/10652/4741

Lee, D.-Y., Elgowainy, A. A., & Dai, Q. (2017). *Life cycle greenhouse gas emissions of by-product hydrogen from chlor-alkali plants*. Argonne National Lab (ANL).

Marcelo, D., & Dell'Era, A. J. I. J. o. H. E. (2008). Economical electrolyser solution. *International Journal of Hydrogen Energy*, *33*(12), 3041–3044.

McNaul, S., White, C., Wallace, R., Warner, T., Matthews, H. S., Ma, J. N., . . . Henriksen, M. (2023). *Strategies for achieving the DOE hydrogen shot goal: Thermal conversion approaches* (No. DOE/NETL-2023/3824). National Energy Technology Laboratory (NETL).

Muradov, N. J. I. J. o. H. E. (2017). Low to near-zero CO_2 production of hydrogen from fossil fuels: Status and perspectives. *International Journal of Hydrogen Energy*, 42(20), 14058–14088.

Palucka, T., & Ingram, B. J. J. M. B. (2019). Materials challenges in the hydrogen cycle. *MRS Bulletin*, 44(3), 164–166.

Philibert, C. J. P. I. E. A. (2017). *Renewable energy for industry* (p. 65). International Energy Agency.

Pivovar, B., Rustagi, N., & Satyapal, S. J. T. E. S. I. (2018). Hydrogen at scale (H2@ Scale): Key to a clean, economic, and sustainable energy system. *The Electrochemical Society Interface*, 27(1), 47.

Pradier, J. P., & Pradier, C.-M. (2014). *Carbon dioxide chemistry: Environmental issues*. Elsevier.

Qazi, U. Y. J. E. (2022). Future of hydrogen as an alternative fuel for next-generation industrial applications; Challenges and expected opportunities. *Energies*, *15*(13), 4741.

Show, K.-Y., Yan, Y., Ling, M., Ye, G., Li, T., & Lee, D.-J. J. B. T. (2018). Hydrogen production from algal biomass – advances, challenges and prospects. *Bioresource Technology*, *257*, 290–300.

Vu, A. T., Gulati, S., Vogel, P.-A., Grunwald, T., Bergs, T. J. I. J. o. H., & Transfer, M. (2021). Machine learning-based predictive modeling of contact heat transfer. *International Journal of Heat and Mass Transfer*, 174, 121300.

Chapter 4

Environmentally friendly and cost-effective hydrogen production strategy

Prospects, challenges, and technology development

Sukanta Roga, Jeet Bhakta, Adesh Singh, and Swapan Bhaumik

4.1 INTRODUCTION

Green hydrogen is a promising solution for decarbonizing applications, complementing renewable electricity. Academic and industrial efforts should focus on green hydrogen as the linchpin economy, enabling a society powered entirely by renewable energy (Oliveira et al., 2021). To tackle climate change and lower greenhouse gas emissions, the whole community is up against an unparalleled challenge. To lessen the effects of climate change, the world is currently going through a major energy transition that will lead to a sustainable and decarbonized future. Hydrogen has become a promising replacement for traditional fossil fuels because of the increased focus on renewable energy sources. A realistic choice for the shift to a low-carbon economy, hydrogen offers a clean and sustainable fuel source for a variety of uses. India is well positioned to make use of the potential of green hydrogen for its energy transition, thanks to its ambitious renewable energy goals and substantial industrial base. With 175 GW of solar, 60 GW of wind, 10 GW of biomass, and 5 GW of ocean energy, the nation has set an ambitious goal of 450 GW of renewable energy capacity by 2030. The National Hydrogen Energy Mission and the Green Hydrogen Fund are two of the programs that India has started to promote the use of hydrogen as a fuel (Sontakke & Jaju, 2021). Water penetrates the cathode, which is typically made of a mixture of Ni and YSZ or a material similar to perovskite, where it is converted to make H_2 and oxygen anions, which travel through a solid solution to the anode (made of perovskites, like lanthanum strontium cobalt ferrite), producing molecular oxygen. Proton-carrying solid-oxide electrolyzers (SOEs) operate similarly to the PEME (Equations 4.1–4.3). Due to the high temperature, SOEs have better thermodynamics and quicker kinetics than AEs and PEMEs, which is evident from Figure 4.1(c). They are more efficient, work at high current densities, produce pure hydrogen, and split water with less electrical power (Raman et al., 2022).

 DOI: 10.1201/9781003537816-4

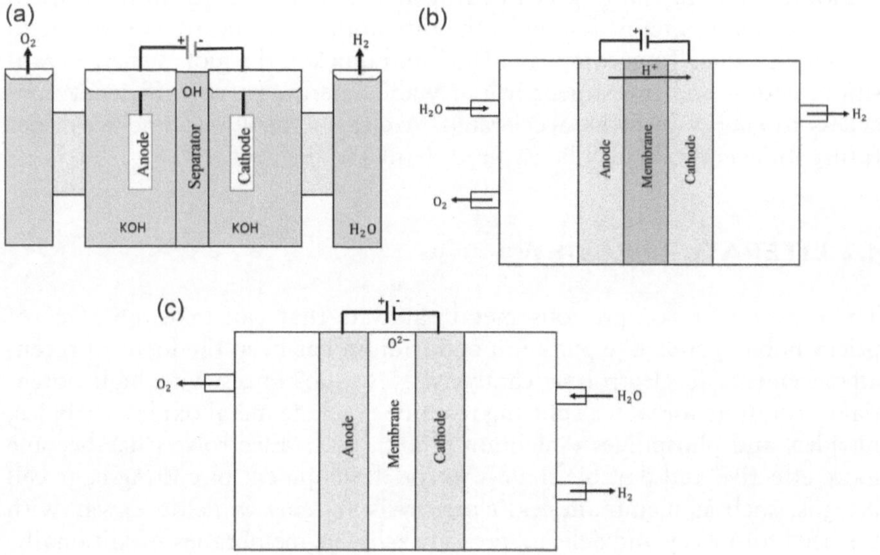

Figure 4.1 Schematic diagram of (a) alkaline electrolyzer, (b) PEM electrolyzer, and (c) solid-oxide electrolyzer.

India must expand its green hydrogen production capabilities and create a stable environment for its transit, storage, and application across many industries. To provide creative solutions for the widespread adoption of green hydrogen, the government, industry, and academics must work together, according to the study. The benefits of using hydrogen as a clean, plentiful, and adaptable fuel source are highlighted in another article titled "Significance of Hydrogen as Economic and Environmentally Friendly Fuel." This chapter does, however, also highlight the difficulties that come with the mass use of hydrogen as a fuel, such as the expense and infrastructural needs. In the transition to a more sustainable energy system, hydrogen can play an important role. The essay also discusses how problems can be overcome through innovation and cooperation (Yang et al., 2023). Sustainable Development Goal 7 (SDG7) must be attained by 2030, and the report "Sustainable Energy for All: Monitoring SDG7—The Energy Progress Report 2021" emphasizes the urgent need for more action and investment in clean energy, energy efficiency, and access to energy. This study demonstrates how the COVID-19 pandemic has had a significant negative influence on the worldwide energy transition, especially in poorer nations. The study emphasizes the urgent need to speed up action and investment in clean energy to achieve SDG7 despite modest success in increasing access to power and clean cooking options (Salehmin et al., 2022). In conclusion, a sustainable and decarbonized future depends on the use of green hydrogen. The reports and materials mentioned above illustrate the potential of green hydrogen as a crucial factor in the shift to a low-carbon economy as well as the difficulties that must be overcome to accomplish this. To accomplish SDG7 and quicken the global energy transition, action and investment in renewable energy, energy efficiency, and access to energy must be accelerated. A cleaner, healthier, and wealthier future for everybody may be ensured thanks to their efforts.

4.2 LITERATURE REVIEW

The creation of non-precious metal catalysts that can take the place of pricey noble metals like platinum and iridium has been the focus of recent advancements in electrolysis catalysts. New substances with high potential as catalysts for water-splitting reactions include metal oxides, carbides, nitrides, and phosphides (Salehmin et al., 2022). Electrolysis has become more effective and durable thanks to the development of cutting-edge cell designs, such as membrane-less electrolysis systems, which do away with the need for pricey and delicate proton exchange membranes. Additionally, these technologies increase process safety and decrease contamination risk. Electrolysis is now being used in novel ways, such as in conjunction with renewable energy sources like solar, wind, and geothermal energy, as well as other technologies like carbon collection and utilization. Particularly, combining electrolysis with carbon capture technology can aid in lowering greenhouse gas emissions and advancing the production of hydrogen sustainably. Even while electrolysis has recently made progress, there are still several issues that must be resolved before this technology for producing hydrogen can be widely used. Among these difficulties are electrolysis' high-energy requirements and expenses, the demand for extensive infrastructure, and the necessity to increase the electrolysis systems' robustness and stability. New materials, cell designs, and electrolysis operating conditions must be developed to overcome these difficulties (Rashid et al., 2015). Electrolysis is increasingly being used in energy storage and renewable energy systems, as well as in conjunction with other technologies like carbon capture and utilization. A potential technology for producing hydrogen is biomass gasification because it is plentiful, environmentally friendly, and renewable (Zhang et al., 2019).

4.3 METHODOLOGY

This current work discusses hydrogen production technologies, including water electrolysis, solid-oxide electrolyzers, alkaline electrolyzers (AE), and proton exchange membranes (PEMs). It compares their performance, benefits, and drawbacks, highlighting their environmental impact. The summary emphasizes the future of sustainable energy and the need for ongoing research to improve efficiency and sustainability in water electrolysis for hydrogen production. It also highlights the importance of environmental sustainability in hydrogen production.

Through a thorough review of the literature and data collection on fuel cell generators, electrolyzers, hydrogen storage systems, and renewable energy sources, the green hydrogen production system model is prepared using Python libraries and HOMER Pro software. It can calculate the capital, operating, and maintenance expenses related to every system component, thanks to the model. After that, a comprehensive sensitivity analysis is carried out to determine how different input parameters would affect the final cost of producing green hydrogen. Through the identification of cost-saving measures and the optimization of system design and operation, the objective was to augment the overall efficiency of the green hydrogen production process. Convincing conclusions regarding the economic viability and feasibility of producing green hydrogen were drawn from the interpretation of the data, especially when considering solar power and fuel cell technology. The study describes economic and sustainable green hydrogen production by providing stakeholders and policymakers with important information.

4.3.1 Production by water electrolysis

The primary way of generating hydrogen from renewable sources is electrolysis, which uses electricity to split water into hydrogen and oxygen. In a typical water electrolyzer, an anode electrode and a cathode electrode are separated by a semipermeable barrier. An external circuit connects the electrodes to the power source. Water enters the electrolyzer where it is subjected to an electric charge, causing it to separate into hydrogen and oxygen. At the cathode, reduction results in the production of H_2, and at the anode, oxidation results in the production of O_2. These two reactions, known as the Oxygen Evolution Reaction (OER) (Sharma et al., 2021) and the Hydrogen Evolution Reaction (HER), respectively, take place in an acidic medium, as shown by the following equations:

$$\text{OER: } H_2O = H_2 + 1/2O_2$$
$$\text{HER: } 2H^{(+)} + 2e^- = H_2$$
$$\text{Overall: } H_2O = 1/2O_2 + 2H^+ + 2e^-$$

Electrocatalysts are needed to reduce the increase in electrochemical processes; they are mainly platinum group metals (Abe et al., 2019). To produce intermediates that facilitate charge transfer in the electrolyzer, they adsorb reactants on their surfaces. These chemical concepts can be used in various electrolyzer designs to produce H_2 from water. The three main technologies of interest for industrial usage are alkaline electrolyzers (AE), PEMEs, and SOEs (James & Menzies, 2023).

In Figure 4.1, it is shown that an anion-permeable membrane (such as asbestos or zirconium) or an anionic polymer membrane is used as a base and electrolyte. Each electrode goes through the following processes:

$$\text{Cathode: } 2H_2O + 2e^- = 2H_2 + 2OH^- \quad (1)$$

$$\text{Anode: } 2OH^- = 2H_2O + \frac{1}{2}O_2 \quad (2)$$

$$\text{Overall: } 2H_2O = H_2 + \frac{1}{2}O_2 \quad (3)$$

Water enters the cathode, where it divides into H_2 and OH^- anions. The divider only allows OH^- anions, which will be oxidized into oxygen, to pass from the cathode to the anode. Water as hydrogen is continuously added to the reservoir to keep it full. During the fission process, oxygen is created. AE has many advantages over PEME because of its simplicity. They are less prone to corrosion than acid electrolyzers, which makes it possible to utilize catalysts other than Pt that are more affordable and accessible, last longer, and need less maintenance, but still have issues, despite being a proven and frequently used technology: Because they are slow in alkaline medium, constrained in current density, low in working pressure, and inefficient energy (70–80%). HER must be altered to perform more dynamically to handle renewable resources with fluctuating energy sources. Recently, options for AE optimization have been proposed to address some of the hardware issues with this technology. These options include electrode porosity and wettability to facilitate bubble separation and surface coverage, improved ionic conductivity of the electrolyte through additives, and separator replacement or tangible separation. Proton-permeable polymer electrolyte membranes (Figure 4.1b) are found in polymer exchange membrane electrolyzers, which are acidic in character. With water penetrating the anode, where the OER takes place, the reactions in the electrodes are the same as in Equations (4.1)–(4.3) with water entering the anode where the OER occurs (Salehmin et al., 2022). Hydrogen cations that reach the cathode, where HER takes place, pass through the membrane and create H_2. Membranes made of perfluorosulfonic acid polymers, such as Nafion, Fumapem, Flemion, and Aciplex, are frequently employed. Due to their larger compactness, higher current densities and efficiency (80–90%), ability to create oxygen and hydrogen with higher purity, and more dynamic operation, PEMEs are better adapted to sustainable energy sources than AEs. The acidic environment of the PEME, however, mandates the employment of expensive noble metal catalysts (Pt/Pd for HER and IrO_2/RuO_2 for OER) (Sharma et al., 2021) whose susceptibility to membrane fouling and anode degradation poses problems with durability and service life. The development of non-noble metal catalysts that meet activity and stability requirements decreased loading catalyst supports, and corrosion-resistant bipolar plates are all ongoing issues for materials. The thick coating of ceramic electrolyte, which is typically made of yttrium-stabilized zirconia, can conduct oxyanions or hydrogen protons, thanks to solid-oxide electrolyzers' ability to work at high temperatures (5000°C to 1000°C).**

The following processes typically take place at each porous electrode:

$$\text{Cathode: } H_2O + 2e^- = H_2 + O^{-2} \tag{4}$$

$$\text{Anode: } O^{-2} = \frac{1}{2}O_2 + 2e^- \tag{5}$$

$$\text{Overall: } H_2O = H_2 + \frac{1}{2}O_2 \tag{6}$$

Despite the potential of this method, the high operating temperatures present significant material durability issues. More information on alternative battery materials and recent developments is provided. For instance, the electrolyte lanthanum strontium gallium magnesium (LSGM) is used for the cathode and a nickel-based material is used for the anode. LSGM lifts the main chain metal catalyst to the surface of the perovskite electrolyte via voltage. Hardware issues have prompted studies into these classes of electrolyzers, as was briefly mentioned. Device effectiveness is impacted by various factors, such as reducing internal impedance, improving membrane electrode assemblies, and choosing separation materials. Considering this, the study on the electrolyte's catalytic components is more active, which this review will analyze from the standpoint of important materials. Electrolysis-produced hydrogen is just as environmentally friendly as the electricity that created it, much like the energy that is stored in battery-powered automobiles (Rashid et al., 2015). The production of electrolytic hydrogen from renewable electricity (wind, solar, hydro, or biomass) has been shown to decrease carbon emissions globally by more than 90%. Given the quantity of electricity generated and the resources required to construct solar panels, wind turbines, and electrolyzers, wind electrolysis has been demonstrated to be the most effective method for producing H_2.

4.4 RESULTS AND DISCUSSION

Hydrogen fuel cells are becoming increasingly popular as an alternative energy source to traditional non-renewable energy sources such as natural gas. In Nagpur, India, a study was conducted to evaluate the cost-effectiveness of producing hydrogen using a solar photovoltaic system of 250 kW and a hydrogen tank with 100 L capacity through the water electrolysis process. The study was conducted using the HOMER software, which is a computerized simulation tool for analyzing the feasibility and cost-effectiveness of renewable energy systems. The study revealed that the most effective method of energy production using solar photovoltaic and hydrogen tank systems was through the use of hydrogen fuel cells. The analysis showed that the hydrogen fuel cell had a leveled cost of electricity (LCOE) of $0.09/kWh, while the LCOE for producing hydrogen using an alkaline electrolyzer was $0.14/kWh.

From Figure 4.2, it is evident that the cost of producing hydrogen using the proton exchange membrane electrolyzer and the solid-oxide electrolyzer was even higher, with LCOEs of $0.22/kWh and $0.34/kWh, respectively. The results indicate that hydrogen fuel cells are the most cost-effective method of energy production using solar photovoltaic and hydrogen tank systems. This is because fuel cells have higher energy conversion rates, lower maintenance costs, and a longer service life compared to the other electrolysis methods. Additionally, fuel cells can operate at a high level of efficiency, making them ideal for use in applications such as transportation and power generation. The study also evaluated the impact of different hydrogen storage capacities on the cost of hydrogen production. The analysis showed that increasing the hydrogen storage capacity from 100 L to 1000 L reduced the LCOE of producing hydrogen using the alkaline electrolyzer from $0.14/kWh to $0.11/kWh.

The LCOE for hydrogen fuel cells remained the same at $0.09/kWh, regardless of the hydrogen storage capacity. Furthermore, the study evaluated the impact of different solar photovoltaic capacities on the cost of hydrogen production. The analysis showed that increasing the solar photovoltaic capacity from 250 kW to 1000 kW reduced the LCOE of producing hydrogen using the alkaline electrolyzer from $0.14/kWh to $0.11/kWh. The LCOE for hydrogen fuel cells remained the same at $0.09/kWh, regardless of the solar photovoltaic capacity. Overall, the study revealed that hydrogen fuel cells are the most effective and cost-efficient method of energy production using solar photovoltaic and hydrogen tank systems according to Figure 4.3. Nonetheless, water electrolysis remains an expensive method of producing hydrogen. The cost of electrolyzers is high, and the use of noble metal catalysts in some types of electrolyzers further increases the cost. In addition to the cost of electrolysis, the study also identified the need for significant investment in hydrogen infrastructure, including hydrogen refueling stations and pipelines. The widespread adoption of hydrogen as a fuel source will require significant investments in infrastructure, which will drive down costs over time as more countries and companies commit to decarbonization. In conclusion, the study demonstrated the cost-effectiveness of hydrogen fuel cells for energy production using the solar photovoltaic and hydrogen tank systems in Nagpur, India.

While the cost of electrolysis and the infrastructure requirements for hydrogen production, storage, and transportation remain significant

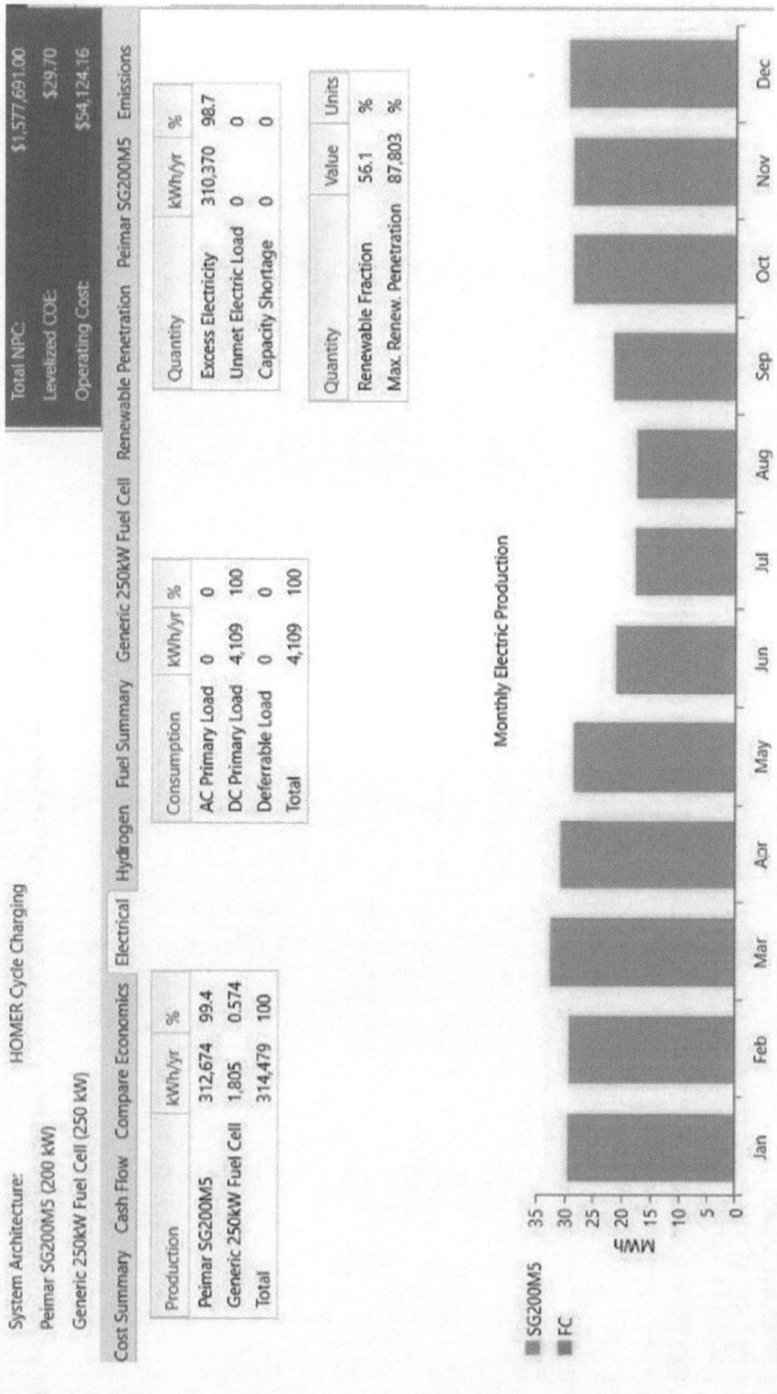

Figure 4.2 Monthly electricity production.

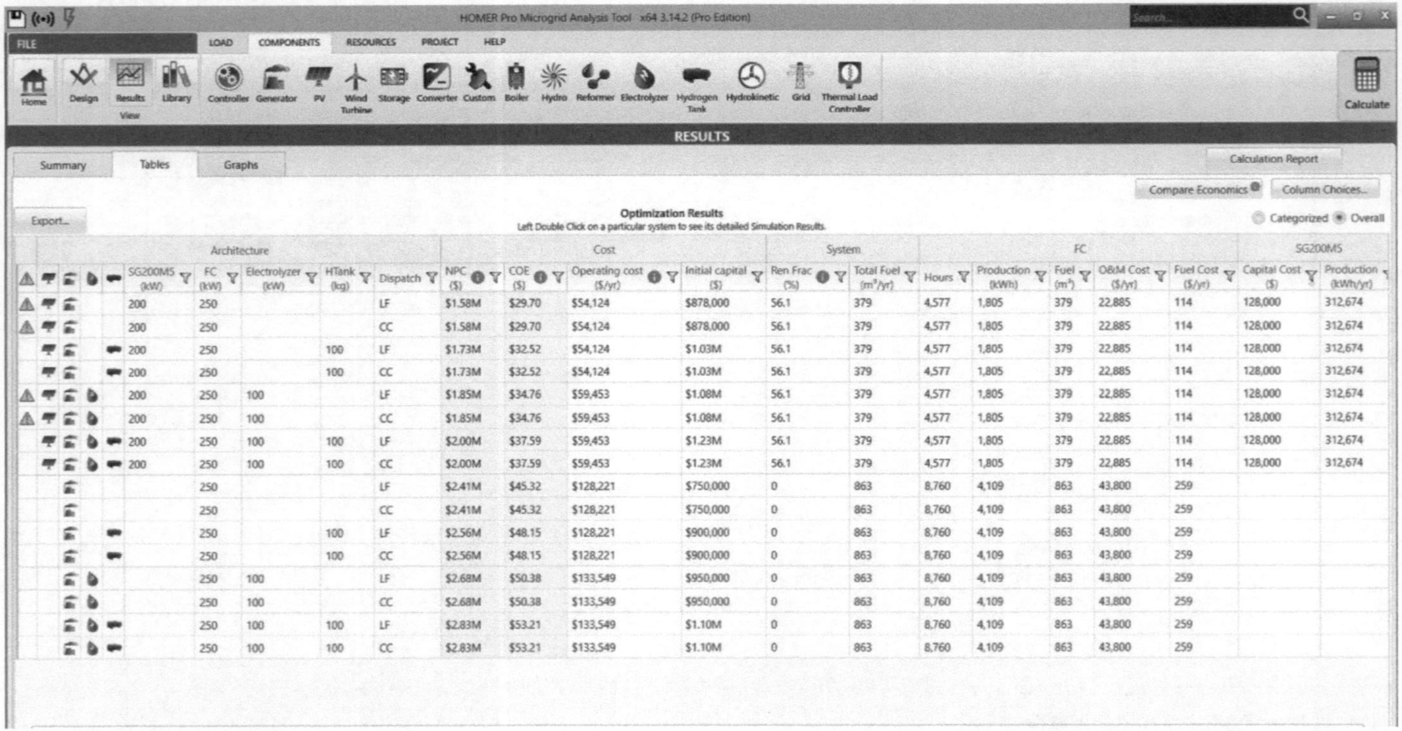

Architecture					Cost				System		FC						SG200MS	
SG200MS (kW)	FC (kW)	Electrolyzer (kW)	HTank (kg)	Dispatch	NPC ($)	COE ($)	Operating cost ($/yr)	Initial capital ($)	Ren Frac (%)	Total Fuel (m^3/yr)	Hours	Production (kWh)	Fuel (m^3)	O&M Cost ($/yr)	Fuel Cost ($/yr)	Capital Cost ($)	Production (kWh/yr)
200	250			LF	$1.58M	$29.70	$54,124	$878,000	56.1	379	4,577	1,805	379	22,885	114	128,000	312,674
200	250			CC	$1.58M	$29.70	$54,124	$878,000	56.1	379	4,577	1,805	379	22,885	114	128,000	312,674
200	250		100	LF	$1.73M	$32.52	$54,124	$1.03M	56.1	379	4,577	1,805	379	22,885	114	128,000	312,674
200	250		100	CC	$1.73M	$32.52	$54,124	$1.03M	56.1	379	4,577	1,805	379	22,885	114	128,000	312,674
200	250	100		LF	$1.85M	$34.76	$59,453	$1.08M	56.1	379	4,577	1,805	379	22,885	114	128,000	312,674
200	250	100		CC	$1.85M	$34.76	$59,453	$1.08M	56.1	379	4,577	1,805	379	22,885	114	128,000	312,674
200	250	100	100	LF	$2.00M	$37.59	$59,453	$1.23M	56.1	379	4,577	1,805	379	22,885	114	128,000	312,674
200	250	100	100	CC	$2.00M	$37.59	$59,453	$1.23M	56.1	379	4,577	1,805	379	22,885	114	128,000	312,674
	250			LF	$2.41M	$45.32	$128,221	$750,000	0	863	8,760	4,109	863	43,800	259		
	250			CC	$2.41M	$45.32	$128,221	$750,000	0	863	8,760	4,109	863	43,800	259		
	250		100	LF	$2.56M	$48.15	$128,221	$900,000	0	863	8,760	4,109	863	43,800	259		
	250		100	CC	$2.56M	$48.15	$128,221	$900,000	0	863	8,760	4,109	863	43,800	259		
	250	100		LF	$2.68M	$50.38	$133,549	$950,000	0	863	8,760	4,109	863	43,800	259		
	250	100		CC	$2.68M	$50.38	$133,549	$950,000	0	863	8,760	4,109	863	43,800	259		
	250	100	100	LF	$2.83M	$53.21	$133,549	$1.10M	0	863	8,760	4,109	863	43,800	259		
	250	100	100	CC	$2.83M	$53.21	$133,549	$1.10M	0	863	8,760	4,109	863	43,800	259		

Figure 4.3 Different combinations of equipment to optimize hydrogen production.

challenges, the study provides valuable insights into the potential for renewable energy systems to provide a sustainable and decarbonized future. Investment and action in renewable energy, energy efficiency, and access to energy must be expedited to achieve this goal, and green hydrogen can play a crucial role in meeting daily uses, as mentioned in Figure 4.4. Water electrolysis is a vital process for generating hydrogen, a clean and sustainable fuel source that offers an alternative to conventional hydrogen production methods that rely on non-renewable energy sources such as natural gas. There are three primary technologies employed in industrial hydrogen production via water electrolysis: alkaline, proton exchange membrane, and solid-oxide electrolyzers. Alkaline electrolyzers are cheaper and require less maintenance, but they have low energy conversion rates and operate slowly in an alkaline medium with low working pressure. Proton exchange membrane electrolyzers are more efficient, have higher current densities and purity levels, and can operate dynamically with renewable energy sources, but they are less durable, require expensive noble metal catalysts, and have a low service life. Solid-oxide electrolyzers are effective and can operate at high temperatures, but they are costly to manufacture, require high operating temperatures, and have low durability.

However, the creation of green hydrogen via water electrolysis offers a promising alternative to conventional hydrogen production methods that rely on non-renewable energy sources such as natural gas. As renewable energy sources become more common, the demand for sustainable hydrogen production will increase, and water electrolysis will play a crucial role in meeting this demand. Hydrogen and oxygen are separated from water using electricity in one method of electrolysis. By passing electric current through water, hydrogen and oxygen ions are formed as a result of the water molecules breaking apart. At the other end of the cathode, hydrogen ions are attracted to oxygen ions. As the hydrogen and oxygen ions combine at the cathode and anode, respectively, hydrogen gas and oxygen gas are produced. Electrolysis can be done using renewable energy sources such as wind and solar power, making the production of green hydrogen possible. The use of renewable energy sources reduces the carbon footprint of the process, making it more environmentally friendly. One advantage of using electrolysis to produce hydrogen is that it can be done on a small scale, making it ideal for decentralized hydrogen production according to Figure 4.5. This is particularly useful for applications such as fuel cell vehicles, where hydrogen is produced on-site using renewable energy sources.

Besides producing high-purity hydrogen, water electrolysis is also useful for fuel cells and other applications. Hydrogen and oxygen are used to generate electricity in fuel cells, and the only byproduct is water. There are a variety of applications for fuel cells, including transportation and energy production. One of the main challenges associated with electrolysis is the cost of the process. Electrolyzers are expensive to manufacture, and the

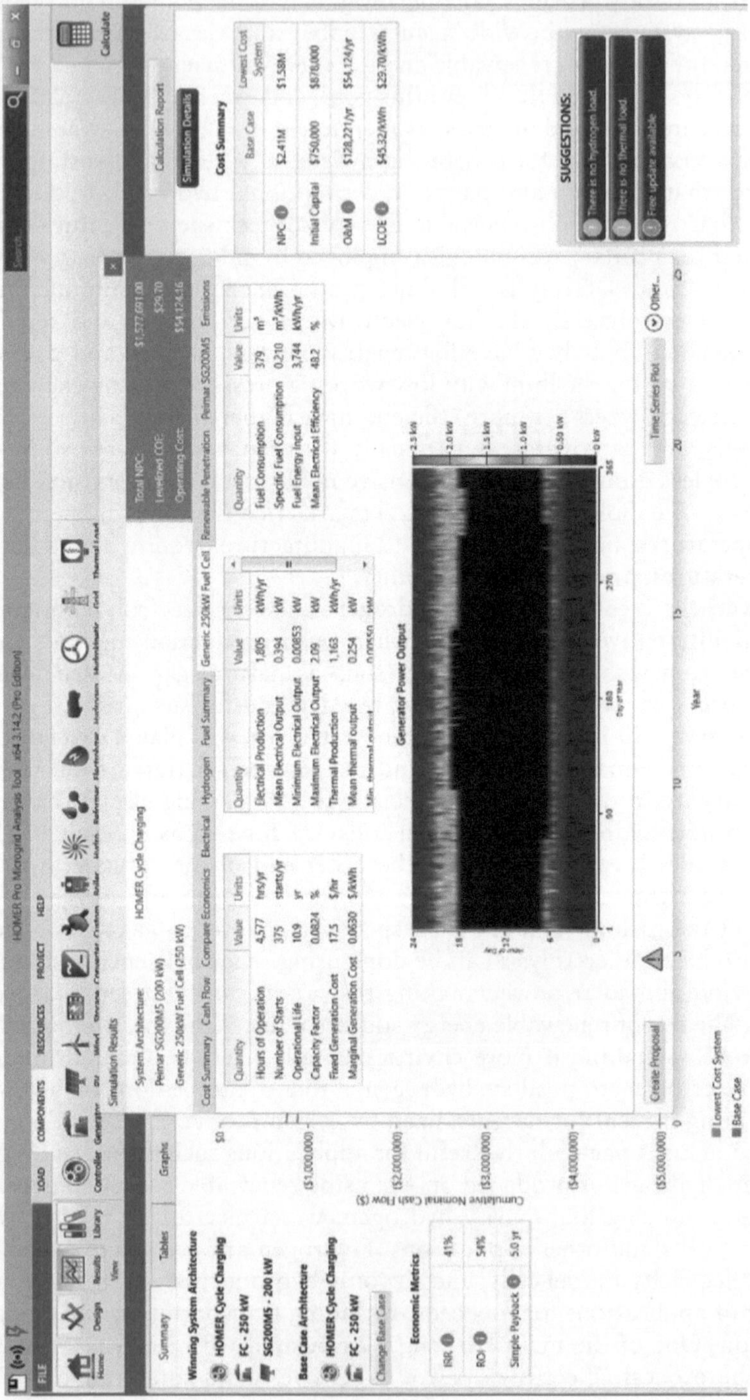

Figure 4.4 Generated power output vs hour of the day vs day of the year.

use of noble metal catalysts in some types of electrolyzers further increases the cost. However, the cost of electrolyzers is expected to decrease as the technology develops and scales up. Another challenge is the infrastructure required for hydrogen production, storage, and transportation. The widespread adoption of hydrogen as a fuel source will require significant investments in infrastructure (Figure 4.5b), including hydrogen refueling stations and pipelines. However, as more countries and companies commit

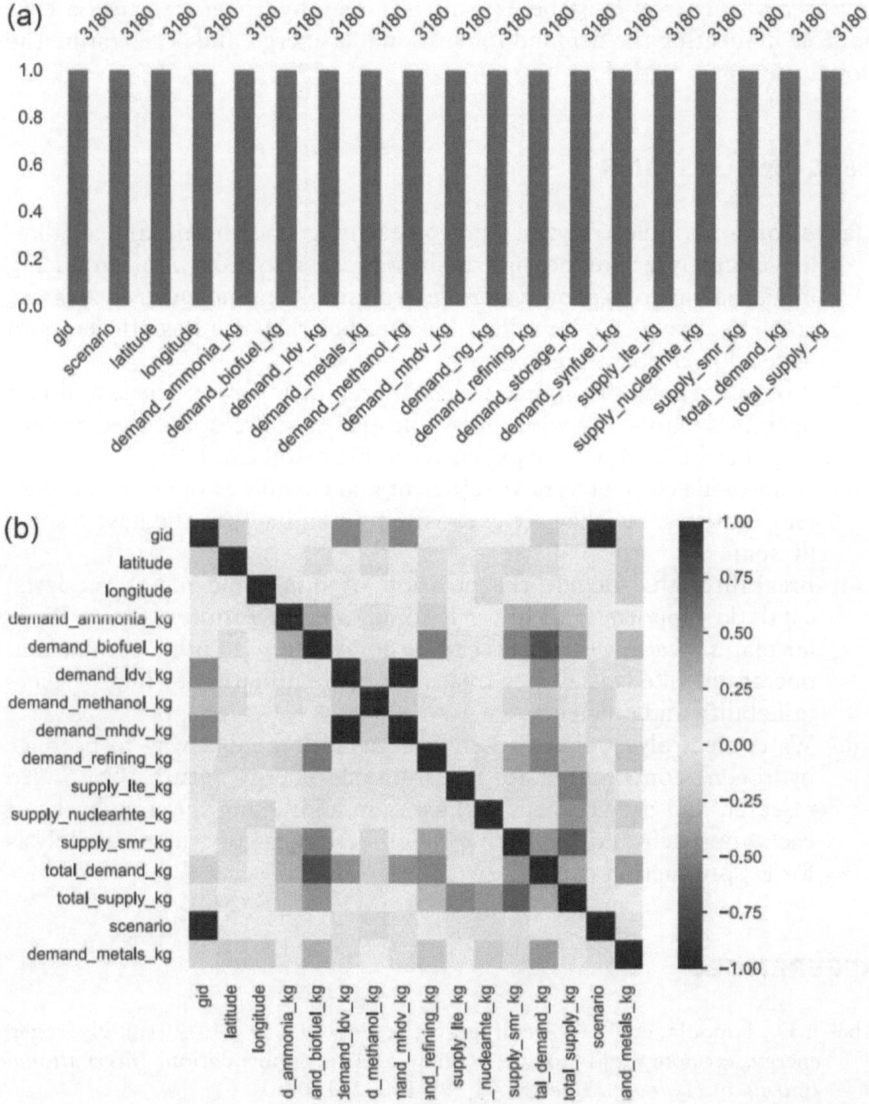

Figure 4.5 (a) Generated a visualization of nullity by fuel demand column and (b) generated a heat map of fuel demand and consumption.

to decarbonization, the demand for hydrogen infrastructure is expected to increase, driving down costs. In conclusion, water electrolysis offers a promising method for producing green hydrogen, a clean and sustainable fuel source. While there are challenges associated with the process, such as the cost of electrolyzers and the infrastructure requirements for hydrogen production, storage, and transportation, these challenges can be overcome with innovation and collaboration. To achieve a sustainable and decarbonized future, investment and action in renewable energy, energy efficiency, and access to energy must be expedited. Green hydrogen can play a crucial role in meeting the demand for sustainable energy and accelerating the global energy transition.

4.5 CONCLUSIONS

(a) A solid-oxide electrolyzer, a proton exchange membrane, and an alkaline electrolyzer are the three main sources of hydrogen for producing industrial hydrogen by water electrolysis. Alkaline electrolyzers are cost-effective and require less upkeep, but they are less efficient and have a lower operating pressure.
(b) Proton exchange membrane electrolyzers are very efficient and can operate dynamically with renewable energy sources, but they are not very durable and require expensive noble metal catalysts.
(c) Solid-oxide electrolyzers are efficient and capable of operating at high temperatures, but they are expensive to manufacture and have a short lifespan.
(d) Breakthroughs include the creation of non-noble metal catalysts, catalytic supports with lower loading, and corrosion-resistant bipolar plates. Water electrolysis can be combined with other commercial operations like wastewater treatment or desalination to improve sustainability and efficiency.
(e) Water electrolysis offers an environmentally benign way to produce hydrogen, contributing to a sustainable energy future. Continued research and development will assist in addressing the drawbacks of each approach and improving the effectiveness of water electrolysis for H_2 production overall.

REFERENCES

Abe, J. O., Popoola, A. P. I., Ajenifuja, E., & Popoola, O. M. (2019). Hydrogen energy, economy and storage: Review and recommendation. *International Journal of Hydrogen Energy*, 44(29), 15072–15086.

James, N., & Menzies, M. (2023). Distributional trends in the generation and end-use sector of low-carbon hydrogen plants. *Hydrogen*, *4*(1), 174–189.

Oliveira, A. M., Beswick, R. R., & Yan, Y. (2021). A green hydrogen economy for a renewable energy society. *Current Opinion in Chemical Engineering*, *33*, 100701.

Raman, R., Nair, V. K., Prakash, V., Patwardhan, A., & Nedungadi, P. (2022). Green-hydrogen research: What have we achieved, and where are we going? Bibliometrics analysis. *Energy Reports*, *8*, 9242–9260.

Rashid, M. D., Al Mesfer, M. K., Naseem, H., & Danish, M. (2015). Hydrogen production by water electrolysis: a review of alkaline water electrolysis, PEM water electrolysis and high temperature water electrolysis. *International Journal of Engineering and Advanced Technology*, 80–93.

Salehmin, M. N. I., Husaini, T., Goh, J., & Sulong, A. B. (2022). High-pressure PEM water electrolyser: A review on challenges and mitigation strategies towards green and low-cost hydrogen production. *Energy Conversion and Management*, *268*, 115985.

Sharma, S., Agarwal, S., & Jain, A. (2021). Significance of hydrogen as economic and environmentally friendly fuel. *Energies*, *14*(21), 7389.

Sontakke, U., & Jaju, S. (2021, November). Green hydrogen economy and opportunities for India. In Dr. Manoj A. Kumbhalkar (Ed.), *IOP Conference Series: Materials Science and Engineering* (Vol. 1206, No. 1, p. 012005). IOP Publishing.

Yang, B., Zhang, R., Shao, Z., & Zhang, C. (2023). The economic analysis for hydrogen production cost towards electrolyzer technologies: Current and future competitiveness. *International Journal of Hydrogen Energy*, *48*(37), 13767–13779.

Zhang, Y., Li, L., Xu, P., Liu, B., Shuai, Y., & Li, B. (2019). Hydrogen production through biomass gasification in supercritical water: A review from exergy aspect. *International Journal of Hydrogen Energy*, *44*(30), 15727–15736.

Chapter 5

Thermochemical conversion of biomass for hydrogen production

Roque Aguado, Simin Anvari, Antonio Escámez, Daniel Sánchez, and David Vera

5.1 INTRODUCTION

Numerous industrial processes require hydrogen as a primary raw material (Binder et al., 2018; Patra et al., 2019). In recent years, hydrogen has been increasingly gaining consideration as a sustainable strategic energy carrier (Binder et al., 2018), providing an alternative to fossil fuels (Lanjekar et al., 2023), particularly in the sectors of bulk energy storage and transportation (Lepage et al., 2021). However, despite being an abundant and widely spread element on Earth, hydrogen is seldom found in its pure form, but rather bound together within heterogeneous molecules (Buffi et al., 2022). Hydrogen can be produced through thermochemical processing of fossil- or bio-derived resources, through biological processes, or through water-splitting processes such as electrolysis, thermolysis, and photo-electrolysis (Buffi et al., 2022; Nikolaidis & Poullikkas, 2017). All these hydrogen production methods can be implemented using renewable energy sources (Binder et al., 2018). The different pathways for renewable hydrogen production are represented in figure 5.1.

Nowadays, the vast majority of hydrogen production worldwide, approximately 96%, is conventionally sourced from fossil resources (Arregi et al., 2018; Binder et al., 2018; Chai et al., 2023; Fermoso et al., 2012; Lepage et al., 2021; Nikolaidis & Poullikkas, 2017; Patra et al., 2019; Singh et al., 2023), offering unmatched affordability if the associated CO_2 emissions are not captured and sequestered, in which case the resulting hydrogen is commonly referred to as "gray hydrogen" (IRENA, 2019). The primary sources of hydrogen production include steam reforming of light hydrocarbons and partial oxidation of heavier hydrocarbons, collectively constituting 78% of the world's hydrogen production, with coal gasification contributing an additional 18% (Arregi et al., 2018; Chai et al., 2023; Dincer & Acar, 2015; Eloffy et al., 2022; Nikolaidis & Poullikkas, 2017). Electrolysis accounts for the bulk of the remaining 4% of global hydrogen production (Chai et al., 2023; Chen et al., 2021), which is obtained as a byproduct of chlorine production (M. Balat, 2008).

 DOI: 10.1201/9781003537816-5

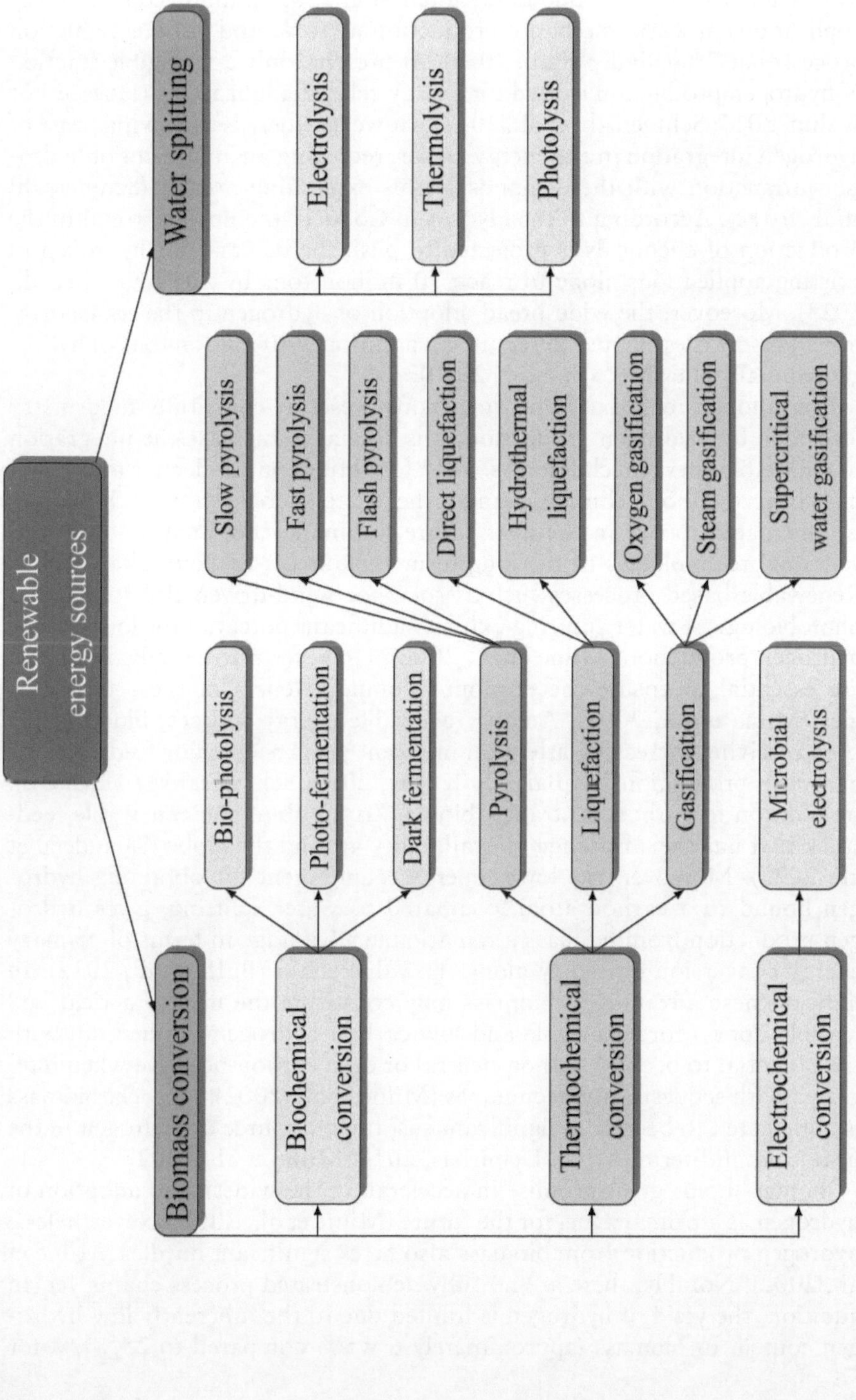

Figure 5.1 Classification of renewable hydrogen production processes, with a particular focus on thermochemical conversion pathways.

The demand for hydrogen on a global scale is predominantly driven by the industrial sector. Ammonia production accounts for roughly 50% of worldwide hydrogen production (Kumar et al., 2023), followed by refinery applications at 22%, methanol production at 14%, and various reduction processes at 7% (Binder et al., 2018). At present, only a negligible fraction of hydrogen production is used for energy-related applications (Hosseini & Wahid, 2016; Schmersahl et al., 2012). However, there is a growing trend of hydrogen integration in the energy sector, requiring an alignment of hydrogen production with the demands of this expanding market (Schmersahl et al., 2012). According to the Hydrogen Council, the upward trend in the production of chemicals is projected to push the demand for hydrogen in existing applications alone to reach 70 million tons by 2050 (Chai et al., 2023). Moreover, the widespread adoption of hydrogen in the transportation industry is estimated to require an additional 40 million tons of hydrogen annually (Levin & Chahine, 2010).

Even though fossil-based hydrocarbons presently constitute the primary feedstock for hydrogen production, it is crucial to increase the integration of renewable energy technologies. With fossil fuels on the decline and growing concerns about climate change, the share of renewable technologies is anticipated to rise in the near future, ultimately dominating over conventional technologies in the long term (Nikolaidis & Poullikkas, 2017). Renewable-based processes such as solar- or wind-driven electrolysis and photobiological water splitting exhibit significant potential for low-carbon hydrogen production (Milne et al., 2002). However, further advancements are essential to ensure the economic competitiveness of these technologies (Milne et al., 2002). Among renewable energy sources, biomass has been receiving increasing attention in recent years as a major feedstock for hydrogen production (H. Balat & Kirtay, 2010; Schmersahl et al., 2012). In addition to carbon neutrality, biomass is an abundant renewable feedstock that has the advantage of availability around the globe (Salaudeen et al., 2023). Moreover, the lower energy requirement for obtaining hydrogen bound to a carbon atom, compared to water splitting, gives hydrogen production from biomass a remarkable advantage in terms of primary energy conversion efficiency along the value chain (Buffi et al., 2022). In light of these advantages, biomass may constitute the most practical and feasible option for renewable and low-carbon hydrogen production, with the potential to become carbon-neutral or even carbon-negative when integrated with sequestration techniques (Milne et al., 2002). As such, biomass is anticipated to become a significant sustainable source of hydrogen in the near- and mid-term (Ayhan Demirbas, 2016; Milne et al., 2002).

Biomass holds great promise in accelerating the widespread adoption of hydrogen as a primary fuel for the future (Milne et al., 2002). Nevertheless, hydrogen production from biomass also faces significant hurdles (Milne et al., 2002). Notably, there are no fully demonstrated process chains yet. In addition, the yield of hydrogen is limited due to the inherently low hydrogen content in biomass (approximately 6 wt.% compared to 25 wt.% for methane), and the energy content is reduced because of the high oxygen content of biomass (around 40 wt.%) (Buffi et al., 2022; Hosseini & Wahid, 2016; Milne et al., 2002). In order to produce low-cost and low-carbon hydrogen, a wide range of biomass feedstocks may be utilized, including lignocellulosic biomass, marine-based biomass, agricultural waste, energy crops, and their byproducts such as biogas, bio-oil, and biochar (Kumar et al., 2023; Singh et al., 2023). The main considerations when selecting biomass as a feedstock for hydrogen production include the types of cultivars available in each region, their growth cycles, and the diverse requirements each cultivar has for other purposes, as well as the associated costs of harvesting, transportation, and pretreatment (Milne et al., 2002; Singh et al., 2023). However, variations and irregularities in the physicochemical properties and morphological structures of different biomass types, including those of the same variety sourced from diverse locations, may alter the resulting distribution of end products and adversely affect the conversion efficiency (Singh et al., 2023).

Hydrogen production from biomass can be accomplished through three different pathways: thermochemical, biochemical, and electrochemical (Eloffy et al., 2022). Each of these processes for biomass-derived hydrogen production varies significantly across several aspects, including the feedstock basis, reaction chemistry, plant design, and plant scales (Schmersahl et al., 2012). Biochemical conversion processes, also referred to as biological conversion processes, and electrochemical conversion processes are both characterized by their lower energy intensity compared to thermochemical conversion processes due to their operation under mild conditions (H. Balat & Kirtay, 2010; Nikolaidis & Poullikkas, 2017). The ability to operate at ambient pressure and temperature is indeed a significant advantage (Binder et al., 2018). However, in both cases, the current state-of-the-art is still limited to the laboratory scale (Binder et al., 2018), and the production of hydrogen at modest rates remains a challenge (H. Balat & Kirtay, 2010; Nikolaidis & Poullikkas, 2017). Moreover, both biochemical and electrochemical conversion processes are quite selective regarding their feedstocks, and lignocellulosic biomass materials cannot be easily processed (Eloffy et al., 2022; Lanjekar et al., 2023). By contrast, thermochemical conversion processes are substantially faster and provide higher stoichiometric yields of hydrogen (Buffi et al., 2022; Megia et al., 2021; Nikolaidis & Poullikkas, 2017), as well as a higher overall efficiency and lower production cost (H. Balat & Kirtay, 2010; Hosseini & Wahid, 2016). Thermochemical pathways such as gasification and steam reforming are already consolidated for industrial-scale hydrogen production from fossil fuels, demonstrating the highest technological maturity with a technology readiness level (TRL) of 9 on a scale of 1 to 9 (Binder et al., 2018). However, renewable hydrogen can also be produced via thermochemical methods using biomass as feedstock (Binder et al., 2018). For these reasons, researchers are increasingly directing their attention toward thermochemical pathways for hydrogen production from biomass (Lanjekar et al., 2023). Actually, thermochemical conversion of biomass is recognized as one of the most promising technologies for renewable hydrogen production (Patra et al., 2019), potentially capable of producing hydrogen from biomass for industrial and commercial applications in an economically viable and environmentally beneficial manner (Nikolaidis & Poullikkas, 2017; Singh et al., 2023). Despite the multiple advantages of biomass-derived hydrogen production, its commercial-scale implementation is currently hindered by various challenges, such as the heterogeneous nature and widespread distribution of biomass, considerations related to catalyst design, pre- and post-treatment requirements, and inherent limitations in thermochemical conversion processes (Chen et al., 2021; Singh et al., 2023).

Thermochemical conversion pathways are focused on breaking down the polymeric structure of biomass into lower molecular weight polymers and hydrogen-rich gases, which are well-suited for generating pure hydrogen alongside various solid, liquid, and gaseous byproducts (Buffi et al., 2022). The resulting intermediates, such as pyrolysis bio-oil or hydrogen-rich gases, can undergo further processing to yield additional hydrogen through the steam reforming and water-gas shift reactions or find value in other markets (Buffi et al., 2022; Megia et al., 2021; Nikolaidis & Poullikkas, 2017). Various thermochemical conversion technologies have been explored and investigated to convert biomass-based feedstocks into a hydrogen-rich gas mixture (Ayhan Demirbas, 2016; Fermoso et al., 2012), which can be broadly categorized into pyrolysis, liquefaction, and gasification processes (Buffi et al., 2022; Eloffy et al., 2022; Jara-Cobos et al., 2023; Megia et al., 2021; V. G. Nguyen et al., 2024; Nikolaidis & Poullikkas, 2017). However, in the context of renewable hydrogen production from biomass, pyrolysis and gasification are the most widely adopted pathways (Kumar et al., 2023; Mishra et al., 2023; V. G. Nguyen et al., 2024). In the upcoming sections, a comprehensive review of the existing thermochemical conversion processes for the production of hydrogen-rich gases from biomass is presented, along with the available approaches to increase the hydrogen yield and the prevailing methods for hydrogen separation and purification.

5.2 HYDROGEN FROM BIOMASS PYROLYSIS

Among thermochemical processes, pyrolysis is a promising pathway to produce hydrogen (Chen et al., 2021). Pyrolysis is a thermochemical conversion process characterized by its endothermic nature, wherein biomass is heated within the temperature range of 400 to 800°C, typically under pressures of up to 5 bar, in the absence of an oxidizing agent (Abdin et al., 2020; Eloffy et al., 2022; Hosseini & Wahid, 2016; Lepage et al., 2021; V. G. Nguyen et al., 2024). However, in some circumstances, a small amount of oxidizer may be introduced to aid in heat generation (Ayhan Demirbas, 2016; Eloffy et al., 2022; Lepage et al., 2021). As shown in Figure 5.2, the distribution

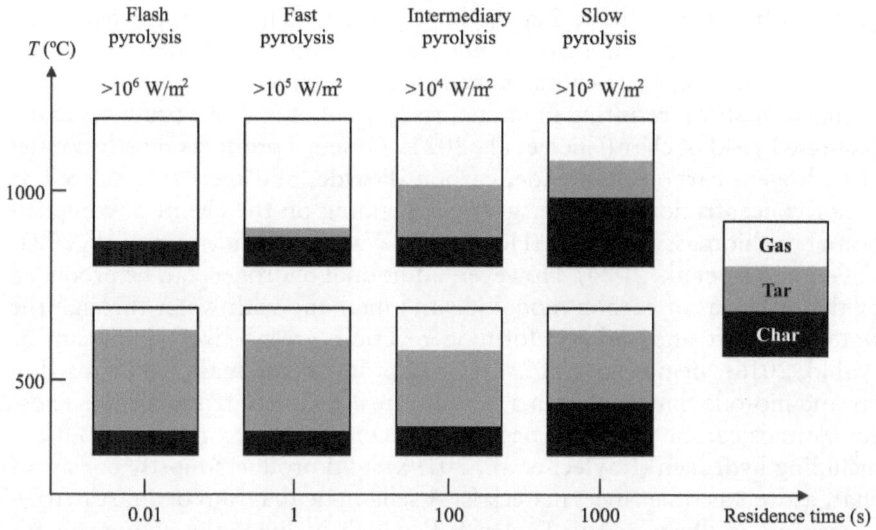

Figure 5.2 **Effect of different pyrolysis conditions on product distribution.**

of pyrolysis products depends on three main operating parameters: heating rate, residence time, and maximum reaction temperature. The products of biomass pyrolysis are solid charcoal, liquid oils, and gaseous compounds. According to the operating temperature and heating rate, this process can be categorized into slow pyrolysis, fast pyrolysis, and flash pyrolysis (Chai et al., 2023; Chen et al., 2021; Eloffy et al., 2022; Hosseini & Wahid, 2016; Lepage et al., 2021; Mishra et al., 2023; V. G. Nguyen et al., 2024).

Slow pyrolysis is operated at temperatures lower than 450°C (Lepage et al., 2021), with a slow heating rate not exceeding 50°C/min and a residence time within the range of 5 to 30 min (Chen et al., 2021). This process yields a substantial amount of charcoal as an end product (Lepage et al., 2021), with longer residence times and slow heating rates decreasing the liquid yield and increasing the gas yield (Chen et al., 2021). However, since the primary products of slow pyrolysis are mostly charcoal, it is generally not considered for hydrogen production (Hosseini & Wahid, 2016; Ni et al., 2006). During the slow pyrolysis process, the hydrogen yield is low, but increases at higher temperatures and heating rates (Chen et al., 2021). For this reason, fast or flash pyrolysis with higher operating temperatures and much shorter residence times are preferred for hydrogen production (Hosseini & Wahid, 2016).

Fast pyrolysis is a high-temperature process whereby the biomass feedstock is rapidly heated in the absence of an oxidizer, resulting in vapor formation that is subsequently condensed into a liquid called bio-oil (Kayfeci et al., 2019; Ni et al., 2006). Fast pyrolysis typically produces a high bio-oil yield of up to 75 wt.% at temperatures between 450 and 600°C with a rapid heating rate of around 300°C/min and a relatively short residence time (Eloffy et al., 2022; Lepage et al., 2021). The products from fast pyrolysis are present in all three phases: gas, liquid, and solid. However, the higher heating rate facilitates the faster depolymerization of biomass to primary volatiles, resulting in an increased yield of gaseous products and a decreased yield of char (Chen et al., 2021). Gaseous products mostly consist of hydrogen, carbon monoxide, carbon dioxide, and methane, as well as minor concentrations of other gases, depending on the chemical composition of the biomass feedstock (Hosseini & Wahid, 2016; Megia et al., 2021; V. G. Nguyen et al., 2024). However, additional hydrogen can be produced by the reactions of carbon monoxide and methane with water through the water-gas shift and steam reforming reactions, respectively (Hosseini & Wahid, 2016; Megia et al., 2021). Liquid products are mainly composed of tar and bio-oils that remain in a liquid state at ambient temperature. These compounds can be further processed and converted into various products, including hydrogen (Kayfeci et al., 2019). Solid products mostly consist of char, as well as other inert materials. A schematic diagram of a fast pyrolysis process is illustrated in Figure 5.3, which includes the equipment for biomass pretreatment, a pyrolysis reactor, a cyclone for char removal, and quench and electrostatic precipitators (ESP) units to increase the liquid yield (Funke & Dahmen, 2020a).

Finally, flash pyrolysis shares similarities with fast pyrolysis but differs in its operation at temperatures above 600°C, faster heating rates exceeding 1000°C/s, and a shorter residence time of less than one second (Lepage et al., 2021; Mishra et al., 2023). The flash pyrolysis process is used to maximize the gas yield. This is because the total yield of hydrogen-rich gaseous products tends to increase as the pyrolysis temperature for the biomass feedstock rises from 500 to 800°C (Ayhan Demirbas, 2016; Li et al., 2004).

Fast or flash pyrolysis, followed by steam reforming, constitutes the standard pathway for hydrogen production from biomass thermochemical processing in the absence of oxygen (Chai et al., 2023). Under fast heating and pyrolysis conditions, an interaction occurs between the biomass pyrolysis and in-situ steam reforming of the pyrolysis intermediate products, which results in a higher yield of hydrogen-rich gas (Li et al., 2004). Nevertheless, the gas yields from both fast and flash pyrolysis are comparatively lower than those from gasification, mostly due to the lower operating temperatures of pyrolysis processes (M. Balat, 2008; Lepage et al., 2021). Actually, the hydrogen concentration in the gaseous products from biomass pyrolysis is still not significant enough to achieve commercial viability (Chen et al., 2021; Singh et al., 2023). Therefore, an effective strategy to increase the hydrogen yield involves increasing the temperature and residence time, as well as applying catalysts in the biomass pyrolysis process (Chen et al., 2021; Singh et al., 2023).

The pyrolysis process is well established at the pilot and commercial scales (TRL 8–9) for bio-oil and char production (Chen et al., 2021). In some of the currently operational plants, the by-produced syngas is also

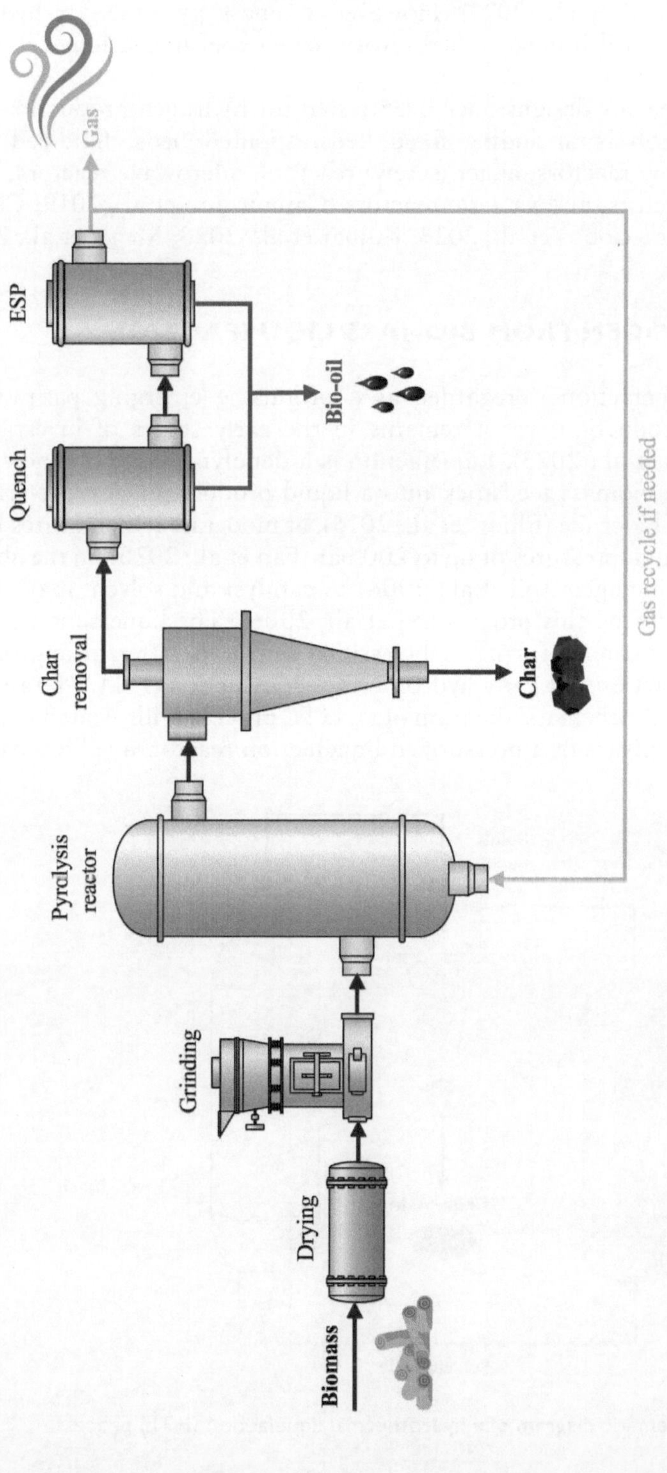

Figure 5.3 Schematic diagram of a fast pyrolysis process.

collected (Chen et al., 2021). However, biomass pyrolysis for hydrogen production is still limited to laboratory-scale experimentation (TRL 1–4) (Chen et al., 2021).

Various reactor designs have been tested for hydrogen production from biomass pyrolysis, including fixed beds, spouted beds, fluidized beds, entrained flow reactors, auger (screw) reactors, microwave reactors, rotating cone reactors, and vacuum reactors (Campuzano et al., 2019; Chen et al., 2021; Jara-Cobos et al., 2023; Kumar et al., 2023; Megia et al., 2021).

5.3 HYDROGEN FROM BIOMASS LIQUEFACTION

Biomass liquefaction is regarded as a promising emerging pathway for hydrogen production, yet it remains in the early stages of investigation (Jara-Cobos et al., 2023). Liquefaction is a depolymerization process that converts the biomass feedstock into a liquid product mixture, referred to as bio-oil or biocrude (Elliott et al., 2015), at moderate temperatures below 400°C and high pressures of up to 200 bar (Fan et al., 2020), in the absence of an oxidizing agent (Ni et al., 2006). A catalyst and solvent may also be introduced during this process (Ni et al., 2006). The liquefaction conversion pathway can be further subclassified into two different approaches: direct liquefaction (DL) and hydrothermal liquefaction (HTL) (Jara-Cobos et al., 2023). A schematic diagram of an HTL process is illustrated in Figure 5.4, which consists of a pressurized liquefaction reactor, a solid separator,

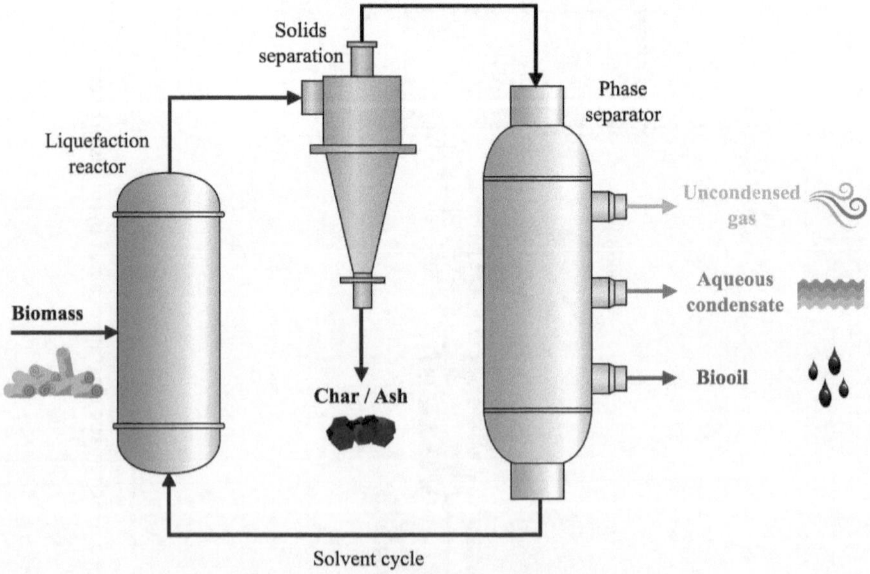

Figure 5.4 Schematic diagram of a hydrothermal liquefaction (HTL) process.

and a three-phase separator (Funke & Dahmen, 2020b). HTL is generally preferred for wet biomass and is typically conducted at moderate temperatures ranging from 250 to 370°C and high pressures between 4 and 22 MPa (Fan et al., 2020), often in the presence of a catalyst to facilitate the breakdown of the polymeric structure of the biomass. The choice of an appropriate catalyst is essential to maximize the bio-oil yield and reduce the temperature and pressure required for this process (Megia et al., 2021). Various catalysts are available, including alkaline, acid, metal-based, and mineral catalysts, with the choice among them depending on the particular substrate used. For instance, alkaline catalysts are usually the preferred option for lignocellulosic and microalgae substrates, whereas zeolites or acid catalysts find preference for alcohols (Basar et al., 2021).

The HTL process consists of three sequential stages: breakdown of the polymeric structure of biomass (depolymerization), decomposition of biomass monomers, and recondensation of reactive intermediates (Chai et al., 2023). The main product of HTL is a liquid bio-oil, along with a gaseous stream, an aqueous phase, and a solid residue byproduct (Chai et al., 2023; Megia et al., 2021). The bio-oil constitutes an organic fraction of nonpolar compounds that can be upgraded through hydrotreating to produce fuels or undergo subsequent reforming to obtain hydrogen, whereas the aqueous phase, with lower value, primarily consists of water but also contains various oxygenated organic compounds (Chai et al., 2023; Elliott et al., 2015; Megia et al., 2021). The aqueous phase can be recirculated back to the HTL reactor to increase the yield of bio-oil or be used for the production of hydrogen-rich syngas through steam reforming (Megia et al., 2021). Nevertheless, the hydrogen yield achieved through this process is thermodynamically constrained by the exothermic nature of the forward water-gas shift reaction, as well as the negative influences of other secondary reactions (Megia et al., 2021).

While HTL may be viewed as a less favorable process for hydrogen production from biomass in comparison to pyrolysis (Chai et al., 2023), it presents several advantages, such as lower operating temperature, reduced tar yield, and higher energy efficiency (Gollakota et al., 2018). Another remarkable advantage over the pyrolysis process is that there is no need for biomass drying pretreatment, resulting in significant energy savings (Jara-Cobos et al., 2023). This makes HTL particularly attractive for processing aquatic biomass, waste, and organic sludge (Jara-Cobos et al., 2023). HTL has experienced tremendous process advancements over the past decades and there are promising initiatives for demonstrating its industrial-scale viability (Funke & Dahmen, 2020b). However, challenges associated with reaching the severe operational conditions and the relatively low hydrogen yield still make biomass liquefaction processes unfavorable for hydrogen production (Jara-Cobos et al., 2023; Ni et al., 2006; Nikolaidis & Poullikkas, 2017).

Biomass liquefaction reactors can operate in batches or continuously (Jara-Cobos et al., 2023; Megia et al., 2021). Despite continuous reactors leading to higher feedstock conversion, most of the investigations have focused on batch conditions (Elliott et al., 2015). Consequently, no commercial-scale liquefaction plant for hydrogen production has been implemented thus far (Jara-Cobos et al., 2023). It is still essential to further investigate reaction mechanisms, thermal and kinetic behavior, process parameter optimization, reactor design, and economic analysis to advance the understanding and potential implementation of this technology.

5.4 HYDROGEN FROM BIOMASS GASIFICATION

The most intensively explored pathway for hydrogen production from biomass is the gasification process (Arregi et al., 2018; Fermoso et al., 2012), which stands out as the most promising thermochemical conversion pathway, considering both economic and environmental factors (Arregi et al., 2018; Milne et al., 2002; Ni et al., 2006; Nikolaidis & Poullikkas, 2017; Parthasarathy & Narayanan, 2014; Schmersahl et al., 2012; Valizadeh et al., 2022). Remarkably, gasification including gas upgrading is the only option for thermochemical hydrogen production from biomass that is nearing commercial readiness (Buffi et al., 2022; Salaudeen et al., 2023), whereas the alternative pathways still need further research and development efforts before scaling up (Buffi et al., 2022; Chai et al., 2023). In fact, all the process equipment required for hydrogen production from biomass is well established commercially, except for the gasifier equipment (H. Balat & Kirtay, 2010; Binder et al., 2018; Hosseini & Wahid, 2016). Moreover, it is worth highlighting that among all the available hydrogen production methods, biomass gasification has a clear advantage in terms of energy and exergy efficiency (Acar & Dincer, 2014; Dincer & Acar, 2015; Lepage et al., 2021).

Gasification is a form of pyrolysis, conducted at higher temperatures in the presence of a small amount of oxygen in order to increase the gas yield (Ayhan Demirbas, 2016). Therefore, it involves the partial oxidation of biomass (Ayhan Demirbas, 2016; Song et al., 2022). One of the key advantages of gasification is the ability to transform biomass into synthesis gas (syngas), which may eventually be used to produce valuable energy carriers, fuels or synthetic chemicals, such as hydrogen, methanol, ethanol, or hydrocarbons (Basu, 2018). Biomass is an attractive feedstock for the production of these compounds due to its little or negligible net carbon dioxide contribution to the atmosphere when grown sustainably (Williams et al., 1995). The gasification process allows most of the energy content of the solid fuel to be transferred to the gas phase. However, a minor fraction of the energy content remains as a carbonaceous solid residue known as biochar, which is discharged as a byproduct. Furthermore, another even smaller liquid fraction constitutes the tar, a highly undesirable black, thick, viscous liquid that may condense downstream in the low-temperature sections of a gasification plant. The presence of impurities such as tar, ash, and suspended particulate matter in the product gas may result in practical issues such as coking, fouling, deposition, sintering, slagging, agglomeration, and catalyst poisoning (Ni et al., 2006; Patra et al., 2019). Therefore, the raw product gas needs thorough cleaning before subsequent use in downstream processes (Patra et al., 2019).

The composition of the product gas from gasification is influenced by several factors, including the choice of biomass feedstock and gasifying agent, as well as the type of gasifier and its internal temperature profile (Patra et al., 2019; Steffen Heidenreich Michael Müller Pier Ugo Foscolo, 2016; Valizadeh et al., 2022). In particular, the type of gasifying agent represents a key decision concerning the composition and applicability of the product gas (V. G. Nguyen et al., 2024). The five major gasifying agents are air, oxygen, steam, carbon dioxide, and hydrogen (Salaudeen et al., 2023; Shayan et al., 2018; Song et al., 2022; Steffen Heidenreich Michael Müller Pier Ugo Foscolo, 2016), as well as mixtures thereof in varying proportions. However, only steam, carbon dioxide, and hydrogen participate in the endothermic gasification reactions, whereas oxygen (or air) is used in the combustion stage to supply the required heat input (Mahinpey & Gomez, 2016). There are several options for hydrogen production through biomass gasification, depending on the choice of gasifying agent. At present, the most favorable approaches from an economic feasibility standpoint involve using oxygen (or air) in an autothermal process, steam, or supercritical water (Lepage et al., 2021). Another important operating parameter is the temperature profile inside the gasifier. It is well known that a higher temperature during the gasification process leads to an increase in hydrogen yield (Salaudeen et al., 2023), because most of the gasification reactions yielding hydrogen (char gasification and methane reforming) are endothermic (Valizadeh et al., 2022). Nevertheless, excessively high gasification reaction temperatures can hinder an increase in hydrogen generation due to the promotion of the reverse water-gas shift reaction (Valizadeh et al., 2022).

Various reactor types are available for biomass gasification processes, including fixed beds (downdraft, updraft, and crossdraft), spouted beds, fluidized beds (bubbling, circulating, and twin bed), entrained flow reactors, rotary kilns, and plasma reactors (Arregi et al., 2018; Basu, 2018; V. G. Nguyen et al., 2024). Based on the available literature, fluidized-bed gasifiers stand out as a major reactor design for hydrogen production through steam gasification of biomass (Binder et al., 2018).

5.4.1 Oxygen gasification

The autothermal operation of a gasifier implies the need to supply oxygen to generate the required heat for the gasification process (Valizadeh et al., 2022). The most frequently used gasifying agent is molecular oxygen, predominantly in the form of air due to its availability and low cost. However, inert components such as nitrogen and, to a much lesser extent, argon dilute the gas mixture, leading to a lean fuel gas with a remarkably high nitrogen content (40–60 vol.% N_2), and thus, a low calorific value (LHV = 4–7 MJ/Nm^3), the term used for which is producer gas (Basu, 2018; Hagos et al., 2014). However, nitrogen dilution also leads to a relatively low hydrogen concentration, which requires a higher effort for hydrogen purification (Steffen Heidenreich Michael Müller Pier Ugo Foscolo, 2016). For this reason, when aiming for hydrogen production, it is advisable to generate a product gas devoid of nitrogen. Therefore, air gasification processes are generally disregarded (Binder et al., 2018).

Gasification with pure oxygen avoids dilution of the product gas with nitrogen and argon, thereby leading to a gaseous product with an increased heating value (LHV = 12–28 MJ/Nm^3) (Basu, 2018; Bridgwater, 1995), which is well-suited for hydrogen production or as synthesis gas for conversion into various products such as methanol, ammonia, methane, or gasoline (Bridgwater, 1995). Indeed, the product gas from oxygen gasification exhibits the highest heating value, with the product gas from steam and air gasification following in descending order of heating value (Basu, 2018). The primary drawback of oxygen-blown gasification often lies in the need for an air separation unit (ASU) (Steffen Heidenreich Michael Müller Pier Ugo Foscolo, 2016). The process of separating oxygen from air is currently accomplished through pressure swing adsorption (PSA), cryogenic distillation, or membrane separation, all of which require substantial energy consumption. Although substituting air with pure oxygen significantly increases the energy density of the product gas, an air separation unit (ASU) is typically not deemed suitable for small-scale applications due to the exorbitant capital and operational costs, which outweigh any potential advantage (Lepage et al., 2021; Singh et al., 2023). Accordingly, PSA is only economically feasible for large-scale oxygen gasification plants (>100 MW_{th}) (Steffen Heidenreich Michael Müller Pier Ugo Foscolo, 2016).

5.4.2 Steam gasification

Steam gasification is recognized as one of the most promising methods for producing hydrogen from biomass (Mishra et al., 2023). Notably, steam gasification allows obtaining syngas with an increased hydrogen concentration and yield (Shayan et al., 2018), as well as a higher H_2/CO molar ratio, mostly attributed to an intensification of the forward water-gas shift reaction (Valizadeh et al., 2022). The hydrogen concentration ranges from 20 to 60 vol.% for steam gasification and only from 5 to 25 vol.% for air gasification (Song et al., 2022). Moreover, steam gasification is capable of processing wet biomass with a moisture content ranging from 5 to 35 wt.%, while air gasification is limited to dry feedstock with a moisture content preferably below 10–15 wt.% (Arregi et al., 2018; Lepage et al., 2021). However, the steam gasification process is allothermal, meaning that it requires an external heat input (Steffen Heidenreich Michael Müller

Pier Ugo Foscolo, 2016; Valizadeh et al., 2022). Steam at a temperature of 700°C is typically required, and high-performance steam generators involve additional costs (Singh et al., 2023). Despite its allothermal nature, steam gasification seems to be the most suitable process for hydrogen production, with a hydrogen concentration in the gas of roughly 40%, a higher H_2/CO ratio of about 5.6, and a reduced presence of impurities compared to air gasification (Lepage et al., 2021). In addition, steam gasification provides more flexibility by allowing the use of wet biomass feedstock without the need for costly oxygen (Lepage et al., 2021). However, the formation of tar in the syngas, with concentrations ranging from 1 to 100 g/Nm^3, represents the primary challenge in steam gasification of biomass at intermediate temperatures (around 700–1100°C), whether in fixed or fluidized beds (Kumar et al., 2023; V. G. Nguyen et al., 2024). The issue of excessive tar production may be addressed either through the use of catalysts or through a thermal cracking process (Kumar et al., 2023).

Figure 5.5 shows a schematic process flow diagram (PFD) for hydrogen production via allothermal biomass gasification using steam as the gasifying agent. After the gasification and standard syngas conditioning process, a reforming stage followed by a water-gas shift stage is generally integrated to achieve a high hydrogen content within the syngas (Steffen Heidenreich Michael Müller Pier Ugo Foscolo, 2016). Incorporating a staged water-gas shift unit is essential to maximize the conversion of carbon monoxide into hydrogen, as otherwise the energy content of unconverted carbon monoxide is wasted (Schmersahl et al., 2012). Rapeseed oil methyl ester (RME) scrubbers and amine scrubbers should also be included downstream of the water-gas shift reactor to separate tar and carbon dioxide from the product gas, respectively (Binder et al., 2018). Pressure swing adsorption or membrane processes may eventually be used to separate and purify hydrogen from the hydrogen-rich syngas (Steffen Heidenreich Michael Müller Pier Ugo Foscolo, 2016). The hydrogen yield from steam gasification is reported to range between 40 and 190 g of hydrogen per kilogram of biomass feedstock (Dincer & Acar, 2015; Lepage et al., 2021).

Steam gasification of biomass constitutes a feasible approach for achieving high yields of hydrogen, enabling the deployment of large-scale hydrogen production plants (Salaudeen et al., 2023). However, the research focus is currently on medium- to small-scale plants, where allothermal gasification processes appear to be more advantageous (Binder et al., 2018). Hydrogen production through steam gasification of biomass using dual-fluidized-bed gasifiers is reported to have reached market maturity (TRL 8) (Binder et al., 2018; Kumar et al., 2023), although the complete process chain has yet to be fully validated in operational environments. Recent advancements in steam gasification have aimed to decrease the process temperatures and the tar and char formation through the use of catalysts to increase the hydrogen yield (Lepage et al., 2021). The main objectives revolve around the development of enhanced catalysts, particularly dolomite and olivine, to decrease the

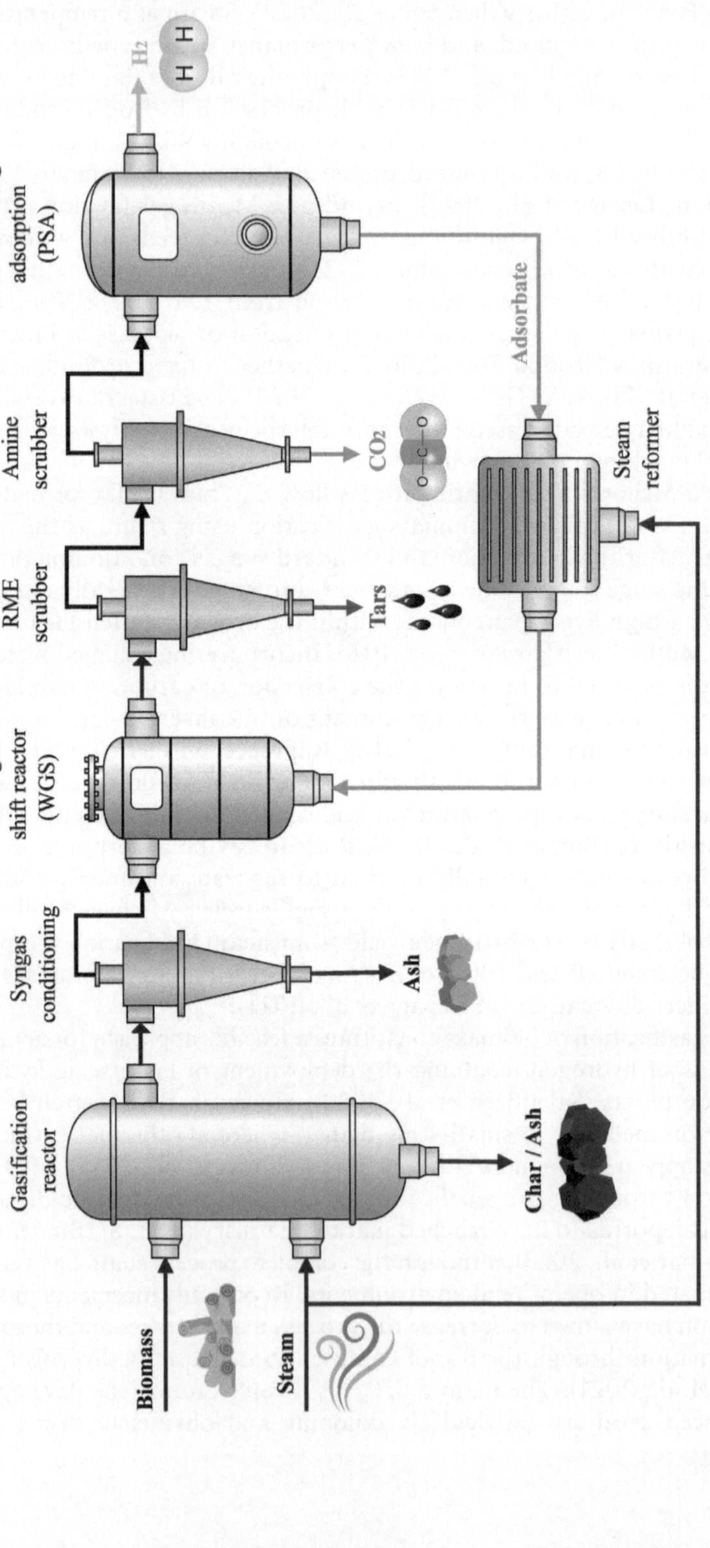

Figure 5.5 Process flow diagram for hydrogen production via steam gasification of biomass.

operating temperature, and consequently, the total energy consumption and tar content, ultimately reducing overall process costs (Lepage et al., 2021).

5.4.3 Supercritical water gasification

Biomass gasification in supercritical water is another pathway with significant potential for renewable hydrogen production that has been attracting increasing interest in recent years (A. Demirbas, 2010; Ayhan Demirbas, 2016; Hosseini & Wahid, 2016; Reddy et al., 2014; Singh et al., 2023). At temperatures and pressures above the critical point (374°C, 221 bar), the dielectric constant of water exhibits a substantial reduction (A. Demirbas, 2010; Eloffy et al., 2022; Lepage et al., 2021; Pinkard et al., 2019; Reddy et al., 2014). Furthermore, the number of hydrogen bonds decreases, and their strength weakens considerably (A. Demirbas, 2010; Ayhan Demirbas, 2016; Eloffy et al., 2022; Lepage et al., 2021). Consequently, upon reaching supercritical conditions, water transitions from a polar to a nonpolar solvent, with exceptional solubility toward organic compounds containing nonpolar groups as well as most permanent gases (A. Demirbas, 2010; Ayhan Demirbas, 2016; Eloffy et al., 2022; Hosseini & Wahid, 2016; Pinkard et al., 2019). Lignocellulosic biomass, mainly composed of lignin, cellulose, and hemicellulose, readily dissolves in supercritical water under high temperature and pressure conditions (Reddy et al., 2014). Cellulose and hemicellulose are broken down into simple sugars, which are subsequently gasified by increasing the reaction temperature and pressure (Reddy et al., 2014). By contrast, lignin is converted into phenolic compounds, which are further reformed into simple gases such as H_2, CO, CO_2, and CH_4 (Reddy et al., 2014).

Unlike biomass gasification using other agents such as air or steam, supercritical water gasification (SCWG) presents the advantage of directly handling wet biomass, even with moisture contents exceeding 50 wt.% without the need for a previous energy-intensive drying step, and achieves higher gasification efficiency at lower temperatures (H. Balat & Kirtay, 2010; A. Demirbas, 2010; Ayhan Demirbas, 2016; Eloffy et al., 2022; Hosseini & Wahid, 2016; Pinkard et al., 2019; Reddy et al., 2014; Yan et al., 2006). Moreover, tar and char formation are substantially lower in biomass gasification under supercritical conditions (Hosseini & Wahid, 2016; Mishra et al., 2023). Supercritical water has a higher density than that of steam, leading to an increased space-time yield (A. Demirbas, 2010). Thus, the residence periods are generally very short (2–6 s) (Eloffy et al., 2022; Lepage et al., 2021). The higher specific heat and thermal conductivity of supercritical water are also advantageous for facilitating the endothermic reforming reactions (A. Demirbas, 2010). Another benefit of SCWG lies in the production of hydrogen at high pressure, which can be directly stored, thereby eliminating the substantial energy costs typically required for compression (A. Demirbas, 2010; Hosseini & Wahid, 2016; Reddy et al., 2014).

The operating temperatures in SCWG can be divided into two temperature ranges, namely, low temperatures (374–500°C) and high temperatures (500–800°C) (Eloffy et al., 2022; Lepage et al., 2021; Reddy et al., 2014), although higher temperatures are conducive to increased hydrogen yields (Reddy et al., 2014). However, sustaining these severe operational conditions demands substantial energy consumption, leading to increased costs and limited scalability (Arregi et al., 2018; Eloffy et al., 2022). Indeed, the operational experience of supercritical steam gasification is at present mostly limited to bench-scale runs conducted in batch reactors and, to a lesser extent, in continuous flow tubular or fluidized-bed reactors (Arregi et al., 2018).

The development of an industrial-scale supercritical gasification system has not been realized thus far, because the high-pressure requirements involve substantial capital and operational expenses (Arregi et al., 2018; Eloffy et al., 2022). Reactors designed for SCWG are currently in a developmental stage, with considerable efforts focused on enhancing their energy efficiency and achieving high hydrogen yields (Reddy et al., 2014). Addressing several technical challenges, such as standardizing reactor design, neutralizing waste, and gaining a comprehensive understanding of the chemical dynamics, is essential for the development of commercially viable SCWG systems (Eloffy et al., 2022; Pinkard et al., 2019). Moreover, the incorporation of catalysts in SCWG is appealing due to their ability to reduce the need for extremely severe operating conditions and increase the hydrogen yields (Reddy et al., 2014; Singh et al., 2023). Various catalysts, including alkali, transition metals, and activated carbon, have been tested. However, common issues such as reactor plugging, corrosion, and fouling, as well as catalyst deactivation and sintering, still need to be overcome (Reddy et al., 2014).

5.4.4 Chemical looping gasification

Chemical looping gasification (CLG) is a novel process for converting carbonaceous materials such as coal, biomass, or other solid fuels into high-quality syngas (N. M. Nguyen et al., 2021), which can then be used to produce high-purity hydrogen or further processed into various valuable chemicals or fuels (Kumar et al., 2023).

In conventional gasification processes, the biomass feedstock is reacted with a gasifying agent (usually air, oxygen, and/or steam) at high temperatures, producing a mixture of syngas and other byproducts. However, CLG differs in that it consists of a cyclic process wherein an oxygen carrier indirectly transfers oxygen from an oxidizing agent to the biomass feedstock, thereby avoiding direct contact between the fuel and air (N. M. Nguyen et al., 2021). Typically used oxygen carriers in CLG processes include transition metal oxides such as Mn_2O_3, Fe_2O_3, NiO, and CuO (Kumar et al., 2023; N. M. Nguyen et al., 2021; Zhao et al., 2017). Among these, nickel-based oxygen carriers exhibit the highest chemical reactivity (N. M. Nguyen et al., 2021), but iron-based oxygen carriers have received more attention due to their cost-effectiveness (Kumar et al., 2023).

The key advantage of chemical looping gasification lies in its potential for higher efficiency and lower CO_2 emissions compared to conventional gasification processes (Goel et al., 2022; N. M. Nguyen et al., 2021). By separating the oxygen carrier from the fuel, CLG avoids dilution of the product gas with atmospheric nitrogen (Goel et al., 2022), leading to a syngas stream with a high concentration of hydrogen, which reduces the complexity of downstream hydrogen separation and purification steps (Zhao et al., 2017). Furthermore, CLG can enable carbon capture and storage (CCS) more effectively by providing a concentrated stream of CO_2 for capture (Zhao et al., 2017). Another benefit of CLG is that certain metal oxide oxygen carriers may exhibit catalytic properties in the cracking of biomass tar (Goel et al., 2022; Kumar et al., 2023; N. M. Nguyen et al., 2021), as well as in the conversion of precursors of polluting by-products such as nitrogen oxides into molecular nitrogen (Goel et al., 2022). At the same time, the inorganic compounds present in biomass ash may also act as effective catalysts for gasification reactions (N. M. Nguyen et al., 2021; Zhao et al., 2017).

Two common reactor types have been proposed for CLG processes: fixed-bed and fluidized-bed reactors, which could be implemented in three configurations, namely a two-reactor system, three-reactor system, and alternating packed or fluidized-bed reactor (N. M. Nguyen et al., 2021). However, the most popular configuration for CLG is the dual reactor system (Goel et al., 2022; Kumar et al., 2023; N. M. Nguyen et al., 2021). As illustrated in Figure 5.6, this configuration consists of two interconnected fluidized-bed reactors: a fuel reactor and an air reactor. In the fuel reactor, the carbonaceous material is exposed to the metal oxide particles (Me_xO_y), which act as the oxygen carrier. At high temperatures, the metal oxide releases oxygen, reacting with the fuel to produce syngas. Most of the reactions occurring in the fuel reactor are endothermic reduction reactions, requiring a large heat supply for stable operation (N. M. Nguyen et al., 2021). The reduced metal oxide (Me_xO_{y-1}) is then transported to the air reactor, where it is oxidized by ambient air, regenerating the metal oxide through exothermic reactions. The released heat can be utilized for various purposes, such as producing steam for power generation or for other industrial processes.

Despite their numerous advantages, chemical looping technologies for hydrogen production are mostly in the research and development phase (TRL 3–5) due to the numerous challenges they still face (Goel et al., 2022; N. M. Nguyen et al., 2021). Over the years, extensive research has been conducted on CLG of biomass, exploring numerous oxygen carriers in pilot-scale gasification plants ranging from a few kilowatts to megawatts to assess their feasibility (N. M. Nguyen et al., 2021). However, further research and development efforts are needed to optimize and scale up CLG technologies for commercial applications.

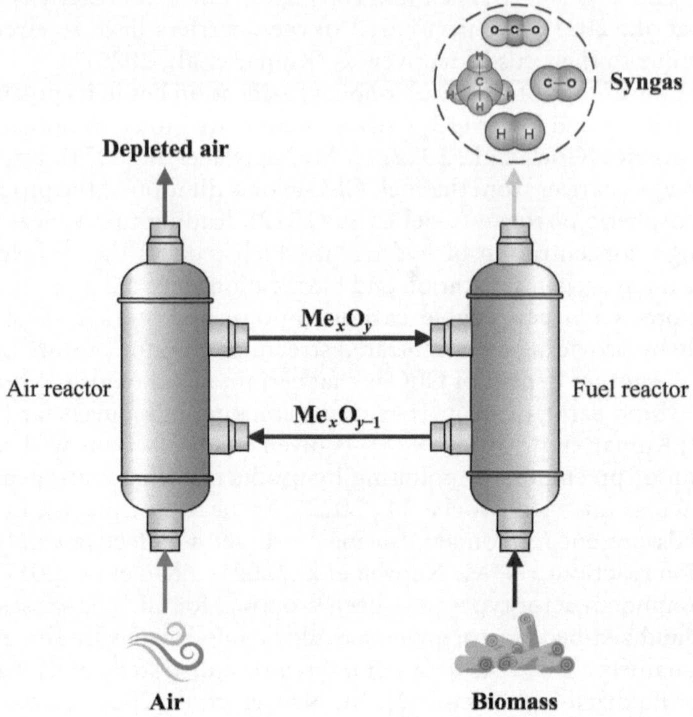

Figure 5.6 Schematic diagram of a dual bed chemical looping gasification (CLG) process.

5.4.5 Sorption enhanced gasification

Sorption enhanced gasification (SEG) is a promising technology for renewable hydrogen production coupled to carbon capture and sequestration, because it can generate hydrogen-rich syngas while simultaneously capturing CO2 in-situ through the use of a solid sorbent (Koppatz et al., 2009; Martínez et al., 2022; Parvez et al., 2021; Pfeifer, 2013). Among various available sorbents, limestone (primarily composed of CaCO3), is extensively used due to its high CO2 capture efficiency and low cost (Koppatz et al., 2009; Müller et al., 2017; Parvez et al., 2021). In addition to fossil fuels, various biomass- and waste-derived feedstocks can also be used in SEG (Parvez et al., 2021; Pfeifer, 2013), potentially achieving carbon negativity if the biogenic CO2 emissions are used for carbon capture, utilization and storage (CCUS).The fundamental principle of the SEG process is illustrated in Figure 5.7a) (Parvez et al., 2021; Pfeifer, 2013). The hot bed material, acting as a heat and CO2 carrier, circulates between two interconnected fluidized bed reactors, namely the gasifier and combustor. Concurrently, the reactive bed material facilitates continuous CO2 removal through the cyclic processes of CaO carbonation and CaCO3 calcination (Koppatz et

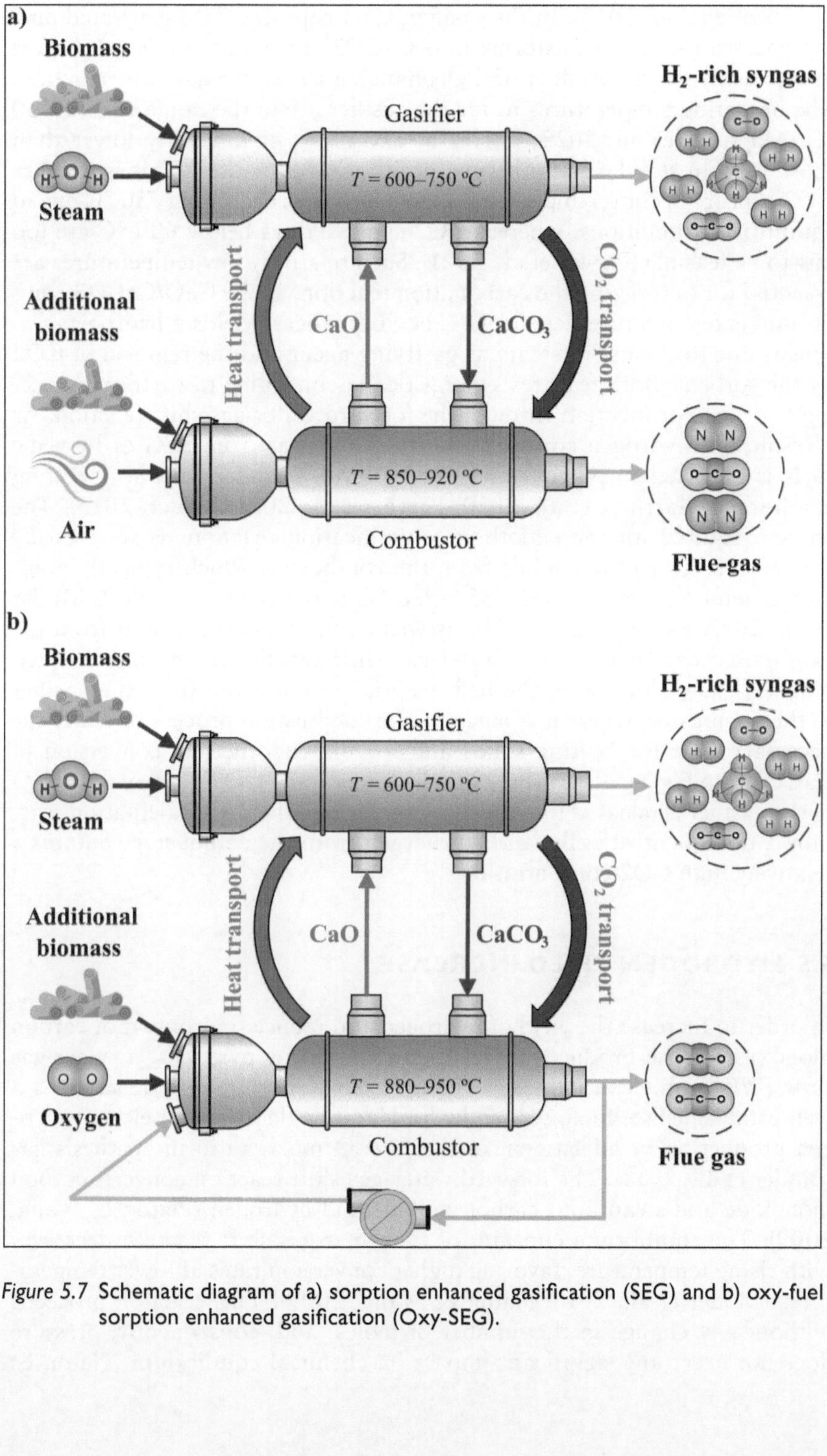

Figure 5.7 Schematic diagram of a) sorption enhanced gasification (SEG) and b) oxy-fuel sorption enhanced gasification (Oxy-SEG).

al., 2009; Pfeifer, 2013). In the gasifier, CaO captures CO2 generated during gasification and transforms into CaCO3 (carbonation). This process promotes hydrogen production by enhancing the water-gas shift reaction. The operating temperatures inside the gasifier are in the range of 600–750 °C (Martínez et al., 2022; Parvez et al., 2021), significantly lower than those used in standard allothermal gasification processes. This is because CO2 capture becomes impractical at temperatures exceeding 750 °C due to equilibrium limitations, whereas CO2 capture rates below 600 °C are too low to be feasible (Parvez et al., 2021). Such relatively low temperatures are essential for optimizing the carbonation reaction, as the CaO/CaCO3 equilibrium is temperature-dependent. The SEG process yields a hydrogen-rich syngas due to the use of steam as gasifying agent and the removal of CO2 by the sorbent. Both features synergistically contribute to further enhancing hydrogen production through the forward water-gas shift reaction. As a result, the hydrogen concentration in the syngas from SEG of biogenic fuels is reported to potentially exceed 70 vol.% under certain operating conditions (Martínez et al., 2020; Parvez et al., 2021; Pfeifer, 2013). The energy required for the endothermic gasification reactions is supplied by the sensible heat of the solids from the combustor, which typically operates at temperatures between 850–920 °C (Koppatz et al., 2009; Müller et al., 2017; Parvez et al., 2021), as well as by the heat released from the exothermic carbonation and water-gas shift reactions. The unconverted biomass char, along with the bed material, is conveyed from the gasifier to the combustor, where it is burned. The combustion process provides the necessary heat for heating solids and for the endothermic conversion of CaCO3 into CaO (calcination). Simultaneously, the CO2 absorbed by CaO in the gasifier is released into the flue gas through the same calcination reaction. Consequently,the flue gas discharged from the combustor contains a relatively high CO2 concentration.

5.5 HYDROGEN YIELD INCREASE

In order to increase the yield of hydrogen and reduce the content of carbon monoxide in the product gas from gasification or pyrolysis, a water-gas shift (WGS) unit can be used. The forward water-gas shift reaction is a well-established technology in industrial large-scale plants for either hydrogen production or adjustment of the H_2/CO molar ratio in synthesis gas (Binder et al., 2018). The forward water-gas shift reaction converts carbon monoxide and steam into carbon dioxide and hydrogen (Platon & Wang, 2009). The equilibrium constant of the water-gas shift reaction decreases with rising temperature, favoring higher conversion rates at lower temperatures (Binder et al., 2018; Platon & Wang, 2009). This reaction proceeds without any change in the number of moles, and, consequently, pressure does not exert any significant impact on chemical equilibrium (Platon &

Wang, 2009). In order to achieve low concentrations of carbon monoxide and high concentrations of hydrogen in the effluent without requiring a substantial excess of steam, operating the reaction at lower temperatures is preferred (Platon & Wang, 2009). However, reaction rates diminish at low temperatures and the reaction becomes kinetically hindered (Platon & Wang, 2009).

On an industrial scale, a WGS unit typically consists of one or more fixed-bed reactors (Binder et al., 2018). For adjustment of the H_2/CO molar ratio in synthesis applications, a high-temperature WGS stage with a relatively small CO conversion is typically enough (Binder et al., 2018). However, for hydrogen production, the syngas stream requires further processing across one or two additional stages with gradually decreasing inlet and outlet temperatures to achieve a sufficiently high CO conversion (Binder et al., 2018). Traditionally, and in most industrial settings, the WGS unit consists of two fixed-bed reactors and only one low-temperature WGS stage is used (Platon & Wang, 2009). The molar steam-to-dry gas ratios in the WGS reactor vary depending on the feed, with a typical range from 0.6 to 2.2, and a minimum ratio of 0.4 (Binder et al., 2018).

In order to reach reasonable reaction rates, different catalysts may be used. Catalysts based on ferrochrome (Fe-Cr) are generally used for the high-temperature water-gas shift stage (Binder et al., 2018). This high-temperature stage operates under adiabatic conditions, with gas inlet temperatures in the range of 350 to 550°C, space velocities between 400 and 1200 h^{-1}, along with operating pressures above 20 bar, which vary according to plant requirements (Binder et al., 2018; Platon & Wang, 2009). The low-temperature stage typically uses Co-Mo or Cu-Zn-based catalysts (Binder et al., 2018) and operates at temperatures around 200°C, leveraging the favorable equilibrium conditions below 250°C (Platon & Wang, 2009).

5.6 HYDROGEN SEPARATION AND PURIFICATION

Hydrogen separation and purification from hydrogen-rich syngas may be accomplished through either pressure swing adsorption (PSA) or membrane-based processes (Edlund, 2009; Kayfeci et al., 2019). However, when a high level of purity (> 99%) is required, a PSA process becomes essential (Binder et al., 2018). Indeed, the use of a PSA process has become a prevalent industrial practice for high-purity hydrogen production from a feed gas containing 70–90% H_2 (Sircar & Golden, 2009).

5.6.1 Pressure swing adsorption

The PSA process relies on the physical binding of gas molecules onto a solid adsorbent material. The interaction between the feed gas and the adsorbent is mostly influenced by factors such as the gas component, its partial

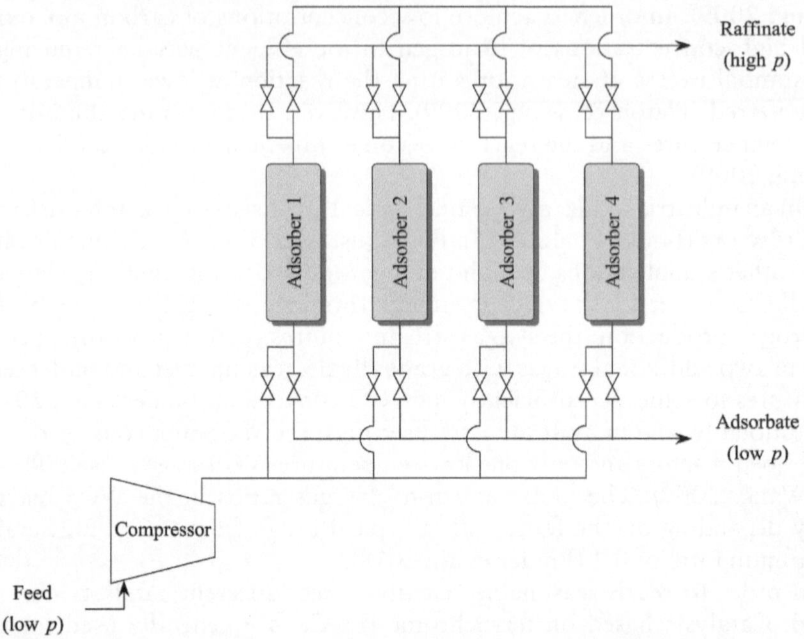

Figure 5.8 Schematic diagram of a pressure swing adsorption (PSA) process.

pressure, the temperature, and the type of adsorbent (Binder et al., 2018). PSA is a mature and widely used process for gas separation at a commercial scale in various industrial applications, such as for air separation, hydrogen production, and biogas upgrading (Binder et al., 2018; Sircar & Golden, 2009). In fact, PSA technology is currently used for hydrogen purification in over 85% of global hydrogen production units, with several hundred PSA process units installed worldwide (Sircar & Golden, 2009).

A simplified process flow diagram of a PSA process is presented in Figure 5.8. The incoming feed gas is compressed and sequentially introduced into various adsorber vessels containing porous solid adsorbents, such as activated carbons, zeolites, silica, or alumina gels (Binder et al., 2018; Kayfeci et al., 2019; Sircar & Golden, 2009). The impurities present in the hydrogen-containing feed gas mixture are selectively adsorbed onto packed columns of adsorbents within the vessels to yield a product gas with high hydrogen purity at feed pressure (Sircar & Golden, 2009). These packed columns are cyclically regenerated through pressure reduction (depressurization) and purging with high-pressure product gas (raffinate) (Binder et al., 2018). The desorbed low-pressure product gas (adsorbate), enriched in the more strongly adsorbed components, contains the impurities from the feed and can be recycled for further utilization within upstream or downstream processes (Binder et al., 2018).

Although the concept of a hydrogen PSA process may seem rather straightforward, designing a functional hydrogen PSA unit can be quite challenging because it generally requires a multicolumn adsorption system operating under cyclic steady-state conditions with a series of sequential, non-isothermal, non-isobaric, and non-steady-state process steps (Sircar & Golden, 2009). These sequential operations include adsorption, desorption, and numerous complementary steps aimed at increasing the hydrogen purity and enhancing recovery in the product gas (Sircar & Golden, 2009).

The affinity for adsorption onto activated carbon of the main gas components of the processed gas can be characterized by the following relationship: $CO_2 > CH_4 > CO > H_2$ (Sircar & Golden, 2009). This indicates that activated carbon has a higher tendency to adsorb CO_2, making it more effective at removing carbon dioxide from the feed gas compared to, for instance, hydrogen (Binder et al., 2018). Therefore, activated carbon is a suitable adsorbent for pure hydrogen production (Binder et al., 2018).

With additional separation of the carbon dioxide prior to the PSA unit and further optimization measures, it is possible to achieve a hydrogen recovery rate of 90% (Binder et al., 2018). Nevertheless, there is an inherent trade-off between hydrogen recovery and its purity (Binder et al., 2018). Accordingly, producing high-grade hydrogen with a purity exceeding 99.999% would require a significantly lower recovery rate (Sircar & Golden, 2009).

5.6.2 Membrane processes

Similar to PSA, membrane-based processes are gas separation processes driven by pressure differences (Binder et al., 2018; Edlund, 2009; Kayfeci et al., 2019). In contrast to PSA, however, membrane processes usually provide a continuous flow of purified product rather than a cyclical one, which simplifies their control (Binder et al., 2018; Edlund, 2009). When choosing between membrane and PSA purification processes, scale considerations are important. Practical experience has demonstrated that PSA can be scaled up economically and may represent the best choice for large-scale stationary applications (Edlund, 2009). Conversely, the economic feasibility of scaling down PSA is generally limited, making membrane-based processes a more favorable choice for small-scale applications (Edlund, 2009).

A membrane constitutes a thin and discrete interface with a molecularly homogeneous or heterogeneous structure that regulates the permeation of the chemical components it comes into contact with. Figure 5.9 illustrates a schematic representation of a membrane separation system for hydrogen purification, wherein a membrane is enclosed within a pressure vessel to form a membrane module (Binder et al., 2018; Edlund, 2009). A feed gas stream, such as syngas from biomass pyrolysis or gasification, is introduced into the membrane system, where it divides into two fractions: permeate

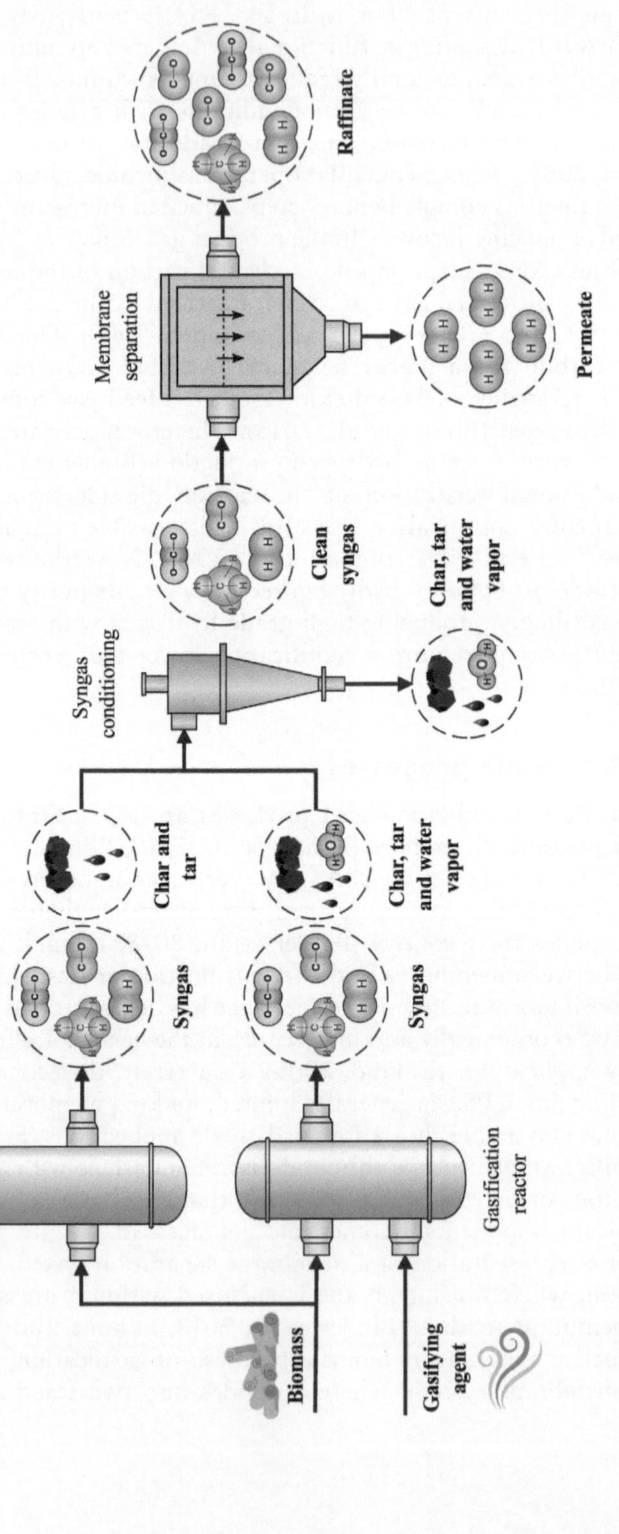

Figure 5.9 Schematic diagram of a membrane for hydrogen separation from syngas.

and raffinate. The permeate fraction contains a higher concentration of gas molecules that more readily penetrate through the membrane, such as hydrogen (Binder et al., 2018).

5.7 CONCLUSIONS

Hydrogen production from thermochemical biomass conversion processes has already been shown to be environmentally and economically attractive and has been demonstrated to be a feasible option once the required technology has reached a state of maturity and becomes commercially available. However, it should be noted that in all thermochemical biomass conversion processes, hydrogen is mixed with several other gas constituents, and thus, downstream unit operations for separation and purification of the hydrogen gas are typically required. Nowadays, several methods, such as water-gas shift reactors, pressure swing adsorption (PSA) units, and membrane separation processes, are technologically proven and available on the market for hydrogen gas concentration, separation, and purification. Nevertheless, thermochemical biomass conversion technologies and their integration with these well-established downstream unit operations remain a bottleneck in achieving market maturity for the overall process. Biomass-to-hydrogen process chains have yet to undergo testing at demonstration scale to achieve the desired configuration in terms of performance and throughput. However, it is anticipated that thermochemical biomass conversion processes, particularly biomass gasification, may soon be available for large-scale hydrogen production. Consequently, continued research and development efforts are still essential to demonstrate the full process chain for successful implementation of the technology. Furthermore, expanding the range of feedstock options would increase the flexibility of thermochemical conversion processes for hydrogen production from biomass and improve their overall economic feasibility.

REFERENCES

Abdin, Z., Zafaranloo, A., Rafiee, A., Mérida, W., Lipiński, W., & Khalilpour, K. R. (2020). Hydrogen as an energy vector. *Renewable and Sustainable Energy Reviews*, *120*, 109620. https://doi.org/10.1016/J.RSER.2019.109620

Acar, C., & Dincer, I. (2014). Comparative assessment of hydrogen production methods from renewable and non-renewable sources. *International Journal of Hydrogen Energy*, *39*(1), 1–12. https://doi.org/10.1016/J.IJHYDENE.2013.10.060

Arregi, A., Amutio, M., Lopez, G., Bilbao, J., & Olazar, M. (2018). Evaluation of thermochemical routes for hydrogen production from biomass: A review. *Energy Conversion and Management*, *165*, 696–719. https://doi.org/10.1016/J.ENCONMAN.2018.03.089

Balat, H., & Kirtay, E. (2010). Hydrogen from biomass – Present scenario and future prospects. *International Journal of Hydrogen Energy*, *35*(14), 7416–7426. https://doi.org/10.1016/J.IJHYDENE.2010.04.137

Balat, M. (2008). Hydrogen-rich gas production from biomass via pyrolysis and gasification processes and effects of catalyst on hydrogen yield. *Energy Sources, Part A*, *30*(6), 552–564. https://doi.org/10.1080/15567030600817191

Basar, I. A., Liu, H., Carrere, H., Trably, E., & Eskicioglu, C. (2021). A review on key design and operational parameters to optimize and develop hydrothermal liquefaction of biomass for biorefinery applications. *Green Chemistry*, *23*(4), 1404–1446. https://doi.org/10.1039/D0GC04092D

Basu, P. (2018). Biomass gasification, pyrolysis and torrefaction: Practical design and theory. In *Biomass gasification, pyrolysis and torrefaction: Practical design and theory*. Elsevier. https://doi.org/10.1016/C2016-0-04056-1

Binder, M., Kraussler, M., Kuba, M., & Luisser, M. (2018). Hydrogen from biomass gasification. *IEA Bioenergy*. https://www.ieabioenergy.com/wp-content/uploads/2019/01/Wasserstoffstudie_IEA-final.pdf

Bridgwater, A. V. (1995). The technical and economic feasibility of biomass gasification for power generation. *Fuel*, *74*(5), 631–653. https://doi.org/10.1016/0016-2361(95)00001-L

Buffi, M., Prussi, M., & Scarlat, N. (2022). Energy and environmental assessment of hydrogen from biomass sources: Challenges and perspectives. *Biomass and Bioenergy*, *165*, 106556. https://doi.org/10.1016/J.BIOMBIOE.2022.106556

Campuzano, F., Brown, R. C., & Martínez, J. D. (2019). Auger reactors for pyrolysis of biomass and wastes. *Renewable and Sustainable Energy Reviews*, *102*, 372–409. https://doi.org/10.1016/J.RSER.2018.12.014

Chai, Y. H., Mohamed, M., Cheng, Y. W., Chin, B. L. F., Yiin, C. L., Yusup, S., & Lam, M. K. (2023). A review on potential of biohydrogen generation through waste decomposition technologies. *Biomass Conversion and Biorefinery*, *13*(10), 8549–8574. https://doi.org/10.1007/s13399-021-01333-z

Chen, W. H., Farooq, W., Shahbaz, M., Naqvi, S. R., Ali, I., Al-Ansari, T., & Saidina Amin, N. A. (2021). Current status of biohydrogen production from lignocellulosic biomass, technical challenges and commercial potential through pyrolysis process. *Energy*, *226*, 120433. https://doi.org/10.1016/J.ENERGY.2021.120433

Demirbas, A. (2010). Hydrogen production from biomass via supercritical water gasification. *Energy Sources, Part A: Recovery, Utilization and Environmental Effects*, *32*(14), 1342–1354. https://doi.org/10.1080/15567030802654038

Demirbas, A. (2016). Comparison of thermochemical conversion processes of biomass to hydrogen-rich gas mixtures. *Energy Sources, Part A: Recovery, Utilization, and Environmental Effects*, *38*(20), 2971–2976. https://doi.org/10.1080/15567036.2015.1122686

Dincer, I., & Acar, C. (2015). Review and evaluation of hydrogen production methods for better sustainability. *International Journal of Hydrogen Energy*, *40*(34), 11094–11111. https://doi.org/10.1016/J.IJHYDENE.2014.12.035

Edlund, D. (2009). Hydrogen membrane technologies and application in fuel processing. In *Hydrogen and syngas production and purification technologies* (pp. 357–384). https://doi.org/10.1002/9780470561256.CH8

Elliott, D. C., Biller, P., Ross, A. B., Schmidt, A. J., & Jones, S. B. (2015). Hydrothermal liquefaction of biomass: Developments from batch to continuous process. *Bioresource Technology*, *178*, 147–156. https://doi.org/10.1016/J.BIORTECH.2014.09.132

Eloffy, M. G., Elgarahy, A. M., Saber, A. N., Hammad, A., El-Sherif, D. M., Shehata, M., Mohsen, A., & Elwakeel, K. Z. (2022). Biomass-to-sustainable biohydrogen: Insights into the production routes, and technical challenges. *Chemical Engineering Journal Advances*, *12*, 100410. https://doi.org/10.1016/J.CEJA.2022.100410

Fan, L., Zhang, H., Li, J., Wang, Y., Leng, L., Li, J., Yao, Y., Lu, Q., Yuan, W., & Zhou, W. (2020). Algal biorefinery to value-added products by using combined processes based on thermochemical conversion: A review. *Algal Research*, *47*, 101819. https://doi.org/10.1016/J.ALGAL.2020.101819

Fermoso, J., Rubiera, F., & Chen, D. (2012). Sorption enhanced catalytic steam gasification process: A direct route from lignocellulosic biomass to high purity hydrogen. *Energy & Environmental Science*, *5*(4), 6358–6367. https://doi.org/10.1039/C2EE02593K

Funke, A., & Dahmen, N. (2020a, July 1). Direct thermochemical liquefaction of biomass – Characteristics, processes and technologies. *IEA Bioenergy*. https://doi.org/10.5445/IR/1000120883

Funke, A., & Dahmen, N. (2020b, July 1). Direct thermochemical liquefaction of biomass – Characteristics, processes and technologies. *IEA Bioenergy*. https://doi.org/10.5445/IR/1000120883

Goel, A., Moghaddam, E. M., Liu, W., He, C., & Konttinen, J. (2022). Biomass chemical looping gasification for high-quality syngas: A critical review and technological outlooks. *Energy Conversion and Management*, *268*, 116020. https://doi.org/10.1016/J.ENCONMAN.2022.116020

Gollakota, A. R. K., Kishore, N., & Gu, S. (2018). A review on hydrothermal liquefaction of biomass. *Renewable and Sustainable Energy Reviews*, *81*, 1378–1392. https://doi.org/10.1016/J.RSER.2017.05.178

Hagos, F. Y., Aziz, A. R. A., & Sulaiman, S. A. (2014). *Trends of syngas as a fuel in internal combustion engines*. Http://Dx.Doi.Org/10.1155/2014/401587. https://doi.org/10.1155/2014/401587

Hosseini, S. E., & Wahid, M. A. (2016). Hydrogen production from renewable and sustainable energy resources: Promising green energy carrier for clean development. *Renewable and Sustainable Energy Reviews*, *57*, 850–866. https://doi.org/10.1016/J.RSER.2015.12.112

IRENA. (2019). *Hydrogen: A renewable energy perspective*. International Renewable Energy Agency. Report. www.irena.org

Jara-Cobos, L., Abril-González, M., & Pinos-Vélez, V. (2023). Production of hydrogen from lignocellulosic biomass: A review of technologies. *Catalysts*, *13*(4), 766. https://doi.org/10.3390/CATAL13040766

Kayfeci, M., Keçebaş, A., & Bayat, M. (2019). Hydrogen production. *Solar Hydrogen Production: Processes, Systems and Technologies*, 45–83. https://doi.org/10.1016/B978-0-12-814853-2.00003-5

Kumar, G., Eswari, A. P., Kavitha, S., Kumar, M. D., Kannah, R. Y., How, L. C., Muthukaruppan, G., & Banu, J. R. (2023). Thermochemical conversion routes of hydrogen production from organic biomass: Processes, challenges and limitations. *Biomass Conversion and Biorefinery*, *13*(10), 8509–8534. https://doi.org/10.1007/s13399-020-01127-9

Lanjekar, P. R., Panwar, N. L., & Agrawal, C. (2023). A comprehensive review on hydrogen production through thermochemical conversion of biomass for energy security. *Bioresource Technology Reports*, *21*, 101293. https://doi.org/10.1016/J.BITEB.2022.101293

Lepage, T., Kammoun, M., Schmetz, Q., & Richel, A. (2021). Biomass-to-hydrogen: A review of main routes production, processes evaluation and techno-economical assessment. *Biomass and Bioenergy*, *144*, 105920. https://doi.org/10.1016/J.BIOMBIOE.2020.105920

Levin, D. B., & Chahine, R. (2010). Challenges for renewable hydrogen production from biomass. *International Journal of Hydrogen Energy*, *35*(10), 4962–4969. https://doi.org/10.1016/J.IJHYDENE.2009.08.067

Li, S., Xu, S., Liu, S., Yang, C., & Lu, Q. (2004). Fast pyrolysis of biomass in free-fall reactor for hydrogen-rich gas. *Fuel Processing Technology*, *85*(8–10), 1201–1211. https://doi.org/10.1016/J.FUPROC.2003.11.043

Mahinpey, N., & Gomez, A. (2016). Review of gasification fundamentals and new findings: Reactors, feedstock, and kinetic studies. *Chemical Engineering Science*, *148*, 14–31. https://doi.org/10.1016/j.ces.2016.03.037

Megia, P. J., Vizcaino, A. J., Calles, J. A., & Carrero, A. (2021). Hydrogen production technologies: From fossil fuels toward renewable sources. A mini review. *Energy and Fuels*, *35*(20), 16403–16415. https://pubs.acs.org/doi/full/10.1021/acs.energyfuels.1c02501

Milne, T., Elam, C., & Evans, R. (2002). *Hydrogen from biomass: State of the art and research challenges* (pp. 1–82). https://doi.org/10.2172/792221

Mishra, K., Singh Siwal, S., Kumar Saini, A., & Thakur, V. K. (2023). Recent update on gasification and pyrolysis processes of lignocellulosic and algal biomass for hydrogen production. *Fuel*, *332*, 126169. https://doi.org/10.1016/J.FUEL.2022.126169

Nguyen, N. M., Alobaid, F., Dieringer, P., & Epple, B. (2021). Biomass-based chemical looping gasification: Overview and recent developments. *Applied Sciences*, *11*(15), 7069. https://doi.org/10.3390/APP11157069

Nguyen, V. G., Nguyen-Thi, T. X., Phong Nguyen, P. Q., Tran, V. D., Ağbulut, Ü., Nguyen, L. H., Balasubramanian, D., Tarelko, W., A. Bandh, S., & Khoa Pham, N. D. (2024). Recent advances in hydrogen production from biomass waste with a focus on pyrolysis and gasification. *International Journal of Hydrogen Energy*, *54*, 127–160. https://doi.org/10.1016/J.IJHYDENE.2023.05.049

Ni, M., Leung, D. Y. C., Leung, M. K. H., & Sumathy, K. (2006). An overview of hydrogen production from biomass. *Fuel Processing Technology*, *87*(5), 461–472. https://doi.org/10.1016/J.FUPROC.2005.11.003

Nikolaidis, P., & Poullikkas, A. (2017). A comparative overview of hydrogen production processes. *Renewable and Sustainable Energy Reviews*, *67*, 597–611. https://doi.org/10.1016/J.RSER.2016.09.044

Parthasarathy, P., & Narayanan, K. S. (2014). Hydrogen production from steam gasification of biomass: Influence of process parameters on hydrogen yield – A review. *Renewable Energy*, *66*, 570–579. https://doi.org/10.1016/J.RENENE.2013.12.025

Patra, T. K., Mukherjee, S., & Sheth, P. N. (2019). Process simulation of hydrogen rich gas production from producer gas using HTS catalysis. *Energy*, *173*, 1130–1140. https://doi.org/10.1016/J.ENERGY.2019.02.136

Pinkard, B. R., Gorman, D. J., Tiwari, K., Rasmussen, E. G., Kramlich, J. C., Reinhall, P. G., & Novosselov, I. V. (2019). Supercritical water gasification: Practical design strategies and operational challenges for lab-scale, continuous flow reactors. *Heliyon*, *5*(2), e01269. https://doi.org/10.1016/J.HELIYON.2019.E01269

Platon, A., & Wang, Y. (2009). Water-gas shift technologies. In *Hydrogen and syngas production and purification technologies* (pp. 311–328). https://doi.org/10.1002/9780470561256.CH6

Reddy, S. N., Nanda, S., Dalai, A. K., & Kozinski, J. A. (2014). Supercritical water gasification of biomass for hydrogen production. *International Journal of Hydrogen Energy*, *39*(13), 6912–6926. https://doi.org/10.1016/J.IJHYDENE.2014.02.125

Salaudeen, S. A., Abhi, T. D., Acharya, B., & Dutta, A. (2023). Valorization of waste biomass for hydrogen-enriched gas production. In *Biomass to bioenergy: Modern technological strategies for biorefineries* (pp. 185–204). https://doi.org/10.1016/B978-0-443-15377-8.00008-4

Schmersahl, R., Klemm, M., Brunstermann, R., & Widmann, R. (2012). Hydrogen from biomass. In *Encyclopedia of sustainability science and technology* (pp. 5116–5133). https://doi.org/10.1007/978-1-4419-0851-3_318

Shayan, E., Zare, V., & Mirzaee, I. (2018). Hydrogen production from biomass gasification; a theoretical comparison of using different gasification agents. *Energy Conversion and Management*, *159*, 30–41. https://doi.org/10.1016/J.ENCONMAN.2017.12.096

Singh, P., Kumar, S., Srivastava, N., Srivastava, R., & Chattopadhyay, J. (2023). Thermochemical hydrogen production. In *Solar-driven green hydrogen generation and storage* (pp. 333–345). https://doi.org/10.1016/B978-0-323-99580-1.00019-4

Sircar, S., & Golden, T. C. (2009). Pressure swing adsorption technology for hydrogen production. In *Hydrogen and syngas production and purification technologies* (pp. 414–450). https://doi.org/10.1002/9780470561256.CH10

Song, H., Yang, G., Xue, P., Li, Y., Zou, J., Wang, S., Yang, H., & Chen, H. (2022). Recent development of biomass gasification for H2 rich gas production. *Applications in Energy and Combustion Science*, *10*, 100059. https://doi.org/10.1016/J.JAECS.2022.100059

Steffen Heidenreich Michael Müller Pier Ugo Foscolo. (2016). *Advanced biomass gasification and product flexibility.* Elsevier. http://www.sciencedirect.com:5070/book/9780128042960/advanced-biomass-gasification

Valizadeh, S., Hakimian, H., Farooq, A., Jeon, B. H., Chen, W. H., Hoon Lee, S., Jung, S. C., Won Seo, M., & Park, Y. K. (2022). Valorization of biomass through gasification for green hydrogen generation: A comprehensive review. *Bioresource Technology*, *365*, 128143. https://doi.org/10.1016/J.BIORTECH.2022.128143

Williams, R. H., Larson, E. D., Katofsky, R. E., & Chen, J. (1995). Methanol and hydrogen from biomass for transportation. *Energy for Sustainable Development*, *1*(5), 18–34. https://doi.org/10.1016/S0973-0826(08)60083-6

Yan, Q., Guo, L., & Lu, Y. (2006). Thermodynamic analysis of hydrogen production from biomass gasification in supercritical water. *Energy Conversion and Management*, 47(11–12), 1515–1528. https://doi.org/10.1016/J.ENCONMAN.2005.08.004

Zhao, X., Zhou, H., Sikarwar, V. S., Zhao, M., Park, A. H. A., Fennell, P. S., Shen, L., & Fan, L. S. (2017). Biomass-based chemical looping technologies: the good, the bad and the future. *Energy & Environmental Science*, *10*(9), 1885–1910. https://doi.org/10.1039/C6EE03718F

Chapter 6

Hydrogen production from biomass

Harnessing nature for sustainable energy

Prashant Sharma, Arun Kumar Sharma, Bhupendra Gupta, Prashant Baredar, and Tushar Choudhary

6.1 INTRODUCTION

Numerous methods may be employed to generate hydrogen from biomass. The use of bioenergy has the capability to create reduced-emission hydrogen concerning very few emission quantities and possibly zero-emission intensities if combined with carbon capture, usage, and its storage. Hydrogen is found in chemical compounds rather than in its pure state in the environment. Therefore, even though it is abundant on the planet, the compounds within which it exists must be converted to generate hydrogen. These days, non-renewable energy resources including coal (18%), natural gas (48%), oil from refineries (30%), and water electrolysis (4%) are among the most prevalent sources of hydrogen (Łukajtis et al., 2018). Hydrogen may be produced using biomass as well as other sustainable energy sources, providing a replacement for more conventional processes. Carbon dioxide, an inevitable outcome of hydrogen production from biomass, may be entirely consumed during photosynthesis in these crops. There are two main ways to turn biomass into hydrogen: thermochemical and biological processes (Sharma et al., 2020; Thammannagowda et al., 2013). Thermochemical techniques convert biomass into hydrogen gas by the application of thermal energy and chemical treatments, while microorganisms like bacteria and algae are used in biological processes to transform biomass into hydrogen gas.

6.2 THERMOCHEMICAL PROCESSES

Thermochemical techniques for producing hydrogen are gaining more attention. There are a few typical thermochemical approaches like biomass gasification, pyrolysis, and liquefaction (Melgar et al., 2007; Vuppaladadiyam et al., 2022).

DOI: 10.1201/9781003537816-6

Gasification: In the gasification process, heating is applied to feedstock in a reducing environment (restricted oxygen) to create a syngas (synthesis gas) blend consisting of various components such as carbon dioxide, carbon monoxide, and hydrogen. After that, the syngas is treated to split and clean the hydrogen. The procedure is carried out at high temperatures (600–1000°C) with restricted availability of oxygen, air, and steam, resulting in partial combustion (Babu & Chaurasia, 2003; Mendiburu et al., 2014). The end outcome is an enormous quantity of gases with minimal carbon. The method is similar to coal gasification, leading to the production of synthesis gas, which contains flammable components, primarily hydrogen, carbon monoxide, and methane, in addition to additional hydrocarbons such as acetylene, ethylene, and inert elements: ammonia, carbon dioxide, and water vapor. In addition, the process output comprises inert elements, such as tar, char, dust, and different contaminants (Badawi et al., 2023; Sharma et al., 2022).

During this procedure, the energy contained in the biomass does not escape instantly; instead, it transforms into the bonding energies of lightweight fuels instead of becoming apparent heat from the residue of combustion. A rigorous series of simultaneous chemical processes takes place in the biomass gasification procedure. Certain of the processes are endothermic, requiring heat to occur, while others are exothermic, emitting heat as they advance. Additionally, the responses are divided into two categories: heterogeneous and homogeneous.

The subsequent physicochemical processes occur in a reactor.

1. Drying (100–150 °C)
2. Pyrolysis (devolatilization) (150–700 °C)
3. Combustion (700–1500 °C)
4. Reduction (800–1100 °C)

Drying, pyrolysis, and reduction are all endothermic reactions that consume heat. Liquid in biomass feed is eliminated during the process of drying. The pyrolysis operation removes non-condensable fluids, gases, and evaporated water, leaving only char. Combustion is an exothermic reaction that oxidizes the feed before reducing the combustion leftovers as gases that ignite in an endothermic approach (Dejtrakulwong & Patumsawad, 2014; Ratnadhariya & Channiwala, 2009). Gasification is occasionally thought of as a three-stage procedure, consisting of decomposition, high-speed methane production, and lower-speed char gasification. The decomposition or pyrolysis step begins gradually at temperatures below 350°C and quickly increases to 700°C. The substances that emerge from this procedure, as well as their chemical constituents, are heavily influenced by the gas composition, temperature, and pressure. Pyrolysis typically yields three materials, notably

1. Gases like hydrogen, carbon dioxide, carbon monoxide, etc.

2. Tar, a black, dense, and caustic fluid made up of complex organic and inorganic compounds.
3. Char, a crystalline residual mostly comprising carbon.

Char ignition is a key process within the gasifier because it supplies all of the heat required to enable endothermic processes to proceed. If oxygen serves as a gasifying agent, it oxidizes the flammable molecules in the bed, producing carbon dioxide and water vapor. These are further reduced when they interact with the char created by devolatilization. Finally, the gasification produces combustible gases like hydrogen, carbon monoxide, and methane via the chain of reactions (Hekimoğlu et al., 2021). It is important to emphasize the potential for various gasification techniques to produce H_2 to show availability, affordability, and broad utility in a range of industries enabling biomass transformation. Thermochemical conversion for various necessities contains varied characteristics or factors such as temperature, ER, and reaction phases such as gasification, pyrolysis, and torrefaction. Pyrolysis constitutes a single step of the gasification process that involves the use of little or no oxygen and produces partially eliminated carbon despite the inclusion of hydrogen. In gasification, oxidizing substances such as steam and air are employed to generate molecular linkages that may liberate energy following oxidation (Pereira & Martins, 2017; Reed, 1985; Xu et al., 2018). Reactions for gasification (Barman et al., 2012; Basu, 2013; Dahmen et al., 2009; Wang & Kinoshita, 1993):

1. *Boudouard reaction*: Within that endothermic reaction, CO_2 produced is reacted with char and generates CO.

$$C + CO_2 \rightarrow 2CO - 172.58 \text{ kJ/mol carbon} \tag{6.1}$$

2. *Water-gas reaction*: In this process, the partial oxidation of carbon occurs in the context of steam. Steam might come through input air, water vapor produced through feedstock, or the pyrolysis process.

$$C + H_2O \leftrightarrow CO + H_2 - 131.38 \text{ kJ / mol carbon} \tag{6.2}$$

Like the gasification medium, steam is provided with or without air or O_2.

3. *Methanation*: CH_4 may be produced in the gasifier as a result of multiple overall reactions:

$$C + 2H_2 \rightarrow CH_4 + 74.90 \text{kJ / mol carbon} \tag{6.3}$$

4. *Shift conversion*: Steam is reduced by CO in this endothermic reaction, and it is desirable as hydrogen has a HHV (higher heating value) compared to carbon monoxide.

$$CO + H_2O \leftrightarrow CO_2 + H_2 - 41.98 \text{ kJ / mol} \tag{6.4}$$

5. *Steam CH_4 reforming reaction*:

$$CH_4 + H_2O \rightarrow CO + 3H_2 \quad (6.5)$$

The gasification process can be classified as air, oxygen, or steam gasification, according to the oxidizer that is utilized. Air is utilized as a gasifying ingredient in air gasification for producing syngas, and it is an uncomplicated method that can be applied to a broad spectrum of biomass. Yet, the initial biomass has comparatively low hydrogen content (3–11 wt%), along with the composition's variability, which results in an uneven and ineffective output of hydrogen and a wide range of contaminants, including tar and char. It is challenging, yet vital, to eliminate all tar from the produced gas (Lepage et al., 2021). Air gasification of biomass produces just syngas containing little hydrogen and lower heating value. Steam gasification generates substantially higher H_2 compared to pyrolysis or air gasification, with a heating energy-to-H_2 transformation rate as high as 52% (Eloffy et al., 2022). Equation (6.1) shows the air gasification reaction.

$$\text{Biomass} + \text{Air} \rightarrow H_2 + CO + CH_4 + N_2 + CO_2 + H_2O + \text{Tar} + \text{Char} \quad (6.6)$$

Several variables, such as temperature, gas flow, pressure, dwell time, equivalency ratio, biomass qualities, and gasifier layout, may have a major effect on the grade of gas generated through biomass gasification. During oxygen gasification, the reactor receives pure oxygen. Oxy-gasification provides a benefit over air gasification since it does not need a separator to extract the nitrogen from the producer gas, which is present within producer gas produced by air gasification, and the tar percentage is lowest in produced syngas. However, pure oxygen must be separated from the air, which requires a significant amount of energy (Mota et al., 2015). In contrast to air gasification, the introduction of pure oxygen causes the gasification to proceed toward an elevated temperature. The production of hydrogen and carbon monoxide rises at elevated temperatures beyond a certain level, whereas the resulting quantity of other gases, including carbon dioxide and methane, decreases (Pandey et al., 2019). Certain procedures that occur primarily in oxy-gasification include devolatilization, tar cracking, Boudouard, water-gas, water-gas shift, and methanation reactions (Sandeep & Dasappa, 2014). As previously stated, pure oxygen promotes elevated temperature gasification, whereas elevated temperatures promote devolatilization and tar cracking reactions, all of which contribute to the production of H_2 and CO. It additionally encourages the Boudouard and water-gas reactions, which generate CO and reduce CO_2 levels in the producing gas. Methanation and the water-gas shift process are both exothermic processes and consequently less dominant at elevated temperatures. The carbon transformation rate improves the composition of syngas owing to the Boudouard reaction. The hydrogen-rich gas might additionally be created by feeding the syngas into the shift reactor. In this process, steam combines with CO across the catalyst to create H_2 and CO_2. The elevated level of nitrogen in the producer gas in air gasification encourages an oxy-gasification procedure, which improves the characteristics and production of producer gases. In oxygen gasification, the quantity of H_2 of the generated gas is determined by a variety of operational factors, including feedstock nature, working temperature, dwell duration, the velocity of delivered oxygen, and reactor type. Increased gasification temperature encourages better generation of carbon monoxide and hydrogen in the oxy-gasification operation. Steam gasification uses steam, air, or pure oxygen as a gasifying agent to generate hydrogen-rich syngas. The application of steam as the gasifying medium improves hydrogen output in comparison to air gasification and produces an intermediate calorific value gas with no nitrogen (12–56 MJ/Nm^3). Subsequently it needs to be mentioned that throughout the gasification, tar is produced along with producing gas as well as additional components. Tar is an unwanted component that is a complicated blend of hydrocarbons. It is being considered a serious barrier to the use of producer gas in energy production applications (Florin & Harris, 2008; Franco et al., 2003). Steam gasification is utilized as a combination of air and oxygen gasification. The gas has the same amount of H_2 as is produced by oxygen gasification, but the method is less expensive. Nonetheless, tar, char, and trace amounts of N_2 are evident. The general steam gasification reaction of biomass is presented in the following equation (Pandey et al., 2019)

$$\text{Biomass} + \text{Steam} \rightarrow H_2 + CO + CH_4 + CO_2 + H_2O + \text{Tar} + \text{Char} \qquad (6.7)$$

Equation (6.2) shows the equation of steam gasification. According to the hydrogen% in the producer gas (40%), greater H_2/CO proportion (1.5), and lower contaminants, steam gasification seems the best procedure for producing hydrogen in comparison with air gasification. Regarding the biomass to be used, steam gasification is viable for moist biomass (water content 5–35 wt%), but air gasification needs dry biomass (Lepage et al., 2021). Following the drying process, the biomass undergoes pyrolysis, resulting in a gas that includes carbon monoxide, carbon dioxide, methane, LHC (C_2H_4), C, and tar. Initial tar comprises a complicated and concentrated combination of polyaromatic hydrocarbons and oxidized compounds such as alcohols, phenolics, furane, and furfuran. With steam gasification temperatures (700–1000°C), homogeneous processes such as pyrolysis and steam methane reforming take place, altering the molecular composition of these oxidized compounds. In contrast to air gasification, steam gasification accelerates the steam reforming process and improves the quantity of hydrogen generated. While in reforming processes, initial tar is broken down to secondary tar, which is subsequently converted to tertiary tar.

Under elevated temperatures (~1300°C), tar can be eliminated. Reforming processes occur on LHCs in an atmosphere of water and carbon dioxide, producing carbon monoxide and hydrogen. Following tar cracking, LHCs (methane and ethane) and carbon oxides (carbon monoxide and carbon dioxide) are produced alongside hydrogen. These substances remain robust during cracking/reforming circumstances. Furthermore, the water-gas shift (WGS) process transforms carbon monoxide in the context of water vapor into a combination of carbon dioxide and hydrogen atoms (Smith et al., 2019). Steam supplied to the gasification procedure improves the reversible reaction by transferring it to hydrogen generation. Yet, the WGS process is exothermic, and following the equilibrium law, rising temperature shifts the reaction to the reacting agents, leading to the creation of carbon monoxide and water. Lastly, heterogeneous processes, like oxidization of carbon to produce carbon dioxide or carbon monoxide, and the Boudouard reaction happen more often than homogeneous reactions. The char produced in the pyrolysis process is predominantly transformed into carbon monoxide and carbon dioxide via gradual heterogeneous reactions throughout steam gasification, necessitating that the char stays in the gasifier for optimal gas transformation. Char may also boost catalyst efficiency by strengthening catalysts. Several elements may be tuned to enhance the hydrogen output generated by gasification. In the completion of steam gasification, the resulting gas is mostly composed of carbon dioxide and hydrogen gas. As a result of this, CO_2-capture mechanisms are being added throughout the entire procedure to raise the amount of hydrogen. Splitting must take place in a similar reactor where the catalysts increase the WGS process, promoting carbon dioxide collection.

Supercritical water gasification (ScWG) serves as a different thermochemical pathway, which was established at laboratory levels and is better suited for many biomass types, including moist biomass (moisture > 35%) like wood and polysaccharides. For the formation of a supercritical fluid, water is required to be heated over 374°C and pressurized beyond 221.2 bars. Under such circumstances, water's relative permittivity and the number of hydrogen bonds diminish (He et al., 2014; Pereira & Martins, 2017).

Supercritical water gasification, also known as hydrothermal gasification, is an alternative to anaerobic breakdown for processing wet substrates like MSW or landfill effluent, which may include as much as 80% moisture. In contrast to anaerobic breakdown, supercritical water gasification is a heating method that finishes the degradation of moist substrates in moments, even in seconds. Organic substances and gases become soluble in supercritical water at elevated temperatures, making them easier to convert. The overall process is endothermic, similar to the process of aqueous phase reforming. Compared to different gasification procedures, dwell periods are often relatively short (<5 s), and the reaction can occur at lower temperatures (650–700°C). Yet, to sustain such conditions, substantial volumes of energy are required, which raises expenses and limits expansion. No industrial framework is currently created due to the extreme pressure needed, which requires significant investment and operating expenses. Several scientific obstacles (waste elimination, reactor consistency, and understanding of reaction kinetics) must be addressed to produce commercially effective ScWG systems (Ghavami et al., 2021). ScWG devices have two temperature areas such as extreme temperatures (usually 500–700°C) and lower temperatures (374–500°C). Pyrolysis processes occur at elevated temperatures prior to gasification, dividing the fuel even without the need for a catalyst. However, alkaline catalysts may be employed to boost hydrogen output by increasing WGS processes. At lower temperatures (374–500°C), hydrolyzed molecules are gasified and decreased via alkaline catalysts, interstitial compounds, and activated charcoal catalysts, which are required for the rapid biomass transformation to producer gas. Furthermore, at lower temperatures, a methanation process starts and absorbs the hydrogen generated, resulting in methane formation, which requires charcoal/carbon-derived catalysts for resolution.

6.3 OPERATING FACTORS/PARAMETERS AFFECTS THE GASIFICATION PROCESS

6.3.1 Biomass characteristics

The characteristics of biomass play a significant role in how much hydrogen is produced during gasification. The amount of hydrogen (H) and oxygen (O) in the biomass directly affects the potential hydrogen yield. Higher H content and lower O content generally lead to more hydrogen production. The fibrous portion of plants, or lignocellulose, that is made up of cellulose, hemicellulose, and lignin, seems to be the greatest noticeable component of biomass and essential to gasification. The cellulose-to-lignin and hemicellulose-to-lignin ratios in an average biomass typically range between 0.5–2.7 and 0.5–2.0, respectively. The amount of lignin in the yield controls its oil proportion, whereas the percentage of cellulose and hemicellulose impacts the gaseous phase's output. Hence, the proportion of cellulose–hemicellulose to lignin determines the gaseous form output. Thus, cellulose and hemicellulose, the main components of plant cell walls, are good sources of hydrogen due to their high carbohydrate content. Lignin, another cell wall component, has a lower hydrogen content but can still contribute. Moisture in biomass is classified as intrinsic and extrinsic, with intrinsic moisture being the water content without weather impact, and extrinsic moisture incorporating weather conditions. Gasifier properties and optimal operation depend on moisture content. Most gasifiers can tolerate 15–30% moisture, but woody and low-moisture herbaceous biomasses with less than 15% water content are suitable for gasification. Biomass with a significant amount of moisture reduces the gasifier's temperature and delays endothermic processes, thus requiring additional energy to vaporize them. As a result, the endothermic process that produces carbon monoxide and hydrogen requires less energy (Rubinsin et al., 2024). As a result, the generation of methane and carbon monoxide involves an increased exothermic process. More dried biomass results in better gasification effectiveness, potentially increasing the hydrogen concentration in the resulting gas. Moisture influences how biomass fuel is handled, stored, and transported. The steam gasification method is suited for biomass with a moisture level below 35% (de Lasa et al., 2011). Air gasification can handle biomass with higher moisture content, typically up to 50% or even slightly more depending on the specific setup. Air gasification does not involve the energy-intensive process of water evaporation like steam gasification. Like steam gasification, oxygen gasification relies on the reactions between the gasifying agent and the biomass components, and excess moisture reduces efficiency. Oxygen gasification, like steam gasification, is suited for biomass with lower moisture content (ideally below 35%). Oxygen gasification produces a syngas with a higher heating value compared to air gasification because it avoids nitrogen dilution. Volatile matter (VM) in biomass has a positive impact on hydrogen yield in all three gasification technologies (steam, air, and oxygen). Flammable gases, including HC, H_2, O, and CO, as well as inert gases, make up volatile matter (VM). Biomass materials have more VM, as much as 87%, while coal has just 40%. VM concentration in the feedstock is further necessary for producing excellent combustible gas production at greater temperatures. Furthermore, a greater temperature above 400°C is beneficial since volatile matter is emitted at this level of heat (Matamba et al., 2023). When the biomass is heated and the volatile content is released, fixed carbon (FC) is the remaining solid residual that remains flammable. The transformation of biomass can be influenced by both FC and VM; however, because volatiles are not as tightly linked in gasification compared to FC, they are more responsive. FC acts as a fuel source during the gasification process, providing the necessary heat for the reactions that convert VM and other components into hydrogen. If the FC content is too low, there might not be enough heat generated to sustain the gasification reactions at the optimal temperature, ultimately reducing hydrogen yield. Steam gasification often requires additional external heat input to maintain the desired reaction temperature. Sufficient FC helps reduce the need for external heat input. Like steam gasification, in air gasification, fixed carbon provides heat for the process. However, since air gasification operates at slightly lower temperatures, the FC content might not be as critical if it is sufficient to sustain the reactions. Due to the higher operating temperatures and absence of nitrogen dilution (unlike air gasification), oxygen gasification can tolerate a wider range of FC content. Even with lower FC content, sufficient external heat can be provided to achieve the desired gasification efficiency (Demirbas, 2009). It comes to light that catalyst loading (CL) has a major impact on the gasification process. A proper loading of the catalyst may decrease the tar percentage in the reactor and raise the amount of hydrogen output. Because the materials are readily accessible and reasonably priced, natural ores including dolomite and olivine as well as alkali and alkaline earth metal catalysts are frequently utilized for biomass gasification (Song et al., 2022). However, there are multiple drawbacks, such as the low catalytic efficiency of natural ores and the volatile, clogging-causing nature of alkali and alkaline earth metals that raise expenditures for operation. Due to its outstanding catalytic activity in generating hydrogen gas and thermal reliability, single-metal catalysts, composite catalysts, and special-structure catalysts containing transition metals at their base are currently garnering a lot of interest. A rise in CL resulted in a decrease in carbon monoxide, rise in methane percentage in produced gas, rise in the proportion of H_2/CO, rise in heating value, rise in hydrogen production, and a decline in tar percentage. After a certain limit, the increase in CL does not result in any further changes in hydrogen yield and tar content due to the limitations in heat and mass transfer.

6.3.2 Temperature and equivalence ratio

The primary factor that has a major influence on gasification is temperature. Gasification reactions are generally endothermic, which means they absorb heat. Elevated pyrolysis and gasification reforming processes are the primary means by which elevated temperatures influence the syngas properties. Higher temperatures increase the reaction rates for processes that convert biomass components, including VM, into hydrogen-rich intermediates and ultimately hydrogen gas by promoting the pyrolysis of raw materials and the secondary cracking of macromolecules (Atnaw et al., 2013; Sharma et al., 2022; Singh Siwal et al., 2020). Furthermore, exothermic reactions' reactants and endothermic reactions' outputs benefit at higher temperatures. Therefore, raising the temperature may boost the endothermic occurrences that occur during the gasification operations. Elevated temperatures increase the generation of hydrogen and carbon monoxide while diminishing the levels of methane, carbon dioxide, and additional compounds in the gasification process. Reducing the tar is similarly linked to a steadily rising gas output, suggesting that elevated temperatures may enhance tar reforming or breaking events. As temperature increases, the ratios of several substances, including H_2/CO and CO/CO_2, H_2/CH_4, CO/CH_4, and H_2/CO_2, might rise; nevertheless, the H_2/CO ratio may drop at a certain elevated temperature, which is approximately 800°C. There are some limitations to the effect of temperature. The residence time for water-gas shift reactions and subsequent reactions is shortened at elevated temperatures. The proportion of fuel to air within stoichiometric circumstances to the identical in actual conditions is known as the equivalency ratio, or ER. Additionally, ER has a major impact on gasification efficiency, product properties, and hydrogen production. The gas yield positively impacts increasing ER up to a certain limit and thereafter it decreases. In most cases, it is optimum at 0.23–0.26 ER, and the minimum is at 0.37 ER.

6.3.3 Gasifiers

Several kinds of gasifiers can be utilized for gasification. Biomass gasifiers that are often used comprise fluidized-bed and fixed-bed reactors. Under more divisions, favored gasifier technologies include fluidized beds, rotary kiln reactors, downdraft, cross-draft, updraft, and plasma technology, among others. Fixed-bed reactors and fluidized-bed reactors have different flow rates of gas within the gasifier. In a fluidized-bed gasifier, the flow rate is greater than in a fixed-bed gasifier. Moreover, the fluidized-bed gasifier maintains a constant temperature in the bed and various regions, When, the fixed-bed gasifier categorized. Each of these kinds of gasifiers has pros and cons of itself (De et al., 2018); Salam et al., 2018). The variety of biomass that is utilized, its structure, the way in which the feedstock and gasifying substance come into collision, its duration, and the speed of heat exchange all influence the selection of gasifier types. In addition, downdraft gasifiers are commonly used in several nations because of the convenience to the biomass (de Souza-Santos, 2004; Pereira & Martins, 2017; Smith et al., 2019).

6.3.4 Steam-to-biomass ratio

The generation of hydrogen during gasification and the content of hydrogen in producer gas are significantly influenced by the steam-to-biomass ratio (S/B). It has been noted that the water-gas shift reaction promotes increased hydrogen generation rates as the S/B ratio increases. An S/B ratio of (1–4) in the middle range (1–4) is ideal; excessive or inadequate can negatively impact the water-gas shift reaction and the grade of the produced gas. In every gasification procedure, the appropriate S/B ratio needs to be considered. It has been shown that, despite being extremely minor in numerous studies, the majority of projects indicate an upsurge in the generation of hydrogen. There aren't many studies that show how a low S/B affects hydrogen generation (Adnan et al., 2017; Cao et al., 2021; Gai et al., 2014).

6.3.5 Pyrolysis

A second thermochemical process that turns biomass into H_2 is called pyrolysis. It is comparable to gasification, except it does not require an oxidizing ingredient and could be used at lower temperatures. In certain situations, a little amount of an oxidizing substance may be used to improve heat generation. Pyrolysis often takes place at pressures as high as 5 bar and temperatures between 400 and 800°C. While producing bio-oil is the primary goal of pyrolysis, flashing or fast pyrolysis with elevated temperatures and suitable residence times can also generate H_2 (Pandey et al., 2019; Roddy & Manson-Whitton, 2012).

The three types of pyrolysis are flash pyrolysis, rapid pyrolysis, and traditional (or slow) pyrolysis, depending on the heating rate and residence period. Because traditional pyrolysis is carried out at temperatures below 450°C, a substantial amount of char is produced. At moderate temperatures (450–600°C), an elevated heating rate (about 300°C/min), and little residence time, rapid pyrolysis provides as much as 75 wt% bio-oil. Like fast pyrolysis, flash pyrolysis is performed at substantially elevated temperatures (over 600°C) and at increased heating rates (>1000°C/s), yet has a smaller residence time (<1 s) to maximize gases produced. Still, in comparison with gasification, the gas outputs via quick and flash pyrolysis are typically smaller, mostly due to the decreased working temperatures (Eloffy et al., 2022).

Following pyrolysis, water, ash, and oxygen-rich substances combine to form an intricate fluid called "bio-oil," which looks like tar. Hydro-insoluble components and hydro-soluble components represent the two categories of bio-oils (Matamba et al., 2023). The fraction that is hydro-insoluble can be separated into various compounds, such as olefins or benzene, toluene, and xylenes, and used in gluing solutions. The procedure may be improved by adding a steam gasification unit to increase the hydro-soluble fraction and the amount of hydrogen generation. Several factors, including the kind of catalyst, operational temperature, biomass, and residence time, are critical to the synthesis of hydrogen by biomass pyrolysis.

Increased pyrolysis temperature, heating rate, and residence time all boost hydrogen production. Fast pyrolysis is more effective at high temperatures than slow pyrolysis. In pyrolysis, based on sources, different types of researches are carried out. Furthermore, excellent syngas production was attained at temperatures between 900 and 2000°C and has been successfully investigated by hot radiofrequency plasma pyrolysis studies. A temperature as elevated as 2000°C may be achieved using solar pyrolysis, and this has been shown to increase the gaseous output. There are some advanced pyrolysis processes, which include autothermal pyrolysis, hydropyrolysis, microwave-assisted pyrolysis, co-pyrolysis, flash pyrolysis, and catalytic pyrolysis. Because traditional pyrolysis is endothermic, it has limitations and requires an outside source with a larger heat flow. This shortfall is a major problem when scaling up processes. With bigger reactors, the total heat requirement of biomass pyrolysis frequently outstrips the heat input. Furthermore, the limited heat transfer ability of biomass might modify processes in various reactors, which may influence the overall performance of the process and the result. To get through the heat transfer restriction in expansion, the autothermal pyrolysis process (APP) was invented. This offers a superior alternative for commercializing pyrolysis innovation. To equalize the heat requirement in an autothermal reactor, some of the heat generated by the feedstock or pyrolysis byproducts is utilized. Hydropyrolysis is a comparatively new technique that transforms biomass into the highest-quality pyrolytic oil by using pressurized hydrogen (>10 MPa). The use of hydropyrolysis has several benefits, such as reducing char production, producing bio-oil with lower oxygen content, and adding a small amount of additional hydrogen to the bio-oil. However, hydropyrolysis itself produces no hydrogen. Usually, the process involves breaking down organic molecules in a hydrogen-deficient atmosphere lacking in hydrogen at elevated temperatures. Instead of pure hydrogen, its decomposition produces char, methane, carbon monoxide, and bio-oil. Certain types of hydropyrolysis involve the addition of steam or hydrogen throughout the procedure. Certain variants, such as hydrothermal liquefaction, can also provide additional compounds in addition to hydrogen. Microwave pyrolysis is a technique that converts organic substances into compounds, fuels, and charcoal by the application of microwaves. It breaks down substances using a thermochemical mechanism that does not require oxygen. Traditional heating methods use convection to transmit heat from elevated temperature fluids to biomass elements, and subsequently disperses via conduction to the core within. Volatiles spread to the exterior because of the temperature differential that is created. But microwaves enter the components of biomass and use the polar chemicals present in the biomass to convert microwave power into heat. This technique provides an improved eco-friendly and effective substitute for conventional heating techniques. In micro-pyrolysis, depending on various factors and catalysts, the range of hydrogen yields is around 5–50 vol% (Vuppaladadiyam et al., 2022; Xie et al., 2015). Co-pyrolysis is a thermochemical process in which multiple substances are heated under an oxygen-free atmosphere (De et al., 2018). Co-pyrolysis differs from conventional pyrolysis primarily because a mixture of initial ingredients is used. Using reciprocal relationships, co-pyrolysis is a possible technique to enhance biomass pyrolysis efficiency. It may be utilized for several biomass types, such as lignocellulosic, algae, or biological waste, and is compatible with current machinery minimal alterations. When using them, nevertheless, one must assess their hydrogen index (Xie et al., 2015).

$$\text{Effective hydrogen index(EHI)}$$
$$= \frac{\text{Hydrogen} - 2 \times \text{Oxygen} - 3 \times \text{Nytrogen} - 2 \times \text{Sulphur}}{\text{Carbon}}$$

The combination is made in such a way that biomass having low EHI is blended with biomass having higher EHI. There is a synergistic effect in which, as heated the blend, the breakdown of one can promote the breakdown of the other. The primary aim of co-pyrolysis is usually to create a usable liquid fuel called bio-oil or syngas (a mixture of gases like CO, H_2, and hydrocarbons) from waste materials. The syngas produced during co-pyrolysis can be further processed using methods like steam reforming to separate out the hydrogen. In this case, co-pyrolysis acts as a preparatory step to increase the yield of hydrogen-rich feedstock. In other ways, the methods to maximize co-pyrolysis for the creation of hydrogen are being investigated. This may entail the use of certain catalysts or the selection of biomass mixtures that facilitate the co-pyrolysis process's degradation of substances into elements that are high in hydrogen. In co-pyrolysis, different variables and catalysts depend on the range of hydrogen yields in about 30–55 vol% (Xie et al., 2015). Pyrolysis accompanied by the application of a catalyst is called catalytic pyrolysis. The chemical processes that take place throughout the decomposition of the substance being utilized are accelerated and directed by this catalyst. Catalytic pyrolysis is commonly used to deoxygenate pyrolytic gases to produce aromatic hydrocarbons, including xylene, phenol, benzene, and toluene. In general, there are usually two different techniques for adding a catalyst during the workflow. The initial procedure is the on-site catalytic procedure, referred to as a catalyst-mixing method (CMM), in which the biomass is mixed with catalyst substances via soaked blending or dry blending and can be employed right away in traditional pyrolysis procedures. On-site catalytic pyrolysis produces fluid and gaseous byproducts while limiting tar deposition. Ex-situ catalytic pyrolysis, on the other hand, is a technique that divides a reactor into two sections, with biomass in one and a catalyst in the other. The raw material is transformed into vapor, which travels via the catalyst and becomes bio-oil. This is thought to be preferable to traditional pyrolysis because of the decreased processing temperatures and power needs. Catalytic pyroly sis may yield a superior-quality fluid substance, yet just between 20 and 30% of biomass carbon is converted to fluid substance, with the other 30% becoming coke and possibly deactivating the catalyst. In catalytic pyrolysis, different variables and catalysts depend on the range of hydrogen yields in about 70–85 vol%.

6.4 HYDROTHERMAL LIQUEFACTION

Hydrothermal liquefaction (HTL) is a thermochemical technique that liquefies excessive water-content substances effectively. This method typically runs at greater pressure (5–20 MPa) and at a lesser (mild) temperature (<400°C), as in contrast to pyrolysis, which works at lesser pressure (0.1–0.5 MPa) and a greater temperature (>450°C) (Xu et al., 2018). Fast pyrolysis is currently the sole commercially viable process for producing bio-oil. Pyrolysis requires essentially dry biomass that restricts its use to biologically moist substances and waste flows containing as much as 90% moisture. The procedure of drying is costly; thus, it might be highly expensive. In hydrothermal liquefaction, biomass is hydrolyzed and depolymerized into monomers and unit structures, then decomposed and degraded to generate intermediaries via chemical processes such as dehydration, dehydrogenation, deoxygenation, cleavage, deamination, decarboxylation, cracking/fragmentation, etc. Fortunately hydrogen exists, groups that have oxygen are removed, and reactionary fragments are rearranged (Xu et al., 2018). As the biomass is heated and pressurized with a solvent, it produces bio-crude oil. The bio-crude oil can then be refined to produce hydrogen.

6.4.1 Aqueous phase reforming

Aqueous phase reforming (APR) is a catalytic process used to convert oxygenated organic compounds, often derived from biomass, into hydrogen. In APR, raw material particles absorb in the liquid phase and interact with water particles at lower temperatures (~275°C) as well as elevated pressures (as much as 50 bar). APR focuses on water-soluble organic molecules like alcohols, carbohydrates, and acidic substances that may be extracted from biomass (plants, algae) or wastewater streams containing organic matter using techniques such as preprocessing or hydrothermal liquefaction (Tian et al., 2024). APR is a high-pressure, high-temperature aqueous method with water as the solvent. A catalyst, commonly valuable metals such as platinum, speeds reformation processes. Water breaks apart organic substances into hydrogen gas (H_2), carbon dioxide (CO_2), and lightweight hydrocarbon particles based on the feedstock and reactivity circumstances. APR is a relatively mild thermochemical process for hydrogen production, operating at lower temperatures (typically around 220–270°C) compared to other methods. It offers high hydrogen yield by efficiently converting organic feedstock into hydrogen, making it an attractive method for hydrogen production. Additionally, APR promotes sustainable biomass utilization for hydrogen generation.

6.4.2 Biological processes

Biohydrogen generation from biomass utilizing microorganisms is a promising method that may generate hydrogen via various renewable sources. Hydrogen is an elemental outcome of microbiological operations that operate in anaerobic circumstances. Microorganisms may generate enzymes which discharge hydrogen through water in response to outside sources of energy like sunshine. The presence of enzymes is essential for the biological processes used to produce hydrogen. Enzymes that facilitate generating hydrogen have centers of activity formed by complicated metalloclusters. The method involves three enzymes: nitrogenase, Fe-hydrogenase, and

NiFe-hydrogenase. Bio-photolysis uses Fe-hydrogenase enzyme, while photo-fermentation uses nitrogenase (Rambabu et al., 2021; Thanigaivel et al., 2023).

There are four main divisions that use biological approaches to produce hydrogen.

6.4.3 Water-splitting photosynthesis

Water-splitting photosynthesis is the initial step of the sunlight-dependent processes in photosynthesis, in which sunlight is utilized to separate water molecules (H_2O) into oxygen (O_2) and hydrogen ions (H+). This mechanism occurs inside the Photosystem II (PSII) protein complex, which is present in the thylakoid membranes of chloroplasts in plants and microorganisms. Water-splitting photosynthesis involves chlorophyll absorbing sunlight and releasing an electron. The electron is subsequently transported to an electron acceptor molecule inside PSII, resulting in a positive potential. To make up for the missing electron, PSII removes one electron from a water molecule and splits it into hydrogen ions and an oxygen atom. The oxygen atom subsequently joins with an additional molecule of oxygen to form oxygen (O_2), which is produced as a result of photosynthesis (Sakurai et al., 2013).

6.4.4 Fermentation

Fermentation is the decomposition of biological substrates by microbes. The transformation of H_2 and H_2-rich gases requires light and can be classified into two categories: dark fermentation (DF) and photo-fermentation (PF) (Shahbaz et al., 2020). Anaerobic bacteria contribute to the transformation of biomass to H_2 through a metabolic pathway in the DF procedure. H_2 is created as the major product by a range of biological processes. This method of biological change is regarded to be incredibly widely recognized and effective. Despite its poor outputs, DF requires basic reactor layouts, making it an intriguing option for H_2 production. In addition, significant study is being conducted to increase H_2 generation yield using blended systems with DF, including microbial electrolysis cells (MECs), PF, anaerobic digester, and DF with a cell-free enzymatic system. The DF route functions at temperatures ranging from 26 to 40°C to over 70°C, with the majority of studies undertaken at 35–37°C because of uncertain advantages beyond mesophilic habitats. The PF procedure is also the fermentative transformation of organic feedstock to bio-H_2 by various photosynthetic bacteria via a three-step cycle of biochemical reactions, like anaerobic transformation. Photosynthetic bacteria like *Rhodospirillum*, *Rhodobacter*, *Rhodobium*, and *Rhodopseudomonas* can produce hydrogen via the function of their MgATP and reductant.

Photosynthetic bacteria use nitrogenous matter to convert biomass into hydrogen during the photo-fermentation process, but only when sunlight is available. Photosynthetic bacteria supply electrons using an oxygen-based mechanism with organic acids or lowering sulfur compounds. The process of H_2 synthesis improves with light intensity, although light conversion efficiency declines. Photosynthetic microbes utilized in photo-fermentation convert light more efficiently than cyanobacteria in secondary bio-photolysis. The technique has disadvantages, including a poor solar energy conversion rate (3–10%), a large anaerobic photo-bioreactor requirement, and substantial energy needs for nitrogenase to be activated. The method uses purple non-sulfur bacteria to transform short-chain organic acids into hydrogen. PF may be done in one or two stages. The single-stage method is of minimal cost, yet has limitations, including excessive demand for energy through nitrogenases and limited solar energy transformation. The two-stage fermentation technique involves dark fermentation initially, followed by PF to create additional hydrogen (Eloffy et al., 2022).

6.5 MICROBIAL ELECTROLYSIS PROCESSING

Microbial electrolysis processing (MEP) is a bio-electrochemical method that uses microbial digestion and electrolysis to generate hydrogen gas (H_2) from organic molecules or sewage. MEP commonly takes advantage of a microbial electrolysis cell (MEC), which is made up of an anode and a cathode separated by a protective barrier. MEP involves immobilizing microorganisms, like bacteria such as *Geobacter* or *Clostridium* species, on the anode of an MEC, which oxidizes organic compounds in wastewater or feedstock, releasing electrons and protons. These ions are moved to the anode, and there they produce electricity by flowing to the cathode via a separate circuit. At the cathode, electrons are consumed in a reduction reaction, forming hydrogen gas (H_2), which can be used as a renewable energy source. The hydrogen gas can be purified and stored for various applications (Aydin et al., 2021; Thammannagowda et al., 2013).

MEP is a renewable energy source that generates hydrogen gas from organic compounds or wastewater, offering a sustainable waste management approach. It also enables the recovery of electrical energy from microbial metabolism, enhancing overall energy efficiency by utilizing organic waste as feedstock.

REFERENCES

Adnan, M. A., Susanto, H., Binous, H., Muraza, O., & Hossain, M. M. (2017). Enhancement of hydrogen production in a modified moving bed downdraft gasifier– A thermodynamic study by including tar. *International Journal of Hydrogen Energy*, *42*(16), 10971–10985.

Atnaw, S. M., Sulaiman, S. A., & Yusup, S. (2013). Syngas production from downdraft gasification of oil palm fronds. *Energy*, *61*, 491–501.

Aydin, M. I., Karaca, A. E., Qureshy, A. M. M. I., & Dincer, I. (2021 [2020, December]). A comparative review on clean hydrogen production from wastewaters. *Journal of Environmental Management*, *279*, 111793.

Babu, B. V., & Chaurasia, A. S. (2003). Modeling, simulation and estimation of optimum parameters in pyrolysis of biomass. *Energy Conversion and Management*, *44*(13), 2135–2158.

Badawi, E. Y., Elkharsa, R. A., & Abdelfattah, E. A. (2023, March). Value proposition of bio-hydrogen production from different biomass sources. *Energy Nexus*, *10*, 100194.

Barman, N. S., Ghosh, S., & De, S. (2012). Gasification of biomass in a fixed bed downdraft gasifier - A realistic model including tar. *Bioresource Technology*, *107*, 505–511.

Basu, P. (2013). *Biomass gasification and pyrolysis: Practical design and theory*. Academic Press.

Cao, Y., Bai, Y., & Du, J. (2021). Air-steam gasification of biomass based on a multi-composition multi-step kinetic model: A clean strategy for hydrogen-enriched syngas production. *Science of the Total Environment*, *753*, 141690.

Dahmen, N., Dinjus, E., & Kruse, A. (2009). Fuels - Hydrogen production | Biomass: Thermochemical processes. In ürgen Garche (Ed.), *Encyclopedia of electrochemical power sources* (pp. 259–267). https://doi.org/10.1016/B978-044452745-5.00313-0 (https://www.sciencedirect.com/science/article/pii/B9780444527455003130).

de Lasa, H., Salaices, E., Mazumder, J., & Lucky, R. (2011). Catalytic steam gasification of biomass: Catalysts, thermodynamics and kinetics. *Chemical Reviews*, *111*(9), 5404–5433.

de Souza-Santos, M. L. (2004). Solid fuels combustion and gasification. In Marcel Dekker A.G. (Ed.), *Solid fuels combustion and gasification* (p. 22). Bassel

Dejtrakulwong, C., & Patumsawad, S. (2014). Four zones modeling of the downdraft biomass gasification process: Effects of moisture content and air to fuel ratio. *Energy Procedia*, *52*, 142–149.

Demirbas, A. (2009). Prediction of higher heating values for vegetable oils and animal fats from proximate analysis data. *Energy Sources, Part A: Recovery, Utilization and Environmental Effects*, *31*(14), 1264–1270.

Eloffy, M. G., Elgarahy, A. M., Saber, A. N., Hammad, A., El-Sherif, D. M., Shehata, M., Mohsen, A., & Elwakeel, K. Z. (2022, September). Biomass-to-sustainable biohydrogen: Insights into the production routes, and technical challenges. *Chemical Engineering Journal Advances*, *12*, 100410.

Florin, N. H., & Harris, A. T. (2008). Enhanced hydrogen production from biomass with in situ carbon dioxide capture using calcium oxide sorbents. *Chemical Engineering Science*, *63*(2), 287–316.

Franco, C., Pinto, F., Gulyurtlu, I., & Cabrita, I. (2003). The study of reactions influencing the biomass steam gasification process. *Fuel*, *82*(7), 835–842.

Gai, C., Dong, Y., & Zhang, T. (2014). Downdraft gasification of corn straw as a non-woody biomass: Effects of operating conditions on chlorides distribution. *Energy*, *71*, 638–644.

Ghavami, N., Özdenkçi, K., Salierno, G., Björklund-Sänkiaho, M., & De Blasio, C. (2021, December). Analysis of operational issues in hydrothermal liquefaction and supercritical water gasification processes: A review. *Biomass Conversion and Biorefinery, 13,* 12367–12394. https://doi.org/10.1007/s13399-021-02176-4

He, C., Chen, C. L., Giannis, A., Yang, Y., & Wang, J. Y. (2014). Hydrothermal gasification of sewage sludge and model compounds for renewable hydrogen production: A review. *Renewable and Sustainable Energy Reviews*, *39*, 1127–1142. https://doi.org/10.1016/j.rser.2014.07.141

Hekimoğlu, G., Sarı, A., Kar, T., Keleş, S., Kaygusuz, K., Yıldırım, N., Tyagi, V. V., Sharma, R. K., & Saleh, T. A. (2021). Carbonized waste hazelnut wood-based shape-stable composite phase change materials for thermal management implementations. *International Journal of Energy Research*, *45*(7), 10271–10284. https://doi.org/10.1002/er.6514

Lepage, T., Kammoun, M., Schmetz, Q., & Richel, A. (2021 [2020, November]). Biomass-to-hydrogen: A review of main routes production, processes evaluation and techno-economical assessment. *Biomass and Bioenergy*, *144*, 105920. https://doi.org/10.1016/j.biombioe.2020.105920

Łukajtis, R., Hołowacz, I., Kucharska, K., Glinka, M., Rybarczyk, P., Przyjazny, A., & Kamiński, M. (2018, March). Hydrogen production from biomass using dark fermentation. *Renewable and Sustainable Energy Reviews*, *91*, 665–694. https://doi.org/10.1016/j.rser.2018.04.043

Matamba, T., Tahmasebi, A., Yu, J., Keshavarz, A., Abid, H. R., & Iglauer, S. (2023, June). A review on biomass as a substitute energy source: Polygeneration influence and hydrogen rich gas formation via pyrolysis. *Journal of Analytical and Applied Pyrolysis*, *175*, 106221. https://doi.org/10.1016/j.jaap.2023.106221

Melgar, A., Pérez, J. F., Laget, H., & Horillo, A. (2007). Thermochemical equilibrium modelling of a gasifying process. *Energy Conversion and Management*, *48*(1), 59–67. https://doi.org/10.1016/j.enconman.2006.05.004

Mendiburu, A. Z., Carvalho, J. A., & Coronado, C. J. R. (2014). Thermochemical equilibrium modeling of biomass downdraft gasifier: Stoichiometric models. *Energy*, *66*, 189–201. https://doi.org/10.1016/j.energy.2013.11.022

Mota, R., Krishnamoorthy, G., Dada, O., & Benson, S. A. (2015). Hydrogen rich syngas production from oxy-steam gasification of a lignite coal - A design and optimization study. *Applied Thermal Engineering*, *90*, 13–22. https://doi.org/10.1016/j.applthermaleng.2015.06.081

De, S., Agarwal, A. K., Moholkar, V. S., & Thallada, B. (2018). *Coal and biomass gasification: recent advances and future challenges*. Preface, V-VII. https://doi.org/10.1007/978-981-10-7335-9

Pandey, B., Prajapati, Y. K., & Sheth, P. N. (2019). Recent progress in thermochemical techniques to produce hydrogen gas from biomass: A state of the art review. *International Journal of Hydrogen Energy*, *44*(47), 25384–25415. https://doi.org/10.1016/j.ijhydene.2019.08.031

Pereira, E. G., & Martins, M. A. (2017). Gasification technologies. In *Encyclopedia of sustainable technologies* (Vol. 4). Elsevier. https://doi.org/10.1016/B978-0-12-409548-9.10133-2

Rambabu, K., Bharath, G., Thanigaivelan, A., Das, D. B., Show, P. L., & Banat, F. (2021 [2020, August]). Augmented biohydrogen production from rice mill wastewater through nano-metal oxides assisted dark fermentation. *Bioresource Technology*, *319*, 124243.

Ratnadhariya, J. K., & Channiwala, S. A. (2009). Three zone equilibrium and kinetic free modeling of biomass gasifier – a novel approach. *Renewable Energy*, *34*(4), 1050–1058. https://doi.org/10.1016/j.renene.2008.08.001

Reed, T. B. (1985). Principles and technology of biomass gasification. In *Advances in solar energy: An annual review of research and development* (Vol. 2, pp. 125–174). https://doi.org/10.1007/978-1-4613-9951-3_3

Roddy, D. J., & Manson-Whitton, C. (2012). Biomass gasification and pyrolysis. In *Comprehensive renewable energy* (Vol. 5). Elsevier Ltd. https://doi.org/10.1016/B978-0-08-087872-0.00514-X

Rubinsin, N. J., Karim, N. A., Timmiati, S. N., Lim, K. L., Isahak, W. N. R. W., & Pudukudy, M. (2024). An overview of the enhanced biomass gasification for hydrogen production. *International Journal of Hydrogen Energy*, *49*, 1139–1164. https://doi.org/10.1016/j.ijhydene.2023.09.043

Sakurai, H., Masukawa, H., Kitashima, M., & Inoue, K. (2013). Photobiological hydrogen production: Bioenergetics and challenges for its practical application. *Journal of Photochemistry and Photobiology C: Photochemistry Reviews*, *17*, 1–25. https://doi.org/10.1016/j.jphotochemrev.2013.05.001

Salam, M. A., Ahmed, K., Akter, N., Hossain, T., & Abdullah, B. (2018). A review of hydrogen production via biomass gasification and its prospect in Bangladesh. *International Journal of Hydrogen Energy*, *43*(32), 14944–14973. https://doi.org/10.1016/j.ijhydene.2018.06.043

Sandeep, K., & Dasappa, S. (2014). Oxy–steam gasification of biomass for hydrogen rich syngas production using downdraft reactor configuration. *International Journal of Energy Research*, *38*, 174–188. https://doi.org/10.1002/er.3019

Shahbaz, M., Al-Ansari, T., Aslam, M., Khan, Z., Inayat, A., Athar, M., Naqvi, S. R., Ahmed, M. A., & McKay, G. (2020). A state of the art review on biomass processing and conversion technologies to produce hydrogen and its recovery via membrane separation. *International Journal of Hydrogen Energy*, *45*(30), 15166–15195. https://doi.org/10.1016/j.ijhydene.2020.04.009

Sharma, P., Gupta, B., & Pandey, M. (2022). Hydrogen rich product gas from air–steam gasification of Indian biomasses with waste engine oil as binder. *Waste and Biomass Valorization*, *13*(6), 3043–3060. https://doi.org/10.1007/s12649-022-01690-4

Sharma, P., Gupta, B., Pandey, M., Singh Bisen, K., & Baredar, P. (2020). Downdraft biomass gasification: A review on concepts, designs analysis, modelling and recent advances. *Materials Today: Proceedings*, *46*(xxxx), 5333–5341. https://doi.org/10.1016/j.matpr.2020.08.789

Singh Siwal, S., Zhang, Q., Sun, C., Thakur, S., Kumar Gupta, V., & Kumar Thakur, V. (2020). Energy production from steam gasification processes and parameters that contemplate in biomass gasifier – A review. *Bioresource Technology*, *297*, 122481.

Smith, J. D., Alembath, A., Al-Rubaye, H., Yu, J., Gao, X., & Golpour, H. (2019). Validation and application of a kinetic model for downdraft biomass gasification simulation. *Chemical Engineering and Technology*, *42*(12), 2505–2519. https://doi.org/10.1002/ceat.201900304

Song, H., Yang, G., Xue, P., Li, Y., Zou, J., Wang, S., Yang, H., & Chen, H. (2022 [2021, December]). Recent development of biomass gasification for H_2 rich gas production. *Applications in Energy and Combustion Science*, *10*, 100059. https://doi.org/10.1016/j.jaecs.2022.100059

Thammannagowda, S., Magnusson, L., Jo, J. H., Maness, P. C., & Seibert, M. (2013). Renewable hydrogen from biomass. In *Encyclopedia of biological chemistry* (2nd ed., pp. 72–75). https://doi.org/10.1016/B978-0-12-378630-2.00302-9

Thanigaivel, S., Rajendran, S., Hoang, T. K. A., Ahmad, A., & Luque, R. (2023 [2022, November]). Photobiological effects of converting biomass into hydrogen – Challenges and prospects. *Bioresource Technology*, *367*, 128278. https://doi.org/10.1016/j.biortech.2022.128278

Tian, Z., Lu, Y., Wang, J., Shu, R., Wang, C., & Chen, Y. (2024). Advances in hydrogen production by aqueous phase reforming of biomass oxygenated derivatives. *Fuel*, *357*(PA), 129691. https://doi.org/10.1016/j.fuel.2023.129691

Vuppaladadiyam, A. K., Vuppaladadiyam, S. S. V., Awasthi, A., Sahoo, A., Rehman, S., Pant, K. K., Murugavelh, S., Huang, Q., Anthony, E., Fennel, P., Bhattacharya, S., & Leu, S. Y. (2022, August). Biomass pyrolysis: A review on recent advancements and green hydrogen production. *Bioresource Technology*, *364*, 128087. https://doi.org/10.1016/j.biortech.2022.128087

Wang, Y., & Kinoshita, C. M. (1993). Kinetic model of biomass gasification. *Solar Energy*, *51*(1), 19–25. https://doi.org/10.1016/0038-092X(93)90037-O

Xie, Q., Addy, M., Liu, S., Zhang, B., Cheng, Y., Wan, Y., Li, Y., Liu, Y., Lin, X., Chen, P., & Ruan, R. (2015). Fast microwave-assisted catalytic co-pyrolysis of microalgae and scum for bio-oil production. *Fuel*, *160*, 577–582. https://doi.org/https://doi.org/10.1016/j.fuel.2015.08.020

Xu, C. C., Liao, B., Pang, S., Nazari, L., Mahmood, N., Tushar, M. S. H. K., Dutta, A., & Ray, M. B. (2018). Biomass Energy. *Comprehensive Energy Systems*, *1–5*, 770–794. https://doi.org/10.1016/B978-0-12-809597-3.00121-8

Chapter 7

Aqueous phase reforming of biomass-derived feedstocks

Toward the design of catalytic systems for the production of hydrogen

Fatima Jalid and Aravi Muzaffar

7.1 INTRODUCTION

Global warming issues and concerns over the depletion of fossil fuel reserves and the pollution caused by continuously increasing energy demands are stimulating researchers around the world to find alternative sources of energy. Under the current circumstances, hydrogen makes an attractive alternative energy source. It is a clean energy carrier producing only water as a product, thereby making it an advantageous clean fuel with tremendous potential for the future. Presently, hydrogen is obtained from petroleum and natural gas by utilizing traditional methods like reforming and gasification, which result in a huge carbon footprint. Besides the conventional energy sources, in principle, hydrogen could also be produced from renewable sources—water or biomass. Generation of hydrogen from water involves splitting it by means of artificial photosynthesis, electrolytic, and plasma processes, which does reduce the carbon emissions. However, these processes involve high energy costs and/or lack an efficient photocatalyst with reliable durability. Nonetheless, if hydrogen production is streamlined from biomass sources, it can possibly change the energy scenario of the world. Previously, it has been observed that the efficient generation of hydrogen from biomass through routes like steam reforming of bio-oil, enzymatic decomposition of sugars, pyrolysis, and gasification results in low production rates of hydrogen, complex reactor processing as well as catalyst requirements, and problems of ash accumulation and formation of tar (Jara-Cobos et al., 2023; Joshi & Vaidya, 2024). A much simpler process in this direction is the aqueous phase reforming (APR) of biomass-based hydrocarbon feedstock—consisting majorly of sugars and alcohols—at low temperatures and high pressure, over transition metal catalysts, producing hydrogen (H_2) and carbon dioxide (CO_2). The process was developed for the first time by Cortright et al. (2002), and is summarized by the following reaction:

$$C_nH_{2y}O_n + nH_2O \rightarrow (n+y)H_2 + nCO_2 \quad (7.1)$$

DOI: 10.1201/9781003537816-7

This single reactor process operating at lower temperatures (~500 K) circumvents the usual prerequisite to vaporize water and oxygenated hydrocarbons, thus lowering the energy requirements. The various undesirable decomposition reactions are also minimized at the operating conditions employing lower temperature. Further, the high pressure (15–50 bar) requirement of the APR process allows effective separation of H_2 by means of membrane separation processes or pressure swing adsorption (R. R. Davda et al., 2005). The operating parameters also enable the water-gas shift (WGS) reaction that consumes carbon monoxide (CO) produced resulting in small amounts of CO in the H_2 produced, thus increasing the H_2 selectivity. This reaction is given by the following equation:

$$CO + H_2O \rightarrow H_2 + CO_2 \tag{7.2}$$

The APR of a number of hydrocarbons obtained from biomass, like ethanol, sorbitol, glucose, glycerol, ethylene glycol, cellulose, etc., was studied experimentally (Chen et al., 2015; Cortright et al., 2002; Vaidya & Lopez-Sanchez, 2017). At the reaction temperature of APR, besides the conversion of hydrocarbon and H_2O to CO_2 and H_2, a side reaction involving their conversion to alkanes and H_2O also takes place, thus making selectivity a key parameter toward the tailoring of such catalytic reaction systems. This necessitates the use of a catalyst that favors the reaction pathway for C-C cleavage, facilitates the release of H_2, and suppresses the formation of alkanes in the process. Most of the experimental studies over APR reactions of different biomass-based molecules are indicative of platinum (Pt) being the most active catalyst (Cortright et al., 2002; Chen et al., 2015; Joshi & Vaidya, 2024). The efficacy of the catalyst has been observed to accentuate upon the use of promoters. Moreover, the addition of another metal to the monometallic supported catalyst has been found to result in enhanced reaction outcomes.

The chapter thus discusses the reaction kinetics and thermodynamics for the APR reaction, mechanistic insights into the various competing reaction pathways, and the impact of catalysts on the reactions as well as the product yield, thereby enhancing the understanding of routes that lead to enhanced H_2 production and selectivity. The chapter also delves into the design of bimetallic alloys and strategies for enhancing the production of H_2 in the process. The production of H_2 from biomass-based sources can also facilitate the development of the biorefinery concept. This would involve the integration of APR toward the production of value-added chemicals using feedstocks originating from biomass. The process can potentially result in higher biomass utilization as a single-pot process where the complete reaction occurs in water. By means of hydrolysis, the cellulose and hemicellulose produce sugars by the action of acid and heat, which is the feedstock for APR reaction.

7.2 FUNDAMENTALS OF AQUEOUS PHASE REFORMING

7.2.1 Mechanism of APR reaction

Oxygenated hydrocarbons obtained from biomass-based feedstock follow a number of reaction pathways in the APR process over the surface of the metal catalyst, resulting in the formation of H_2, CO_2, and alkanes, as depicted in Figure 7.1. The bio-based substrate initially undergoes a dehydrogenation reaction forming adsorbed intermediates over the catalyst surface. The interactions between the metal catalyst and adsorbates may result in the C-C or C-O bond scission, forming syngas and alcohols, respectively. As the reaction takes place in an aqueous medium, WGS reaction of the syngas will occur, producing CO-free H_2 (Grenoble et al., 1981). At the reaction conditions, CO and CO_2 from these reactions can react with H_2 forming alkanes and H_2O by Fischer–Tropsch as well as methanation reactions (Iglesia et al., 1992), depicted by the following equations.:

$$3H_2 + CO \rightarrow CH_4 + H_2O \quad (7.3)$$

$$4H_2 + CO_2 \rightarrow CH_4 + 2H_2O \quad (7.4)$$

$$nCO + 2nH_2 \rightarrow [\text{—}CH_2\text{—}]_n + nH_2O \quad (7.5)$$

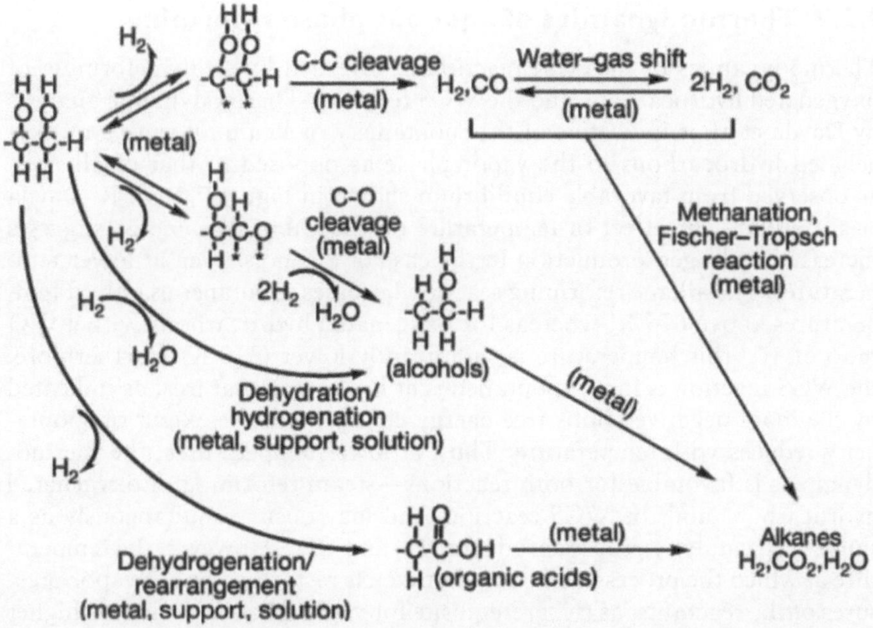

Figure 7.1 Reaction pathways for APR of oxygenated hydrocarbons derived from biomass-based feedstock (Cortright et al., 2002).

The reactions pose a series-selectivity challenge considering H_2 is the desired product in the APR reaction. Parallelly, the alcohols formed upon C—O scission also undergo catalytic hydrogenation reactions forming undesirable alkanes, which constitute a parallel-selectivity challenge. Furthermore, dehydration via C—O bond cleavage of the precursor hydrocarbon can take place owing to acidic sites in the catalyst support or catalyzed by protons present in the aqueous media, following which the dehydrogenation reaction takes place, thereby aggravating the issue of parallel-selectivity (Cortright et al., 2002). Lastly, organic acids produced by dehydrogenation and rearrangement reactions also result in the formation of alkanes. The organic acids being stable do not undergo reforming, and thus may potentially deactivate the catalyst (R. R. Davda et al., 2003). The product distribution of APR is thus a function of the interactions of the molecule with the active site of the catalyst, the support, and the catalyst–support interface. The reaction mechanism gives insight into various selectivity challenges and desired catalyst properties, which are quintessential toward the design of the catalyst. In order to maximize H_2 formation, the reaction conditions as well as the catalyst and support should be such that they facilitate dehydrogenation, C—C bond scission, activation of H_2O, and WGS reaction; and inhibits C—O bond scission, dehydration, and Fischer–Tropsch reaction (Pipitone et al., 2022).

7.2.2 Thermodynamics of aqueous phase reforming

Thermodynamics of the APR reaction is governed by steam reforming of oxygenated hydrocarbons and the WGS reaction. Thermodynamic analysis by Davda et al. is indicative of the spontaneity of steam reforming of oxygenated hydrocarbons in the vapor phase as opposed to that of alkanes, as observed from favorable equilibrium shown in Figure 7.2 (R. R. Davda et al., 2005). The effect of temperature on the Gibbs free energy suggests increased hydrogen production in the case of alcohols, even at lower temperatures. The alkane reforming reaction becomes spontaneous only at temperatures above 675 K, whereas for oxygenated hydrocarbons with a C:O ratio of 1:1, this temperature is significantly lower (320 K). Furthermore, the WGS reaction is more spontaneous at lower temperatures, as indicated by the more negative Gibbs free energy change, and the extent of spontaneity reduces with temperature. Thus, at lower temperatures, the thermodynamics is favorable for both reactions—steam reforming of oxygenated hydrocarbons and the WGS reaction, and may occur simultaneously as a single-step catalytic process producing H_2 and CO_2. However, the temperature at which the process will be undertaken is restrained by the vapor pressures of the reactants, as the prerequisite for vapor-phase reactions is higher

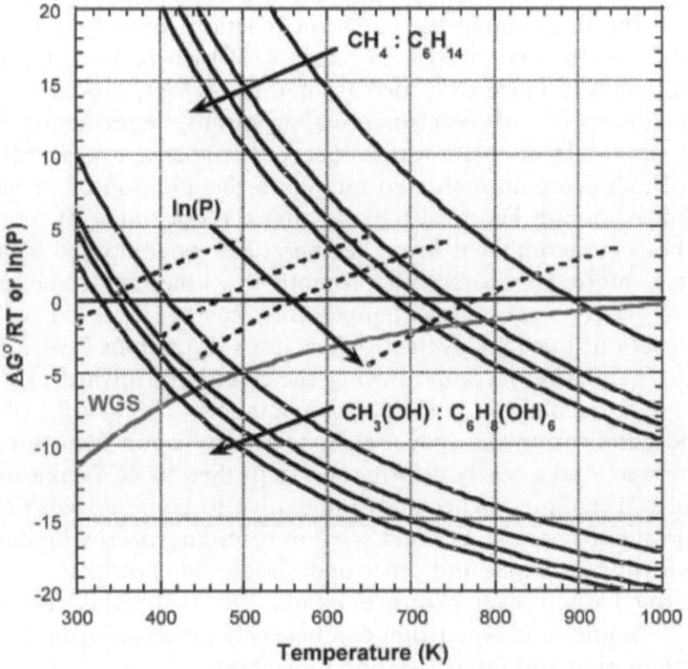

Figure 7.2 Gibbs free energy change versus temperature plot for alcohols and small alkanes (black) and WGS (gray); and vapor pressure versus temperature for alcohols (dotted lines) (R. R. Davda et al., 2005).

temperatures. Nevertheless, by employing APR, the need to vaporize the substrate feedstock is eliminated, thereby facilitating the one-pot process at lower temperatures.

A study based on APR of ethylene glycol, xylitol, xylose, and glycerol using Aspen Plus software revealed that when the reaction is conducted in the liquid state, WGS reaction becomes endothermic as opposed to being exothermic in the vapor state (Ripken et al., 2017). As a consequence of this, the thermodynamics of APR becomes slightly unfavorable; however, the fact that steam reforming incurs the heat of vaporization makes APR thermodynamically more viable. The endothermicity of WGS in the liquid phase makes it compete for energy available for chemical transformations with APR. On the contrary, for the reaction involving reforming of hydrocarbons in the vapor phase, WGS is exothermic and hence facilitates the reforming reaction. Thus, the phase of water plays a pivotal role in dictating the energetics of these reactions.

Detailed thermodynamic analysis conducted by Luo et al. for APR of polyols showed that CO formation could be lowered by employing pressure marginally higher than the vapor pressure of H_2O. Additionally, lower temperatures also inhibited the production of CO; however, it was observed to promote the undesirable methanation reaction forming alkanes (Luo et al., 2007). Similar studies based on APR of hydroxylated C_6 aliphatic compounds showed more CO_2 formation and less CH_4 production with a rise in the number of hydroxyl functional groups of the precursor molecule (Jun et al., 2013). H_2 formation was observed only at temperatures greater than 500 K, showing an increased rate upon the addition of excess water in the feed. Although Figure 7.2 shows that the reforming of oxygenated hydrocarbons is favorable at lower temperatures as compared to alkanes, nonetheless, higher temperatures promote the kinetics of the reaction. Conducting the reaction at lower temperatures ensures inhibition of decomposition reactions but enables the side reactions (Equations 7.3–7.5), which are exothermic, thereby compromising the selectivity toward H_2. Similar observations were made in a few thermodynamic studies of APR of various compounds, indicating the reduced H_2 selectivity upon incorporation of methanation as well as carbon formation steps (Seretis & Tsiakaras, 2015; Xie et al., 2011). Thus, it becomes imperative to come up with catalysts that would suppress the C-O bond scission reaction, thereby hindering the reactions forming alkanes and other undesirable side products, as the CO_2 and H_2 formed at lower temperatures exhibit thermodynamic instability as opposed to alkanes and water that can be easily produced from the subsequent methanation and Fischer–Tropsch reactions.

7.3 CATALYTIC SYSTEMS FOR AQUEOUS PHASE REFORMING

A number of biomass-based oxygenated hydrocarbons have been explored for H_2 production by the APR process, employing a wide range of catalytic systems. The nature of the metal catalyst as well as the catalyst support is observed to influence the outcome of the APR reaction. Based on the reactivity of monometallic catalysts and their downsides, numerous bimetallic alloy catalysts have also been studied. Insights into these facets can help facilitate the fabrication of the optimum catalyst for the reaction by tuning various parameters that will eventually maximize the production of H_2.

7.3.1 Monometallic catalysts

The most optimum catalyst for the APR process will primarily be dictated by the synergy of C-C and C-O bond cleavage over the surface of the catalyst and its propensity to facilitate the WGS. Of all the transition metal catalysts, copper (Cu) shows the highest rate for the WGS reaction; however, it is unsuitable for the reforming of heavier oxygenated hydrocarbons owing to its low reactivity toward C-C bond cleavage (Grenoble et al., 1981;

Sinfelt & Yates, 1967). The rate of C-C bond scission over transition metal catalysts in ethane hydrogenolysis showed the following trend: nickel (Ni) > ruthenium (Ru) > iridium (Ir) ~ rhodium (Rh) > iron (Fe) ~ cobalt (Co) > platinum (Pt) ~ palladium (Pd) (Sinfelt & Yates, 1967). Although Pt and Pd exhibit the lowest C-C scission rates, these are the catalysts exhibiting comparable rates of WGS and C-C scission rates and much lower methanation rates compared to other catalysts (Grenoble et al., 1981; Vannice, 1977). DFT studies based on ethanol decomposition over Pt(111) and Pd(111) have suggested faster C-C bond scission rates as opposed to C-O bond scission owing to lower activation barriers in the former (Alcalá et al., 2003; Li et al., 2010). Furthermore, they noted the impact of the level of dehydrogenation on the scission rates, with more dehydrogenated species facilitating the C-C cleavage.

Dumesic et al. studied the APR of ethylene glycol over silica-supported metal catalysts, wherein they observed the activity trend to be: Pt ~ Ni > Ru > Rh ~ Pd > Ir at 483 K. Despite the high reactivity of Ni, Ru, and Rh, it was observed that these are selective toward alkane formation (R. R. Davda et al., 2003). This experimental study revealed Pt and Pd to be the most promising candidates for the formation of H_2, thereby corroborating the theoretical trends, as is evident from Figure 7.3, whereby the selectivity trends for APR are the most desirable on the surface of Pd; however, the reaction throughput is lower compared to Pt. The experimental results indicate Pt to be the most optimum catalyst for the APR reaction of

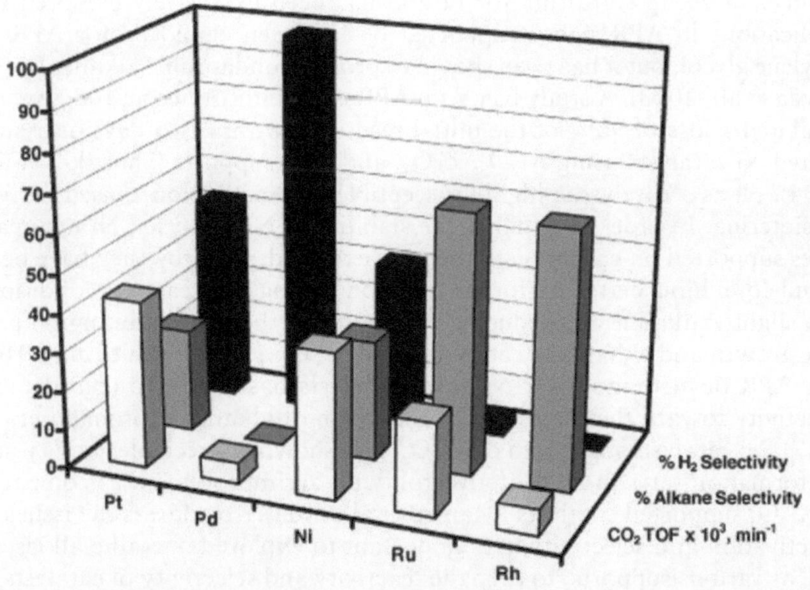

Figure 7.3 Performance of transition metal catalysts for the APR of ethylene glycol (R. R. Davda et al., 2003).

ethylene glycol. Furthermore, the first ever reported APR reaction employed Pt/Al_2O_3 as the catalyst, showing high selectivity toward H_2 and low alkane formation, with the selectivity order being: glucose < sorbitol < glycerol < ethylene glycol < methanol (Cortright et al., 2002).

A number of studies based on different oxygenated substrates have shown Pt to be very efficient for the production of H_2 in APR. Wen et al. observed the activities of catalysts based on alumina support for APR of glycerol to follow the trend as—Pt > Cu > Ni > Co, with Pt exhibiting high stability and Ni and Co undergoing significant deactivation (Wen et al., 2008a). An experimental study over APR of glycerol revealed that with an increase in particle size of the catalyst, the selectivity toward H_2 for the APR of glycerol enhanced; however, the conversion of the reactant remained constant, thus establishing the structure sensitivity of APR reactions (Lehnert & Claus, 2008). This implies that C-C scission is favored at face atoms as an increase in the particle size results in increased face atoms and a decrease in edge and corner atoms. Similar results were observed in the APR of sorbitol over Pt/Al_2O_3 where H_2 production increased with an increase in Pt loading (Duarte et al., 2016). As compared to sorbitol, xylitol has shown increased activity as well as selectivity toward H_2 formation over Pt/Al_2O_3 (Kirilin et al., 2012). Another comparative study over the same catalyst showed sorbitol to be a more promising APR substrate as opposed to glucose (Rupali R. Davda & Dumesic, 2004).

A downside of using a Pt catalyst is its high price, and thus it becomes incumbent to look for alternatives. Ni, a generally active metal catalyst, could be a cheap substitute for Pt and has been extensively explored for applications in APR. Silica-supported Ni has been employed for APR of ethylene glycol, but it has been shown to produce undesirable alkanes (R. R. Davda et al., 2003). A study based on APR of sorbitol, glucose, and glycerol exhibited a loss of 90% of the initial reactivity within two days over supported Ni catalysts using Al_2O_3, ZrO_2, and SiO_2 supports (Shabaker et al., 2004). Ni as a catalyst is highly susceptible to deactivation caused mainly by sintering. In order to enhance the stability of Ni catalysts, Ni nanoparticles supported on carbon nanofiber are explored, whereby they have been found to exhibit better performance upon tuning the reaction conditions to a slightly alkaline and reducing environment, thereby inhibiting Ni particle growth and enhancing catalytic activity (van Haasterecht et al., 2014). The APR of methanol over Ni-based catalysts has exhibited considerable selectivity toward the undesirable production of alkanes (Coronado et al., 2017). Ni catalysts supported on CeO_2 have shown appreciable activity and H_2 formation rates in APR of glycerol, with an improvement as compared to Al_2O_3 supported catalysts (Manfro et al., 2011). To address the issues of deactivation and selectivity, it is important to explore bimetallic alloys, as well as various supports, to tune the reactivity and selectivity of catalysts in line with higher hydrogen production in APR reactions.

7.3.2 Bimetallic catalysts

As bimetallic catalysts provide great scope for accentuating the activity, selectivity, and stability of catalysts, as well as the process economics when compared to their monometallic counterparts, a number of alloy combinations have been explored for the APR reaction of a variety of biomass-based feedstock molecules. Pt, being the most active metal catalyst as well as the most extensively used for the APR of oxygenated hydrocarbons, is also one of the most expensive. Alloying Pt with Fe, Co, and Ni for reforming reaction of ethylene glycol in the aqueous phase enhances the production of H_2, due to the lowering of the d-band center that decreases the heat of adsorption of CO and H_2, thereby increasing the surface available for reaction. Furthermore, the WGS reaction was suggested to be the rate-limiting step in the entire process (George W. Huber et al., 2006). Another study based on Pt alloys for the APR of glycerol revealed PtFe to exhibit better performance than PtCo, PtNi, and PtCu, where Fe was shown to play an essential role by facilitating the dissociation of H_2O toward WGS and freeing up active sites (Guo et al., 2019). Pt-Fe bimetallic catalyst has also shown improved catalytic performance toward the APR of polyols—ethylene glycol, xylitol, sorbitol, and glycerol, with the activity following the order: ethylene glycol > glycerol > xylitol > sorbitol (M.-C. Kim et al., 2016). The formation of Pt-Co bimetallic alloys has improved the H_2 selectivity as well as the reaction turnover in glycerol APR reactions (Dietrich, Akatay et al., 2014; Dietrich, Sollberger et al., 2014). The use of the Pt-Cu bimetallic catalyst supported over Mg(Al)O for glycerol reforming has also exhibited enhancement in selectivity toward H_2 production (Boga et al., 2013). The use of zinc (Zn) with a Pt nanocatalyst has also enhanced the reactivity and H_2 selectivity in the APR reaction of 1-propanol (Lei et al., 2016). The bimetallic alloy of Pt with manganese (Mn) has also shown increased production of H_2 as well as higher conversion (Bossola et al., 2017). A number of studies based on APR over Pt–Re have shown that the addition of rhenium (Re) greatly enhances the reaction rate but shifts the reaction toward undesirable alkane formation (Ciftci, Ligthart, Sen et al., 2014; Godina et al., 2018; King et al., 2010). Similar results are observed upon alloying Pt with molybdenum (Mo), leading to a loss in H_2 selectivity in the APR of glycerol (Dietrich et al., 2012, 2013). Overall, it can be seen that the formation of bimetallic alloy catalysts of Pt with other transition metals, except Re, and Mo has shown better APR performance.

Apart from the Pt catalyst, Ni has also been widely explored for APR reactions. The importance of Ni as a catalyst was explained earlier; however, the major drawback faced upon employing the Ni catalyst is its low stability, which can be altered by using bimetallic alloys. Upon alloying Ni with the most active catalyst for APR—Pt, the activity, selectivity, and stability have been observed to improve greatly. From Figure 7.4, it is clear that with 1Pt-6Ni, the highest H_2 production as well as selectivity for APR

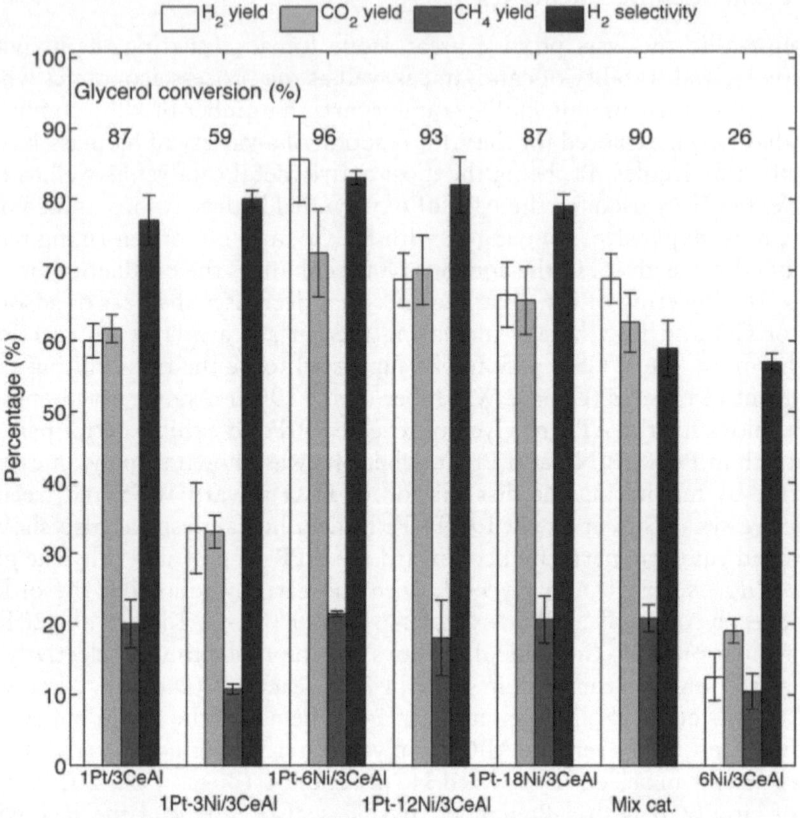

Figure 7.4 The yield, selectivity, and conversion of glycerol over xPt–yNi/3CeAl in glycerol APR (Rahman et al., 2014).

of glycerol is the highest (Rahman et al., 2014). Further, the addition of tin (Sn) to Ni catalyst has been found to significantly improve its performance toward APR of glycerol, ethylene glycol, and sorbitol, whereby the decoration of Ni defect sites by Sn has been found to decrease the turnover of CH_4 formation by inhibiting C—O bond scission while retaining high turnover frequency of C—C bond scission (G. W. Huber et al., 2003). This non-precious catalyst is a cheaper alternative to Pt for H_2 production as it shows comparable rates. Cu catalyst, known for its potential for WGS, has also been observed to enhance the conversion of Ni catalyst for the glycerol APR reaction in a Cu–Ni bimetallic catalyst by prompting WGS and inhibiting methanation compared to monometallic Ni (Wu et al., 2022). Yet another study based on Cu–Ni bimetallic catalyst supported over carbon nanotubes for the APR of glycerol revealed similar results where the synergy between Ni and Cu is proposed to boost the Ni reducibility and the dispersion of metal, with an increase in Cu content furthering the stability against sintering (Rahman, 2015). The inclusion of Co in the Ni catalyst has also been evaluated for the APR reaction, whereby the Co–Ni bimetallic catalyst showed increased selectivity toward H_2 as opposed to the pure Ni catalyst; however, the resistance against sintering was only found to improve with the addition of cerium (Luo et al., 2010). The ability to combat the sintering issue of the Ni catalyst by formulating Ni-based bimetallic alloys opens up the possibility of developing novel alloy combinations that will be more economical compared to Pt, with enhanced reaction performance as well.

In an experimental study of ethylene glycol APR over metal catalysts, Pd has shown the maximum selectivity toward H_2 production, however, with low conversion rates (R. R. Davda et al., 2003). To augment the performance of Pd, alloys of Pd with Fe and Zn have been evaluated experimentally. In this direction, Dumesic and his group assessed the activity of the Pd–Fe bimetallic catalyst, which showed 46 times higher turnover frequency of H_2 production compared to monometallic Pd (George W. Huber et al., 2006). This activity was further improved by adding Fe_2O_3 support that aided the WGS reaction. The addition of Zn to the Pd catalyst led to a tenfold enhancement in the production of H_2 in the APR of ethanol; however, the yield still remained considerably lower when compared to the stoichiometric value (Xiong et al., 2015). The Pt–Pd bimetallic catalyst has been studied for the APR of ethylene glycol, erythritol, sorbitol, and glycerol, whereby the catalyst has consistently shown high H_2 production rates with further augmentation in H_2 turnovers upon the addition of formic acid (Alvear et al., 2020). Nonetheless, the alloys of Pt and Ni are more active and selective compared to those of Pd.

In addition to the previously mentioned alloy combinations, there are also reports of the use of Re to improve the activity of Rh and Ir sup ported on TiO_2 (Nozawa et al., 2015). Another study showed the Re–Rh bimetallic catalyst to outperform the monometallic Rh catalyst as well as the bimetallic Re–Pt catalyst (Ciftci et al., 2014). Investigations of more than two metal catalysts for APR have also been studied in the literature. Pt-Ir-ReOx/SiO_2 was found to be an effective catalyst for APR of glycerol as compared to the addition of Ru and Pd to Ir–ReOx catalyst (S. Liu et al., 2018). Amalgamations of NiFeCo have also been studied in the APR of ethylene glycol whereby the role of each metal was clearly elucidated—Ni aided in conversion of ethylene glycol, Fe assisted with WGS, and Co helped to enhance adsorption/desorption kinetics over the catalyst surface (Tao et al., 2020). The usage of bimetallic alloy catalysts for APR reactions paves the path for finding more active, stable, and selective catalysts that may be employed in the establishment of the biorefinery concept, toward the integration of fermentation-based processes forming biomass-derived platform molecules with catalytic conversions of these molecules producing utility chemicals.

7.3.3 Effect of catalyst support

An essential part of a catalytic system is the support over which the catalyst is dispersed, as that affects the overall performance of the catalyst in heterogeneous reactions. Dumesic and his group explored the role of support in the APR of ethylene glycol, testing Pt supported on SiO_2-Al_2O_3, SiO_2, TiO_2, ZrO_2, Al_2O_3, CeO_2, carbon, and ZnO for the reforming reaction (Shabaker et al., 2003). The results of this study reveal the most enhancement in H_2 production over TiO_2 at both temperatures evaluated. The activity of support toward the reaction at 498 K is found to be as: TiO_2 > carbon > Al_2O_3 > ZrO_2 > Al_2O_3 sintered ~ SiO_2-Al_2O_3 > ZnO > CeO_2 > SiO_2. However, Pt supported on SiO_2-Al_2O_3, TiO_2, and carbon forms liquid-phase compounds and gaseous alkanes, thereby forming alkanes at higher conversions. Thus, Al_2O_3 is proposed to be the best support for the APR reaction, showing the highest H_2 selectivity. Similar observations were made in the APR of glycerol on supported Pt catalysts, where they observed the supports to follow the activity trends as: SAPO-11 < active carbon < HUSY < SiO_2 < MgO < Al_2O_3, with activity as well as selectivity showing a direct relationship with the support basicity (Wen et al., 2008b). Thus, the support plays a dominant role in altering the selectivity toward H_2 production attributable to its acid–base properties. In the reforming reaction of glycerol, the basic properties of the support were investigated in detail to unravel the interplay of APR and WGS, employing Pt supported over SiO_2, TiO_2, CeO_2, MgO, and Al_2O_3 (Guo et al., 2012). It was noted that the basic sites of the support facilitate the WGS, thereby enhancing the rate of APR. The basicity of the supports was ranked as MgO > Al_2O_3 > CeO_2 > TiO_2 > SiO_2, which was the same as the support activity obtained in APR of glycerol, thus elucidating the direct relationship. Yet another study of Pt supported over Al_2O_3, MgO, CeO_2, and ZrO_2 revealed MgO and ZrO_2 to be potential candidates showing high activity as well as APR selectivity (Menezes et al., 2011).

The support effect has also been explored for bimetallic alloys. The roles of SiO_2, Al_2O_3, CMK-3, and activated carbon were investigated for the Pt–Re catalytic system for APR of polyols, whereby the activity was observed to increase in the order as Al_2O_3 < SiO_2 < activated carbon < CMK-3 (H.-D. Kim et al., 2012b). The highest activity of ordered mesoporous carbon was attributed to higher dispersion and optimum rates of adsorption and desorption. Ordered mesoporous carbon has also been found to enhance the activity of Pt-Mn/CMK-3 and Pt–Fe/CMK-9 for APR of polyols (M.-C. Kim et al., 2016) and ethylene glycol (H.-D. Kim et al., 2012a), respectively. Supported Ni–Pt bimetallic alloys were also explored for APR of sorbitol with the supports showing H_2 production activity as alumina nanofiber > ZrO_2 > ceria–zirconia–silica (Tanksale et al., 2009). Besides these supports, various other studies have also been undertaken to unravel the effect of boehmite, MoS_2, graphene, as well as oxide supports based on La, Zr, and Ce for APR reactions at large, which have shown promising results (Koichumanova et al., 2018; Y. Liu et al., 2022; Stekrova et al., 2018).

7.4 CONCLUSIONS

APR of biomass-based hydrocarbon feedstock is an energy-efficient single-step catalytic approach for H_2 formation that operates at low temperatures and higher pressures, enabling the WGS reaction, thus minimizing the production of CO with H_2. The APR process is presented with challenges of selectivity—parallel as well as series, as the reactants may undergo Fischer–Tropsch reaction, methanation, and other decomposition reactions forming undesirable alkanes. This necessitates the design of the catalyst for the process such that the catalyst enables the scission of C-C bonds, inhibiting the scission of C-O bonds forming alkanes. In this direction, a number of experimental studies have been undertaken, revealing Pt to be the catalyst of choice for the APR process for a large number of oxygenated substrates. Pd has also been found to be highly selective toward H_2 production, however, with the turnovers being low. Besides Pt, Ni has also been widely explored, but it faces the issues of rapid deactivation owing to sintering. In order to tailor the activity and selectivity of these catalysts and to explore cost-effective catalysts, bimetallic alloy catalysts have also been tested for the APR reaction. The addition of Sn to Ni has been observed to combat the issues of sintering as Sn preferably sits at the defect sites of Ni, inhibiting the scission of the C-O bond that takes place at these sites. Pt–Fe and Pd–Fe have been found to show enhanced production of H_2 as compared to monometallic Pt and Pd, as the presence of Fe in the catalyst specifically aids the WGS reaction, and makes available the catalytic sites for the reaction, which otherwise remain covered by adsorbed species. Furthermore, the role of supports has also been very crucial in the design of catalytic systems for these reactions as they can potentially modify the selectivity trends due to their basic properties. A number of studies have shown that basic sites make way for WGS, which is an essential reaction in the APR process, converting water to H_2. Alumina, zirconia, and titanium dioxide have shown the most enhancement as supports in aiding the APR reactions, increasing the production of H_2. With the proper design of catalytic systems, the turnover of APR reactions of oxygenated hydrocarbons as well as the H_2 selectivity can be tailored to meet the requirements of post-processing of hydrogen such that it can be employed in various other applications.

REFERENCES

Alcalá, R., Mavrikakis, M., & Dumesic, J. A. (2003). DFT studies for cleavage of C-C and C-O bonds in surface species derived from ethanol on Pt(111). *Journal of Catalysis*, *218*(1), 178–190. https://doi.org/https://doi.org/10.1016/S0021-9517(03)00090-3

Alvear, M., Aho, A., Simakova, I. L., Grénman, H., Salmi, T., & Murzin, D. Y. (2020). Aqueous phase reforming of alcohols over a bimetallic Pt-Pd catalyst in the presence of formic acid. *Chemical Engineering Journal*, *398*, 125541. https://doi.org/https://doi.org/10.1016/j.cej.2020.125541

Boga, D. A., Oord, R., Beale, A. M., Chung, Y.-M., Bruijnincx, P. C. A., & Weckhuysen, B. M. (2013). Highly selective bimetallic Pt-Cu/Mg(Al)O catalysts for the aqueous-phase reforming of glycerol. *ChemCatChem*, *5*(2), 529–537. https://doi.org/https://doi.org/10.1002/cctc.201200112

Bossola, F., Pereira-Hernández, X. I., Evangelisti, C., Wang, Y., & Dal Santo, V. (2017). Investigation of the promoting effect of Mn on a Pt/C catalyst for the steam and aqueous phase reforming of glycerol. *Journal of Catalysis*, *349*, 75–83. https://doi.org/https://doi.org/10.1016/j.jcat.2017.03.002

Chen, G. yi, Li, W. qing, Chen, H., & Yan, B. bei. (2015). Progress in the aqueous-phase reforming of different biomass-derived alcohols for hydrogen production. *Journal of Zhejiang University: Science A*, *16*(6), 491–506. https://doi.org/10.1631/jzus.A1500023

Ciftci, A., Ligthart, D. A. J. M., & Hensen, E. J. M. (2014). Aqueous phase reforming of glycerol over Re-promoted Pt and Rh catalysts. *Green Chemistry*, *16*(2), 853–863. https://doi.org/10.1039/C3GC42046A

Ciftci, A., Ligthart, D. A. J. M., Sen, A. O., van Hoof, A. J. F., Friedrich, H., & Hensen, E. J. M. (2014). Pt-Re synergy in aqueous-phase reforming of glycerol and the water–gas shift reaction. *Journal of Catalysis*, *311*, 88–101. https://doi.org/https://doi.org/10.1016/j.jcat.2013.11.011

Coronado, I., Stekrova, M., García Moreno, L., Reinikainen, M., Simell, P., Karinen, R., & Lehtonen, J. (2017). Aqueous-phase reforming of methanol over nickel-based catalysts for hydrogen production. *Biomass and Bioenergy*, *106*, 29–37. https://doi.org/https://doi.org/10.1016/j.biombioe.2017.08.018

Cortright, R. D., Davda, R. R., & Dumesic, J. A. (2002). Hydrogen from catalytic reforming of biomass-derived hydrocarbons in liquid water. *Nature*, *418*(6901), 964–967. https://doi.org/10.1038/nature01009

Davda, R. R., & Dumesic, J. A. (2004). Renewable hydrogen by aqueous-phase reforming of glucose. *Chemical Communications*, *1*, 36–37. https://doi.org/10.1039/B310152E

Davda, R. R., Shabaker, J. W., Huber, G. W., Cortright, R. D., & Dumesic, J. A. (2003). Aqueous-phase reforming of ethylene glycol on silica-supported metal catalysts. *Applied Catalysis B: Environmental*, *43*(1), 13–26. https://doi.org/10.1016/S0926-3373(02)00277-1

Davda, R. R., Shabaker, J. W., Huber, G. W., Cortright, R. D., & Dumesic, J. A. (2005). A review of catalytic issues and process conditions for renewable hydrogen and alkanes by aqueous-phase reforming of oxygenated hydrocarbons over supported metal catalysts. *Applied Catalysis B: Environmental*, *56*, 171–186. https://doi.org/10.1016/j.apcatb.2004.04.027

Dietrich, P. J., Akatay, M. C., Sollberger, F. G., Stach, E. A., Miller, J. T., Delgass, W. N., & Ribeiro, F. H. (2014). Effect of Co loading on the activity and selectivity of PtCo aqueous phase reforming catalysts. *ACS Catalysis*, *4*(2), 480–491. https://doi.org/10.1021/cs4008705

Dietrich, P. J., Lobo-Lapidus, R. J., Wu, T., Sumer, A., Akatay, M. C., Fingland, B. R., … Miller, J. T. (2012). Aqueous phase glycerol reforming by ptmo bimetallic nano-particle catalyst: Product selectivity and structural characterization. *Topics in Catalysis*, *55*(1), 53–69. https://doi.org/10.1007/s11244-012-9775-5

Dietrich, P. J., Sollberger, F. G., Akatay, M. C., Stach, E. A., Delgass, W. N., Miller, J. T., & Ribeiro, F. H. (2014). Structural and catalytic differences in the effect of Co and Mo as promoters for Pt-based aqueous phase reforming catalysts. *Applied Catalysis B: Environmental, 156–157*, 236–248. https://doi.org/https://doi.org/10.1016/j.apcatb.2014.03.016

Dietrich, P. J., Wu, T., Sumer, A., Dumesic, J. A., Jellinek, J., Delgass, W. N., … Miller, J. T. (2013). Aqueous phase glycerol reforming with Pt and PtMo bimetallic nanoparticle catalysts: The role of the mo promoter. *Topics in Catalysis*, *56*(18), 1814–1828. https://doi.org/10.1007/s11244-013-0115-1

Duarte, H. A., Sad, M. E., & Apesteguía, C. R. (2016). Aqueous phase reforming of sorbitol on Pt/Al2O3: Effect of metal loading and reaction conditions on H_2 productivity. *International Journal of Hydrogen Energy*, *41*(39), 17290–17296. https://doi.org/https://doi.org/10.1016/j.ijhydene.2016.07.071

Godina, L. I., Kirilin, A. V, Tokarev, A. V, Simakova, I. L., & Murzin, D. Y. (2018). Sibunit-supported mono- and bimetallic catalysts used in aqueous-phase reforming of xylitol. *Industrial & Engineering Chemistry Research*, *57*(6), 2050–2067. https://doi.org/10.1021/acs.iecr.7b04937

Grenoble, D. C., Estadt, M. M., & Ollis, D. F. (1981). The chemistry and catalysis of the water gas shift reaction: 1. The kinetics over supported metal catalysts. *Journal of Catalysis*, *67*(1), 90–102. https://doi.org/https://doi.org/10.1016/0021-9517(81)90263-3

Guo, Y., Azmat, M. U., Liu, X., Wang, Y., & Lu, G. (2012). Effect of support's basic properties on hydrogen production in aqueous-phase reforming of glycerol and correlation between WGS and APR. *Applied Energy*, *92*, 218–223. https://doi.org/https://doi.org/10.1016/j.apenergy.2011.10.020

Guo, Y., Liu, X., & Wang, Y. (2019). Catalytic and DRIFTS studies of Pt-based bimetallic alloy catalysts in aqueous-phase reforming of glycerol. *Industrial & Engineering Chemistry Research*, *58*(8), 2749–2758. https://doi.org/10.1021/acs.iecr.8b05774

Huber, G. W., Shabaker, J. W., & Dumesic, J. A. (2003). Raney Ni-Sn catalyst for H_2 production from biomass-derived hydrocarbons. *Science*, *300*(5628), 2075–2077. https://doi.org/10.1126/science.1085597

Huber, G. W., Shabaker, J. W., Evans, S. T., & Dumesic, J. A. (2006). Aqueous-phase reforming of ethylene glycol over supported Pt and Pd bimetallic catalysts. *Applied Catalysis B: Environmental*, *62*(3), 226–235. https://doi.org/https://doi.org/10.1016/j.apcatb.2005.07.010

Iglesia, E., Soled, S. L., & Fiato, R. A. (1992). Fischer-Tropsch synthesis on cobalt and ruthenium. Metal dispersion and support effects on reaction rate and selectivity. *Journal of Catalysis*, *137*(1), 212–224. https://doi.org/https://doi.org/10.1016/0021-9517(92)90150-G

Jara-Cobos, L., Abril-González, M., & Pinos-Vélez, V. (2023). Production of hydrogen from lignocellulosic biomass: A review of technologies. *Catalysts*. https://doi.org/10.3390/catal13040766

Joshi, A. N., & Vaidya, P. D. (2024). Recent studies on aqueous-phase reforming: Catalysts, reactors, hybrid processes and techno-economic analysis. *International Journal of Hydrogen Energy*, *49*, 117–137. https://doi.org/10.1016/j.ijhydene.2023.06.314

Jun, J.-W., Suh, Y.-W., Suh, D. J., & Lee, Y.-K. (2013). Thermodynamic analysis of steam and aqueous reforming of hydroxylated C6 aliphatic compounds. *Journal of Industrial and Engineering Chemistry*, *19*(6), 2072–2078. https://doi.org/https://doi.org/10.1016/j.jiec.2013.03.020

Kim, H.-D., Park, H. J., Kim, T.-W., Jeong, K.-E., Chae, H.-J., Jeong, S.-Y., … Kim, C.-U. (2012a). Hydrogen production through the aqueous phase reforming of ethylene glycol over supported Pt-based bimetallic catalysts. *International Journal of Hydrogen Energy*, *37*(10), 8310–8317. https://doi.org/https://doi.org/10.1016/j.ijhydene.2012.02.160

Kim, H.-D., Park, H. J., Kim, T.-W., Jeong, K.-E., Chae, H.-J., Jeong, S.-Y., … Kim, C.-U. (2012b). The effect of support and reaction conditions on aqueous phase reforming of polyol over supported Pt–Re bimetallic catalysts. *Catalysis Today*, *185*(1), 73–80. https://doi.org/https://doi.org/10.1016/j.cattod.2011.08.012

Kim, M.-C., Kim, T.-W., Kim, H. J., Kim, C.-U., & Bae, J. W. (2016). Aqueous phase reforming of polyols for hydrogen production using supported PtFe bimetallic catalysts. *Renewable Energy*, *95*, 396–403. https://doi.org/https://doi.org/10.1016/j.renene.2016.04.020

King, D. L., Zhang, L., Xia, G., Karim, A. M., Heldebrant, D. J., Wang, X., … Wang, Y. (2010). Aqueous phase reforming of glycerol for hydrogen production over Pt–Re supported on carbon. *Applied Catalysis B: Environmental*, *99*(1), 206–213. https://doi.org/https://doi.org/10.1016/j.apcatb.2010.06.021

Kirilin, A. V., Tokarev, A. V., Kustov, L. M., Salmi, T., Mikkola, J.-P., & Murzin, D. Y. (2012). Aqueous phase reforming of xylitol and sorbitol: Comparison and influence of substrate structure. *Applied Catalysis A: General*, *435–436*, 172–180. https://doi.org/https://doi.org/10.1016/j.apcata.2012.05.050

Koichumanova, K., Vikla, A. K. K., Cortese, R., Ferrante, F., Seshan, K., Duca, D., & Lefferts, L. (2018). In situ ATR-IR studies in aqueous phase reforming of hydroxyacetone on Pt/ZrO_2 and Pt/AlO(OH) catalysts: The role of aldol condensation. *Applied Catalysis B: Environmental*, *232*, 454–463. https://doi.org/https://doi.org/10.1016/j.apcatb.2018.03.090

Lehnert, K., & Claus, P. (2008). Influence of Pt particle size and support type on the aqueous-phase reforming of glycerol. *Catalysis Communications*, *9*(15), 2543–2546. https://doi.org/https://doi.org/10.1016/j.catcom.2008.07.002

Lei, Y., Lee, S., Low, K. B., Marshall, C. L., & Elam, J. W. (2016). Combining electronic and geometric effects of ZnO-promoted Pt nanocatalysts for aqueous phase reforming of 1-propanol. *ACS Catalysis*, *6*(6), 3457–3460.

Li, M., Guo, W., Jiang, R., Zhao, L., & Shan, H. (2010). Decomposition of ethanol on Pd(111): A density functional theory study. *Langmuir*, *26*(3), 1879–1888. https://doi.org/10.1021/la902641t

Liu, S., Tamura, M., Shen, Z., Zhang, Y., Nakagawa, Y., & Tomishige, K. (2018). Hydrogenolysis of glycerol with in-situ produced H2 by aqueous-phase reforming of glycerol using Pt-modified Ir-ReOx/SiO2 catalyst. *Catalysis Today*, *303*, 106–116. https://doi.org/https://doi.org/10.1016/j.cattod.2017.07.025

Liu, Y., Li, S., & Zhu, S. (2022). Novel Pt/MoS2 nanosheet catalyst for hydrogen production via aqueous-phase reforming of methanol. *Reaction Kinetics, Mechanisms and Catalysis*, *135*(5), 2579–2589. https://doi.org/10.1007/s11144-022-02275-x

Luo, N., Cao, F., Zhao, X., & Xiao, T. (2007). Thermodynamic analysis of aqueous-reforming of polylols for hydrogen generation. *Fuel Processing Technology*, *86*, 1727–1736. https://doi.org/10.1016/j.fuel.2006.12.016

Luo, N., Ouyang, K., Cao, F., & Xiao, T. (2010). Hydrogen generation from liquid reforming of glycerin over Ni–Co bimetallic catalyst. *Biomass and Bioenergy*, *34*(4), 489–495. https://doi.org/https://doi.org/10.1016/j.biombioe.2009.12.013

Manfro, R. L., da Costa, A. F., Ribeiro, N. F. P., & Souza, M. M. V. M. (2011). Hydrogen production by aqueous-phase reforming of glycerol over nickel catalysts supported on CeO_2. *Fuel Processing Technology*, *92*(3), 330–335. https://doi.org/https://doi.org/10.1016/j.fuproc.2010.09.024

Menezes, A. O., Rodrigues, M. T., Zimmaro, A., Borges, L. E. P., & Fraga, M. A. (2011). Production of renewable hydrogen from aqueous-phase reforming of glycerol over Pt catalysts supported on different oxides. *Renewable Energy*, *36*(2), 595–599. https://doi.org/https://doi.org/10.1016/j.renene.2010.08.004

Nozawa, T., Yoshida, A., Hikichi, S., & Naito, S. (2015). Effects of Re addition upon aqueous phase reforming of ethanol over TiO_2 supported Rh and Ir catalysts. *International Journal of Hydrogen Energy*, *40*(11), 4129–4140. https://doi.org/https://doi.org/10.1016/j.ijhydene.2015.01.152

Pipitone, G., Zoppi, G., Pirone, R., & Bensaid, S. (2022). A critical review on catalyst design for aqueous phase reforming. *International Journal of Hydrogen Energy*, *47*(1), 151–180. https://doi.org/10.1016/j.ijhydene.2021.09.206

Rahman, M. M. (2015). H_2 production from aqueous-phase reforming of glycerol over Cu–Ni bimetallic catalysts supported on carbon nanotubes. *International Journal of Hydrogen Energy*, *40*(43), 14833–14844. https://doi.org/https://doi.org/10.1016/j.ijhydene.2015.09.015

Rahman, M. M., Church, T. L., Variava, M. F., Harris, A. T., & Minett, A. I. (2014). Bimetallic Pt–Ni composites on ceria-doped alumina supports as catalysts in the aqueous-phase reforming of glycerol. *RSC Advances*, *4*(36), 18951–18960. https://doi.org/10.1039/C4RA00355A

Ripken, R. M., Meuldijk, J., Gardeniers, J. G. E., & LeGac, S. (2017). Influence of the water phase state on the thermodynamics of aqueous-phase reforming for hydrogen production. *ChemSusChem*, *10*(24), 4909–4913. https://doi.org/https://doi.org/10.1002/cssc.201700189

Seretis, A., & Tsiakaras, P. (2015). A thermodynamic analysis of hydrogen production via aqueous phase reforming of glycerol. *Fuel Processing Technology*, *134*, 107–115. https://doi.org/10.1016/j.fuproc.2015.01.021

Shabaker, J. W., Huber, G. W., Davda, R. R., Cortright, R. D., & Dumesic, J. A. (2003). Aqueous-phase reforming of ethylene glycol over supported platinum catalysts. *Catalysis Letters*, *88*(1), 1–8. https://doi.org/10.1023/A:1023538917186

Shabaker, J. W., Huber, G. W., & Dumesic, J. A. (2004). Aqueous-phase reforming of oxygenated hydrocarbons over Sn-modified Ni catalysts. *Journal of Catalysis*, *222*(1), 180–191. https://doi.org/https://doi.org/10.1016/j.jcat.2003.10.022

Sinfelt, J. H., & Yates, D. J. C. (1967). Catalytic hydrogenolysis of ethane over the noble metals of Group VIII. *Journal of Catalysis*, *8*(1), 82–90. https://doi.org/https://doi.org/10.1016/0021-9517(67)90284-9

Stekrova, M., Rinta-Paavola, A., & Karinen, R. (2018). Hydrogen production via aqueous-phase reforming of methanol over nickel modified Ce, Zr and La oxide supports. *Catalysis Today*, *304*, 143–152. https://doi.org/https://doi.org/10.1016/j.cattod.2017.08.030

Tanksale, A., Zhou, C. H., Beltramini, J. N., & Lu, G. Q. (2009). Hydrogen production by aqueous phase reforming of sorbitol using bimetallic Ni–Pt catalysts: metal support interaction. *Journal of Inclusion Phenomena and Macrocyclic Chemistry*, *65*(1), 83–88. https://doi.org/10.1007/s10847-009-9618-6

Tao, J., Hou, L., Yan, B., Chen, G., Li, W., Chen, H., … Lin, F. (2020). Hydrogen production via aqueous-phase reforming of ethylene glycol over a nickel–iron alloy catalyst: Effect of cobalt addition. *Energy & Fuels*, *34*(2), 1153–1161. https://doi.org/10.1021/acs.energyfuels.9b02149

Vaidya, P. D., & Lopez-Sanchez, J. A. (2017). Review of hydrogen production by catalytic aqueous-phase reforming. *ChemistrySelect*, *2*(22), 6563–6576. https://doi.org/10.1002/slct.201700905

van Haasterecht, T., Ludding, C. C. I., de Jong, K. P., & Bitter, J. H. (2014). Toward stable nickel catalysts for aqueous phase reforming of biomass-derived feedstock under reducing and alkaline conditions. *Journal of Catalysis*, *319*, 27–35. https://doi.org/https://doi.org/10.1016/j.jcat.2014.07.014

Vannice, M. A. (1977). The catalytic synthesis of hydrocarbons from H_2-CO mixtures over the Group VIII metals: V. The catalytic behavior of silica-supported metals. *Journal of Catalysis*, *50*(2), 228–236. https://doi.org/https://doi.org/10.1016/0021-9517(77)90031-8

Wen, G., Xu, Y., Ma, H., Xu, Z., & Tian, Z. (2008a). Production of hydrogen by aqueous-phase reforming of glycerol. *International Journal of Hydrogen Energy*, *33*(22), 6657–6666. https://doi.org/10.1016/j.ijhydene.2008.07.072

Wen, G., Xu, Y., Ma, H., Xu, Z., & Tian, Z. (2008b). Production of hydrogen by aqueous-phase reforming of glycerol. *International Journal of Hydrogen Energy*, *33*(22), 6657–6666. https://doi.org/https://doi.org/10.1016/j.ijhydene.2008.07.072

Wu, K., Dou, B., Zhang, H., Liu, D., Chen, H., & Xu, Y. (2022). Aqueous phase reforming of biodiesel byproduct glycerol over mesoporous Ni-Cu/CeO_2 for renewable hydrogen production. *Fuel*, *308*, 122014. https://doi.org/https://doi.org/10.1016/j.fuel.2021.122014

Xie, J., Su, D., Yin, X., Wu, C., & Zhu, J. (2011). Thermodynamic analysis of aqueous phase reforming of three model compounds in bio-oil for hydrogen production. *International Journal of Hydrogen Energy*, *36*(24), 15561–15572. https://doi.org/10.1016/j.ijhydene.2011.08.103

Xiong, H., DeLaRiva, A., Wang, Y., & Datye, A. K. (2015). Low-temperature aqueous-phase reforming of ethanol on bimetallic PdZn catalysts. *Catalysis Science & Technology*, *5*(1), 254–263. https://doi.org/10.1039/C4CY00914B

Lei, Y., Lee, S., Low, K. B., Marshall, C. L., & Elam, J. W. (2016). Combining electronic and geometric effects of ZnO-promoted Pt nanocatalysts for aqueous phase reforming of 1-propanol. *ACS Catalysis*, *6*(6), 3457-3460.

Chapter 8

Thermochemical hydrogen production methods

A review

Sangharatna M. Ramteke, Sachin Akoji Meshram, H. Chelladurai, and Akhilesh Kumar Choudhary

8.1 INTRODUCTION

Hydrogen, often praised as the "future fuel," holds significant promise as a clean energy carrier. It is a plentiful element that can be generated through diverse sustainable means, like electrolysis employing renewable energy sources. When utilized in fuel cells or combustion engines, hydrogen emits solely water vapor as a byproduct, making it a zero-emission energy option (Ibrahim et al., 2021). Its versatility extends to various sectors, such as transportation, industry, and energy storage, where it can play a pivotal role in minimizing greenhouse gas emissions and helping to mitigate climate change (Siddik et al., 2021). While challenges like efficient production and infrastructure development remain, hydrogen's potential to decarbonize our energy systems and foster a more sustainable future is undeniably significant (Siddik and Hangun et al., 2022).

Thermochemical hydrogen production is a promising technology for producing hydrogen from a range of feedstocks, including water, biomass, and fossil fuels. This method uses heat and chemical reactions to split water molecules into hydrogen and oxygen or to release hydrogen from other feedstocks (Funk, 2001; Sato, 1979). This method typically requires high temperatures (500–2000°C), which can be generated using a variety of energy sources, mainly fossil fuels, nuclear energy, or renewable energy sources like solar and geothermal energy (Rosen, 2010; Lee et al., 2022). One of the main advantages of this method is that it can produce hydrogen from various feedstocks. This is important because this method can be customized to the specific resources and needs of a particular region or country. For example, countries with abundant biomass resources could use this method to produce hydrogen from biomass, while countries with abundant solar resources could use it to generate hydrogen from water using solar energy (Karaca et al., 2023). Another advantage is that it can be efficient, with overall efficiencies of up to 90%. This means that this method can produce hydrogen with relatively low energy inputs (Beghi, 1986; Z. Ma et al., 2022). Furthermore, this method encompasses a range of sophisticated

 DOI: 10.1201/9781003537816-8

chemical reactions and cycles, each with its unique set of advantages and challenges. Whether through the high-temperature electrolysis of water, intricate sulfur–iodine or copper–chlorine cycles, or the efficient use of renewable heat sources like concentrated solar power, this method holds promise as a sustainable and environmentally friendly means of satisfying the growing global demand for clean hydrogen fuel, with potential applications spanning from transportation to industrial processes and energy storage. As the technology continues to develop, thermochemical hydrogen production is expected to play an increasingly important role in the global hydrogen economy (Abanades & Flamant, 2006; Kodama et al., 2005).

One of the primary obstacles in thermochemical hydrogen production is reducing the cost of generating high-temperature heat. Another challenge is developing thermochemical water-splitting processes that can operate at lower temperatures and with higher efficiencies. There are a number of opportunities to improve thermochemical hydrogen production methods. For example, researchers are developing new materials that can be used to reduce the operating temperature of thermochemical water-splitting reactions (Onuki et al., 1994; Diver et al., 2008; Budama et al., 2023). Other researchers are developing new ways to integrate thermochemical hydrogen production methods with other industrial processes, such as power generation and refining (Khan et al., 2023; Pregger et al., 2009; Izanloo et al., 2021; Likkasit et al., 2018). Thermochemical hydrogen production methods have the potential to play a major role in the future of clean energy. By enhancing the efficiency and cost-effectiveness of thermochemical hydrogen production methods, we can make hydrogen a more viable fuel for a variety of applications.

To summarize, the hydrogen industry faces significant hurdles, but the prospects it holds for decarbonization, versatility, and renewability are substantial. Governments, corporations, and the public are increasingly recognizing its potential, leading to growing investments and support. The journey toward a cleaner energy future is underway, and hydrogen is set to be a crucial player in this transformation. Overcoming the challenges and capitalizing on the opportunities will be instrumental in unlocking hydrogen's complete potential as a fundamental element of a sustainable energy infrastructure.

8.2 HYDROGEN PRODUCTION METHODS

This section will review the different hydrogen production methods that are currently available, particularly thermochemical processes. We will discuss the advantages and disadvantages of each method in terms of efficiency, cost, environmental impact, and scalability.

8.2.1 Thermochemical processes

Thermochemical processes are the most widely used methods for producing hydrogen today. These processes use heat and/or pressure to convert fossil fuels or other hydrocarbons into hydrogen and other products. Thermochemical processes are typically energy-intensive, and some of them also emit greenhouse gases (Manna et al., 2021; Aziz et al., 2021). However, they are currently the most economically efficient methods for hydrogen production in large quantities. In the following, the three most common thermochemical processes for producing hydrogen are reviewed.

8.2.1.1 Steam methane reforming

Steam methane reforming (SMR) stands as the most commonly employed method for hydrogen production in the world. It involves reacting methane with steam at high temperatures and pressures in the presence of a catalyst to generate hydrogen, along with carbon monoxide and a minor quantity of carbon dioxide. SMR is a highly efficient process, with a typical methane conversion rate of over 90%. However, it is also a very energy-intensive process, requiring a significant amount of heat to drive the reaction. Additionally, SMR generates greenhouse gas emissions, including carbon dioxide and methane (Razi et al., 2023; Basile et al., 2015). The schematic diagram of an SMR process (Figure 8.1) outlines a series of essential steps in the generation of hydrogen from natural gas. This process begins with desulfurization, where the incoming natural gas feedstock undergoes treatment to eliminate sulfur compounds, preventing catalyst poisoning. Subsequently, the preheating phase raises the temperature of the natural gas and steam, enhancing the efficiency of the reforming process within the reformer. Within the reformer itself, a high-temperature, high-pressure environment, along with a nickel catalyst, facilitates the chemical reaction

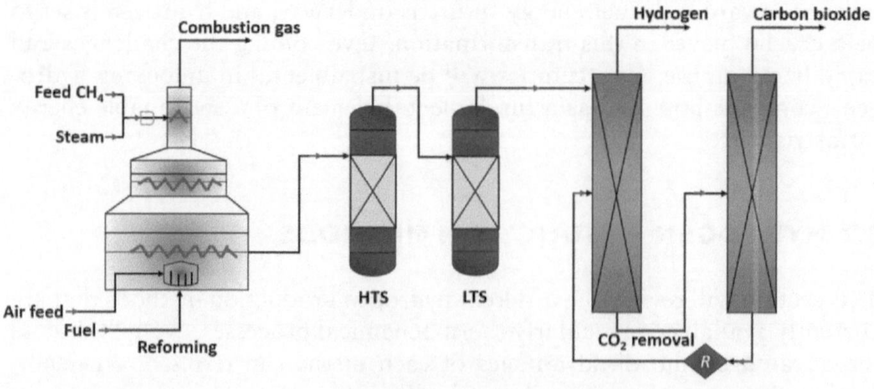

Figure 8.1 Schematic diagram of an SMR process. Redrawn and adapted from Khan et al., 2023.

between natural gas and steam. This reaction generates a blend of hydrogen, carbon monoxide, carbon dioxide, and unreacted methane. The shift converter, acting as a subsequent stage, converts a portion of the carbon monoxide in the reformer product gas into additional hydrogen and carbon dioxide, employing a metal oxide catalyst. The process then transitions to cooling and purification, where the reformer product gas is cooled and thoroughly purified to eliminate water, carbon dioxide, and other impurities. The end result is a high-purity hydrogen gas stream, ready for various industrial applications. Key components in this SMR process include the reformer itself, responsible for the primary chemical conversion, the shift converter for carbon monoxide conversion, and the cooling and purification system, which ensures the production of clean and high-quality hydrogen. This well-established process serves as a crucial method for industrial hydrogen production, contributing to a range of sustainable energy applications (Basile et al., 2015).

In the study by Abd et al., (2023), an innovative approach for biomethane and hydrogen production was explored, involving two key steps: pressure swing adsorption (PSA) for biogas upgrading and SMR for methane conversion. The integration of these processes offered advantages in relation to efficiency, cost-effectiveness, and environmental impact. The PSA phase efficiently separated methane from carbon dioxide within biogas, providing a high-purity methane stream for direct input into the SMR process, eliminating the need for additional purification (Figure 8.2a). Furthermore, the PSA phase could be powered by renewable energy sources, diminishing dependence on fossil fuels and decreasing overall production costs. Importantly, the integrated process generated no greenhouse gas emissions, reducing its environmental footprint. However, scalability remained a challenge due to the complexity and cost of the PSA phase, limiting its application on a larger scale. Salahi et al. (2023), investigated the influence of several factors on hydrogen production efficiency through SMR utilizing a Y-promoted Ni/Al_2O_3 catalyst. They applied response surface methodology (RSM) to optimize key parameters, including Ni loading, Y loading, reaction temperature, and the steam-to-methane ratio. RSM, a statistical technique, was employed to model and optimize the impact of these parameters on hydrogen yield and methane conversion. The optimized conditions for hydrogen production were identified as follows: Ni loading at 16.2%, Y loading at 2.7%, a reaction temperature of 690°C, and a steam-to-methane ratio of 3.4. Under these conditions, the authors achieved an impressive hydrogen yield of 97.4% and a methane conversion rate of 93.8% (Figure 8.2b). The methodology offers several advantages, including its efficiency in optimizing various parameters and its reliability due to its statistically rigorous nature. However, it also comes with disadvantages, such as complexity and potential time-consuming implementation, as well as the requirement for a substantial amount of data to train the

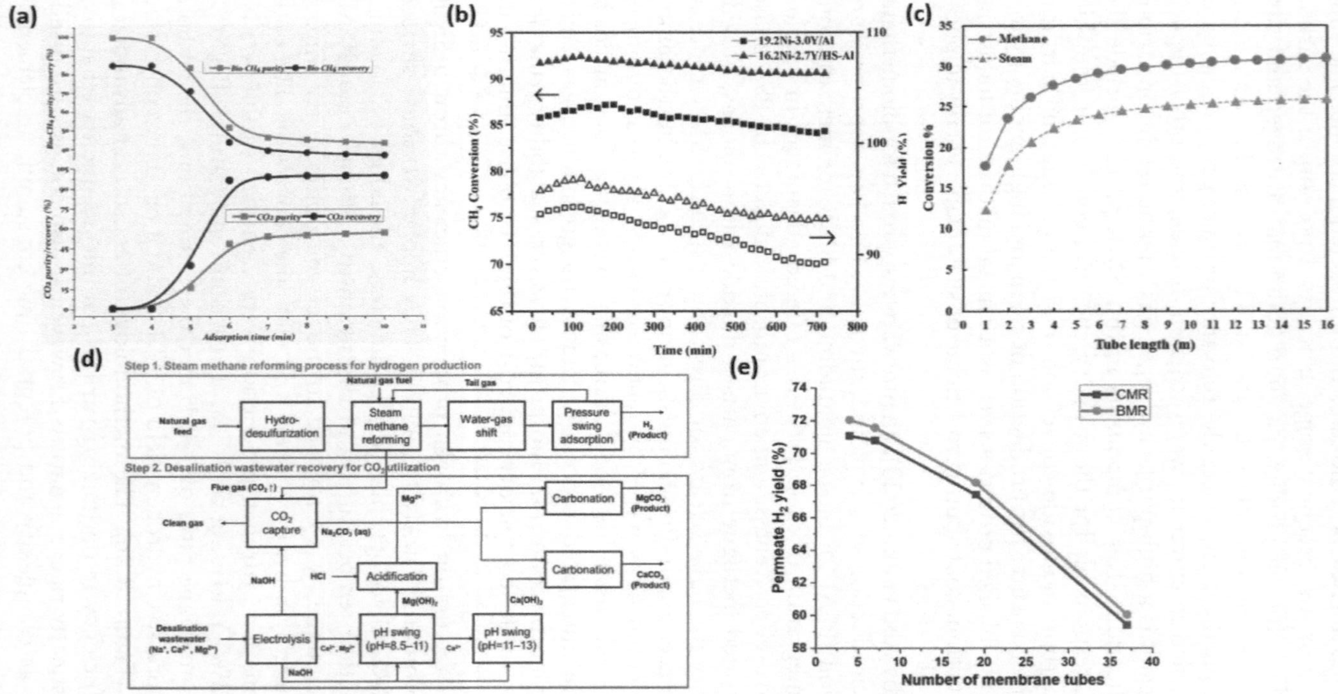

Figure 8.2 (a) The impact of adsorption time on Bio-CH_4/CO_2 purity and recovery in the product stream, using a four-bed, two-vessel, in-series configuration under a pressure of 200 kPa. Reprinted and adapted from Abd et al. (2023) with permission. (b) Assessing the activity performance of 19.2Ni-3.0Y/Al and 16.2Ni-2.7Y/HS-Al catalysts during a 12-hour SMR reaction. Reprinted and adapted from Salahi et al. (2023) with permission. (c) Analyzing the fluctuation in methane and steam conversion concerning the length of the reformer tube. Reprinted and adapted from Salahi et al. 2023) with permission. (d) A streamlined illustration depicting the proposed process. Reprinted and adapted from Salahi et al. (2023) with permission. (e) Comparison of permeate H_2 yield between the CMR and the BMR as a function of the number of membrane tubes. Reprinted and adapted from Choi et al. (2022) with permission.

model. Amini et al. (2023) employed computational fluid dynamics (CFD) to simulate the SMR process within an industrial-scale side-fired reformer. This CFD simulation aimed to analyze the impact of various parameters on hydrogen production efficiency, including reformer tube length, reformer tube diameter, steam-to-methane ratio, and surface heat flux. The optimal conditions for hydrogen production were determined as follows: reformer tube length of 11 m (Figure 8.2c), reformer tube diameter of 0.172 m, steam-to-methane ratio of 1.6, and a surface heat flux of 95 kW/m^2, with a predicted potential increase in hydrogen production efficiency of up to 20%. The methodology's advantages lie in CFD's capability to effectively simulate complex systems, enabling the investigation of numerous parameters and the identification of areas for system improvement. However, CFD simulations can be computationally expensive and time-consuming, necessitate accurate input data, such as fluid properties and system geometry, and require validation against experimental data to ensure accuracy. Lim et al.'s (2023) groundbreaking approach to carbon-neutral hydrogen production was investigated by integrating SMR with CO_2 utilization from desalination wastewater (Figure 8.2d). This innovative process aims to produce hydrogen efficiently while mitigating its environmental impact. The potential advantages of this methodology include enhanced efficiency through waste CO_2 utilization, reduced carbon emissions for a more sustainable production process, and the repurposing of CO_2 from desalination wastewater. However, the investigation lacks specific data on efficiency gains, cost-effectiveness, and scalability, leaving important practical implications unanswered. Technical complexities, energy requirements, and economic viability also need further exploration for a comprehensive evaluation of this novel approach. Choi et al. (2022) employed CFD to assess the performance of a baffled membrane reactor (BMR) for hydrogen production via catalytic membrane reactor (CMR). They discovered that the BMR could significantly enhance hydrogen production efficiency while reducing carbon dioxide emissions compared to conventional CMR reactors (Figure 8.2e). Moreover, the authors developed a scale-up procedure for the BMR, offering the potential to design and construct large-scale BMRs for commercial hydrogen production.

To summarize, in the examination of innovative approaches for biomethane and hydrogen production, various studies contributed insights into efficiency, cost-effectiveness, environmental impact, and scalability. Integrated processes that combined PSA and SMR showed promise, with PSA efficiently purifying biogas for direct input into SMR, reducing costs and environmental footprint. The optimization of SMR for hydrogen production revealed effective parameters using RSM, achieving high yields. CFD simulations demonstrated potential efficiency gains in SMR, though challenges such as computational costs and data accuracy were noted. Carbon-neutral hydrogen production, integrating SMR with CO_2

utilization, presented environmental benefits, but scalability and specific performance data required further exploration. The use of CFD to assess a baffled membrane reactor showcased improved efficiency and reduced emissions, offering a scalable approach for commercial hydrogen production. These studies collectively provided valuable insights, emphasizing the need to address challenges to enhance the overall feasibility and sustainability of hydrogen production methods.

8.2.1.2 Partial oxidation of methane

Partial oxidation of methane (POM) is a thermochemical process that converts methane into a blend of hydrogen and carbon monoxide, known as syngas (York et al., 2003). It is a promising technology for hydrogen production, as it can be used to generate hydrogen from a variety of feedstocks, including natural gas, biogas, and landfill gas (Fleys et al., 2006). The POM process is typically conducted at high temperatures (650–1100°C) and pressures (1–10 atm) in the presence of a catalyst. The catalyst helps to activate the methane molecules and promote the desired reactions. The most common catalysts used for POM are nickel (Ni)-based catalysts, but other catalysts such as rhodium and platinum are also being investigated. The POM reaction can be described by the following chemical equation (8.1) (Abdulrasheed et al., 2019). However, in practice, the POM reaction is more complex and produces a variety of byproducts, including carbon dioxide, water, and soot. The composition of the product gas depends on the reaction conditions and the type of catalyst used.

$$CH_4 + \frac{1}{2} O_2 \rightarrow CO + 2H_2 \tag{8.1}$$

Fazlikeshteli et al. (2023) investigated the use of palladium (Pd)–Ni bimetallic catalysts supported on ceria (CeO_2) for the POM to syngas. Syngas, a blend of hydrogen and carbon monoxide, has versatile applications in fuel and chemical production. The study demonstrated that Pd–Ni/CeO_2 catalysts could achieve syngas production at lower temperatures than monometallic Pd or Ni catalysts, offering potential energy savings and reduced costs (Figure 8.3a). The catalysts were prepared using a mechanochemical method, which was relatively simple and cost-effective. They also exhibited stability, making them suitable for extended use. However, challenges included achieving low metal loadings and the need for specialized equipment. The mechanochemical preparation method was scalable, but further research was needed to optimize it for even lower reaction temperatures. Overall, the methodology presented a promising and environmentally friendly approach to methane-derived syngas production. Furthermore, Yang et al. (2023) investigated the utilization of Pd-decorated anatase TiO_2 for the photocatalytic POM to ethane. The study revealed that decorated

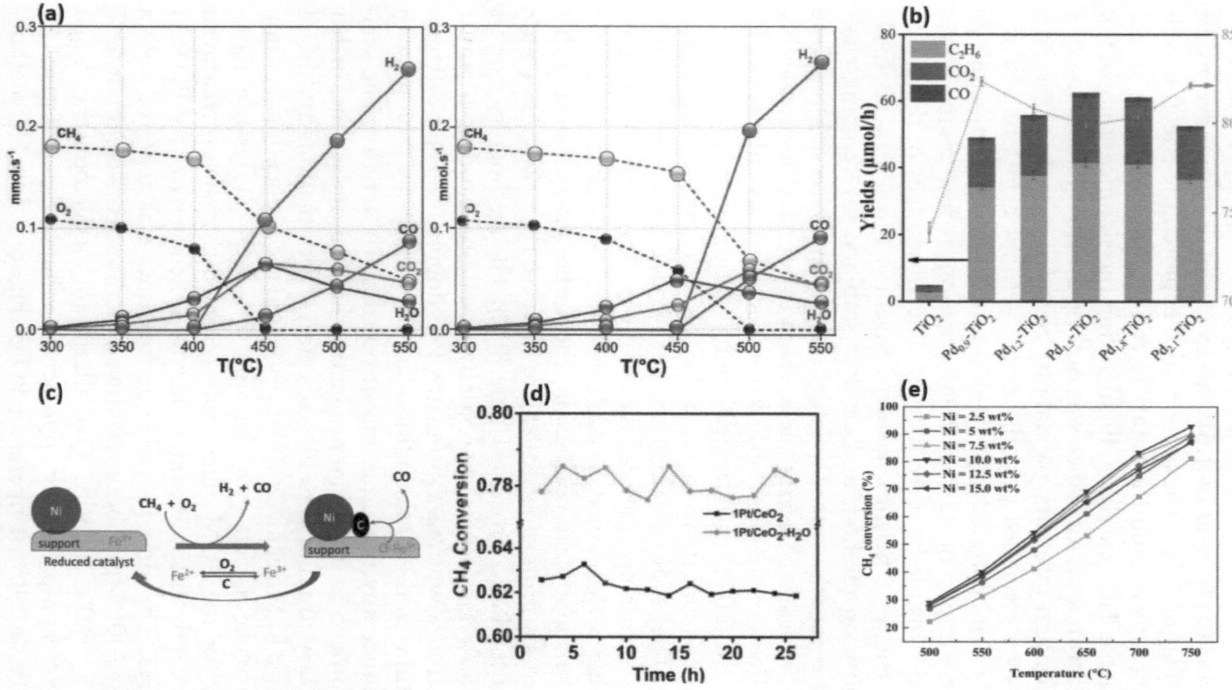

Figure 8.3 (a) Molar flowrates measured for bimetallic catalysts, specifically $Pd_{(1)}$-$Ni_{(0.5)}/CeO_2$, prepared using both BM (ball milling) at 50 Hz for 20 min and IWI (incipient wetness impregnation) methods. Reprinted and adapted from Choi et al. (2022) with permission. (b) Investigated product distribution and selectivity toward C_2H_6 over TiO_2 and Pd*x*-TiO_2 (where $x = 0.9, 1.2, 1.5, 1.8$, and 2.1). Reprinted and adapted from Choi et al. (2022) with permission. (c) A schematic depiction illustrating the redox cycle facilitated by iron ions and their involvement in the removal of coke. Reprinted and adapted from Choi et al. (2022) with permission. (d) Methane conversion over time on stream in POM for two different catalyst types at 7000°C. Reprinted and adapted from Choi et al. (2022) with permission. (e) Methane conversion of the x%Ni/Al_2O_3-10%CeO_2 catalysts. Reprinted and adapted from Choi et al. (2022) with permission.

anatase TiO_2 catalysts could efficiently generate ethane (C_2H_6) at low temperatures and pressures when exposed to visible light irradiation. The most effective performance was achieved with a catalyst prepared by depositing 1 wt% Pd on anatase TiO_2 using a photodeposition method. This catalyst demonstrated a C_2H_6 yield of 1.5% at 25°C and 1 atm under visible light irradiation (Figure 8.3b). The authors also explored the effect of various preparation parameters on the catalytic activity and selectivity of the Pd-decorated anatase TiO_2 catalysts, highlighting the significant impact of Pd loading. The optimal performance was observed with a Pd loading of 1 wt%. The investigation into catalyst stability revealed that these catalysts remained stable over extended periods of use without a loss in activity. Overall, the study demonstrated that Pd-decorated anatase TiO_2 catalysts hold promise for the photocatalytic POM reaction at low temperatures and pressures under visible light irradiation. The advantages of this methodology include energy efficiency, cost-effectiveness through a relatively simple photodeposition method, and catalyst stability. Overcoming challenges associated with achieving low Pd loadings and addressing the need for a light source is crucial. Nevertheless, there is potential for scalability in industrial production. Further research is necessary to enhance C_2H_6 yields. Khaleel et al. (2023) investigated the utilization of iron (Fe)-doped γ-alumina (γ-Al_2O_3) supports to increase the coking resistance of Ni catalysts in the POM. The study revealed that the introduction of Fe^{2+} ions into the γ-Al_2O_3 support notably improved the coking resistance of Ni catalysts. The most favorable performance was achieved with a catalyst containing 3.2 wt% Fe in the support. In conclusion, the authors demonstrated that Fe^{2+} doping presents a promising strategy to increase the coking resistance of Ni catalysts in the POM process (Figure 8.3c). The methodology offers simplicity and cost-effectiveness, potentially extending catalyst lifetimes and reducing reactor downtime. Optimizing Fe^{2+} doping levels for different catalyst formulations and POM conditions presents challenges, along with the need to minimize any potential impact on catalytic activity. The scalability of this approach to industrial production holds promise, but further research is necessary to tailor it to diverse catalyst formulations and POM reaction conditions. L. Ma et al. (2023) investigated the application of steam treatment to introduce defects into platinum (Pt)/CeO_2 catalysts, enhancing their activity for the POM. The study revealed that steam treatment of Pt/CeO_2-H_2O catalysts significantly improved their activity for POM at lower temperatures. The most favorable performance was achieved with a catalyst treated at 700°C for 1 hour, yielding a methane conversion of 62% and a syngas yield of 60% at 550°C. Overall, the authors demonstrated that steam treatment offers a promising approach to enhance the activity of Pt/CeO_2-H_2O catalysts for low-temperature POM (Figure 8.3d). The methodology is simple and cost-effective, potentially reducing energy consumption and POM process costs. Challenges include optimizing steam treatment conditions for different catalyst formulations and POM conditions while minimizing potential impacts on catalyst stability. Babakouhi et al. (2023) investigated the impact of different support promoters and Ni loadings on the performance of mesoporous (M) nanostructured Ni/M-Al_2O_3 catalysts for the combined CO_2 reforming and partial oxidation of methane (CRM-POM). The study revealed that the addition of support promoters like CeO_2, lanthanum oxide (La_2O_3), and magnesium oxide (MgO) to the Ni/Al_2O_3 catalysts significantly enhanced their activity and stability in CRM-POM. The most favorable performance was achieved with a Ni/CeO_2-Al_2O_3 catalyst containing 10 wt% Ni, exhibiting a methane conversion of 90% and a syngas yield of 85% at 700°C (Figure 8.3e). Overall, the authors demonstrated that incorporating support promoters and optimizing nickel loading significantly enhances the performance of mesoporous nanostructured Ni/M-Al_2O_3 catalysts in CRM-POM. The methodology offers the potential for improved energy efficiency, reduced costs, and relatively straightforward implementation. Addressing challenges involves optimizing support promoter and nickel loading for different catalyst formulations and CRM-POM conditions. Additionally, there is a need to consider potential cost impacts associated with support promoters.

In summary, Pd–Ni bimetallic catalysts on CeO_2 demonstrated the potential for energy-efficient syngas production at lower temperatures, utilizing a cost-effective mechanochemical preparation method. Pd-decorated anatase TiO_2 catalysts efficiently generated C_2H_6 under visible light, showcasing stability and cost-effectiveness. Fe-doped γ-alumina provided a straightforward strategy to extend the lifespan of Ni catalysts by enhancing coking resistance. Steam treatment of Pt/CeO_2 catalysts emerged as a promising, cost-effective method for improving POM activity at lower temperatures, potentially reducing energy consumption. Mesoporous Ni/M-Al_2O_3 catalysts for CRM-POM exhibited enhanced activity and stability with support promoters, suggesting potential improvements in energy efficiency and reduced costs. Despite the encountered challenges, these studies collectively indicate viable pathways for advancing cost-effective and energy-efficient methane conversion processes.

8.2.1.3 Coal gasification

Coal gasification involves a thermochemical process converting coal into a gas mixture called syngas, comprising hydrogen, carbon monoxide, carbon dioxide, and methane. The syngas can then be used to produce electricity, hydrogen, or other fuels and chemicals. Coal gasification is a complex process that involves a series of chemical reactions. The first step is to heat the coal to a high temperature (1000–1500°C) in the presence of a gasifying agent, such as oxygen, steam, or air. This causes the coal to decompose and release volatile matter, such as methane, tar, and water vapor. The remaining solid material, known as char, then reacts with the gasifying agent to produce syngas. The composition of the syngas depends on the type of coal used, the gasifying agent used, and the operating conditions. For example, using steam as a gasifying agent produces a syngas that is richer in hydrogen (Bell et al., 2010). Coal gasification is a promising technology for clean coal utilization. It can be used to produce electricity and other products from coal with lower emissions than traditional coal combustion technologies. Additionally, coal gasification can be used to generate hydrogen, which is a clean and versatile energy carrier (Liu et al., 2010).

Yin et al. (2023) conducted a series of experiments using three different types of coal: lignite, long flame coal, and bituminous coal. The experiments were conducted in a simulated underground coal gasification (UCG) reactor at a temperature of 1000°C and a pressure of 1 MPa. The authors found that the hydrogen yield from UCG varied depending on the type of coal used. The highest hydrogen yield was obtained from lignite, followed by long flame coal and bituminous coal. The authors attributed the higher hydrogen yield from lignite to its higher volatile matter content. The authors also found that the hydrogen concentration in the product gas was higher for the lignite and long flame coal experiments than for the bituminous coal experiment. This suggests that the type of coal used can affect the composition of the product gas as well as the hydrogen yield. The authors' findings suggest that the type of coal used in UCG can have a significant impact on hydrogen production. Lignite and long flame coals are more suitable for UCG than bituminous coal due to their higher hydrogen yields and higher hydrogen concentrations in the product gas (Figure 8.4a). Gao et al. (2023) developed a new microwave absorber material using coal gasification fine slag residual carbon (CGRC) adorned with hollow-spherical ferrous-ferric oxide (Fe_3O_4) nanoparticles. The authors used a simple hydrothermal method to synthesize the hollow-spherical Fe_3O_4 nanoparticles. The nanoparticles were then deposited on the surface of the CGRC using a physical adsorption method. The authors characterized the morphology and structure of the CGRC/Fe_3O_4 composite material using scanning electron microscopy (SEM), transmission electron microscopy (TEM), and X-ray diffraction (XRD). The microwave absorption performance of the composite material was measured using a vector network analyzer (VNA). The authors found that the CGRC/Fe_3O_4 composite material exhibited excellent microwave absorption performance. The composite material achieved a minimum reflection loss of 55.2 dB at a thickness of 1.8 mm and a frequency of 11.8 GHz. Figure 8.4b illustrates the possible microwave absorption mechanisms exhibited by the CGRC/Fe_3O_4 composites. The enhanced absorption capability of these composites is ascribed to the synergistic effects of dielectric, conduction, and magnetic losses. Jiang et al. (2023) proposed a numerical model to simulate UCG in deep coal seams and investigated its implications for carbon storage. The authors used their model to simulate UCG in a deep coal seam with a thickness of 100 m and a depth of 1000 m. The authors found that the UCG process (Figure 8.4c) was able to convert more than 70% of the coal into syngas. The authors also found that the UCG process produced

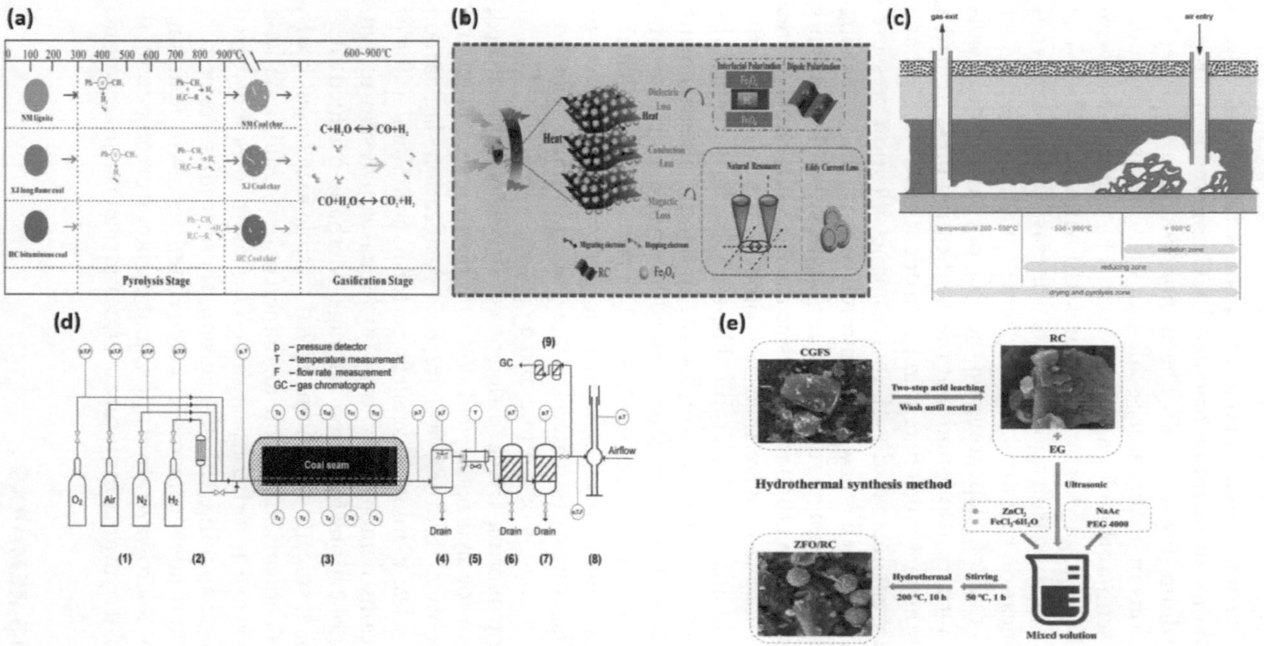

Figure 8.4 (a) A schematic illustration depicting the process of hydrogen production through the pyrolysis and gasification of three different coals. Reprinted and adapted from Choi et al. (2022) with permission. (b) The potential electromagnetic wave (EMW) absorbing mechanisms of CGRC/Fe_3O_4 (Fe_3O_4 nanoparticles embedded in reduced graphene oxide) composites. Reprinted and adapted from Gao et al. (2023) with permission. (c) A simplified schematic representation of the UCG process. Reprinted and adapted from Jiang et al. (2023) with permission. (d) Diagram outlining the configuration of the ex situ high-pressure UCG installation: (1) compressed reagents, (2) hydrogen preheater, (3) gasification chamber, (4) wet scrubber for gas cleaning, (5) air cooler, (6,7) gas separators, (8) thermal combustor, and (9) gas treatment module prior to GC analysis. Reprinted and adapted from Jiang et al. (2023) with permission. (e) A visual representation detailing the fabrication process of the ZFO (zinc ferrite)/RC (reduced graphene oxide) composite. CGFS, coal gasification fine slag; EG, ethylene glycol; RC, residual carbon; ZFO, $ZnFe_2O_4$. Reprinted and adapted from Jiang et al. (2023) with permission.

a significant amount of heat, which caused the coal seam temperature to increase by up to 500°C. The authors then investigated the implications of UCG for carbon storage. They found that the UCG process could store a significant amount of carbon dioxide in the coal seam. Kapusta et al. (2023) investigated the feasibility of methane-oriented UCG employing hydrogen as the gasification medium (Figure 8.4d). The authors conducted a series of experimental simulations in a laboratory-scale UCG reactor using two different types of coal: semi-anthracite and hard coal. The experiments were carried out at two different gasification pressures: 20 bar and 40 bar. The authors found that the hydrogasification process was feasible for both types of coal and both gasification pressures. The methane content in the product gas ranged from 20% to 27%, depending on the type of coal and the gasification pressure. The authors also found that the gasification pressure had a significant impact on the methane content in the product gas. The methane content in the product gas increased with increasing gasification pressure. Zhang et al. (2023), developed a new ultra-thin microwave absorber using zinc-ferrites ($ZnFe_2O_4$) nanospheres decorated residual CGRC. The authors used a simple hydrothermal method to synthesize the $ZnFe_2O_4$ nanospheres (Figure 8.4e). The nanospheres were then deposited on the surface of the CGRC using a physical adsorption method. The authors characterized the morphology and structure of the $ZnFe_2O_4$/CGRC composite material using SEM, TEM, and XRD. The microwave absorption performance of the composite material was measured using a VNA. The authors found that the $ZnFe_2O_4$/CGRC composite material exhibited excellent microwave absorption performance. The composite material achieved a minimum reflection loss of −49.2 dB at a thickness of 1.5 mm and a frequency of 10.2 GHz.

To summarize, in exploring UCG, different coal types were analyzed, revealing lignite's superior hydrogen yield due to its higher volatile matter content. A new microwave absorber, utilizing coal gasification slag carbon with Fe_3O_4 nanoparticles, exhibited excellent performance. Simulations of UCG in deep coal seams demonstrated over 70% coal conversion to syngas, with potential implications for carbon storage. Methane-oriented UCG using hydrogen as the gasification medium proved feasible for various coals, showing pressure-dependent impacts on methane content. Additionally, a novel ultra-thin microwave absorber using $ZnFe_2O_4$ nanospheres showcased remarkable absorption capabilities. In conclusion, these studies underscore diverse applications and advancements in coal-related research, spanning hydrogen production, microwave absorption, and carbon storage potential, highlighting their broad relevance for future energy and material innovations.

8.3 CONCLUDING REMARKS

The reviewed thermochemical processes (Table 8.1), including SMR, POM, and coal gasification, each offer unique advantages and challenges in the

Table 8.1 Comparison of hydrogen production methods

Methods	*Efficiency*	*Cost*	*Environmental impact*	*Scalability*	*References*
SMR	High methane conversion (>90%)	Cost-effective at large scales	Greenhouse gas emissions (CO_2, CH_4)	Established technology, scalable	Abd et al. (2023); Amini et al. (2023); Choi et al. (2022); Lim et al. (2023); Salahi et al. (2023)
POM	Variable, influenced by reaction conditions and catalysts	Costs influenced by feedstock, catalysts, and process complexity	Formation of byproducts (CO_2, CO)	Challenges in scalability due to process complexity	Babakouhi et al. (2023); Fazlikeshteli et al. (2023); Khaleel et al. (2023); L. Ma et al. (2023); Yang et al. (2023)
Coal gasification	Depends on coal type, with variable syngas composition	Costs can be competitive, influenced by coal type and gasification method	Potential for cleaner coal utilization, with CO_2 storage potential	Scalable but may face challenges in coal availability and infrastructure	Gao et al. (2023); Jiang et al. (2023); Kapusta et al. (2023); Yin et al. (2023); Zhang et al. (2023)

quest for efficient, cost-effective, and environmentally friendly hydrogen production. SMR, being the most widely employed method, boasts high efficiency but raises concern about energy intensity and greenhouse gas emissions. POM presents a promising technology, with its versatility in utilizing various feedstocks, yet the complexity of the process and the formation of byproducts necessitate careful consideration. Coal gasification emerges as a complex yet potentially cleaner alternative, particularly in terms of hydrogen production from coal. Innovations, such as bimetallic catalysts, novel absorber materials, and numerical simulations, contribute to advancements within these methods.

The future of thermochemical hydrogen production holds exciting possibilities for further improvements in efficiency, cost-effectiveness, and environmental sustainability. Researchers continue to explore innovative catalysts, materials, and process optimization techniques to enhance the performance of existing methods. Additionally, the integration of renewable energy sources and advancements in process intensification could play a pivotal role in shaping the future landscape of hydrogen production. Further research is needed to address challenges related to scalability, byproduct formation, and the environmental impact of these methods. Additionally, exploring alternative thermochemical cycles, such as solar thermochemical water-splitting and biomass gasification, could provide alternative pathways for sustainable hydrogen production. As the demand for clean and renewable energy sources grows, thermochemical hydrogen production methods are likely to play a crucial role in meeting these demands. Collaborative efforts between academia, industry, and policymakers will be essential to drive research, development, and deployment of advanced thermochemical hydrogen production technologies.

REFERENCES

Abanades, S., & Flamant, G. (2006). Thermochemical hydrogen production from a two-step solar-driven water-splitting cycle based on cerium oxides. *Solar Energy*, *80*(12), 1611–1623.

Abd, A. A., Othman, M. R., Majdi, H. S., & Helwani, Z. (2023). Green route for biomethane and hydrogen production via integration of biogas upgrading using pressure swing adsorption and steam-methane reforming process. *Renewable Energy*, *210*, 64–78.

Abdulrasheed, A., Jalil, A. A., Gambo, Y., Ibrahim, M., Hambali, H. U., & Hamid, M. Y. S. (2019). A review on catalyst development for dry reforming of methane to syngas: Recent advances. *Renewable and Sustainable Energy Reviews*, *108*, 175–193.

Amini, A., Sedaghat, M. H., Jamshidi, S., Shariati, A., & Rahimpour, M. R. (2023). A comprehensive CFD simulation of an industrial-scale side-fired steam methane reformer to enhance hydrogen production. *Chemical Engineering and Processing-Process Intensification*, *184*, 109269.

Aziz, M., Darmawan, A., & Juangsa, F. B. (2021). Hydrogen production from biomasses and wastes: A technological review. *International Journal of Hydrogen Energy*, *46*(68), 33756–33781.

Babakouhi, R., Alavi, S. M., Rezaei, M., Akbari, E., & Varbar, M. (2023). Combined CO_2 reforming and partial oxidation of methane over mesoporous nanostructured Ni/M-Al_2O_3 catalyst: Effect of various support promoters and nickel loadings. *Journal of CO_2 Utilization*, *70*, 102427.

Basile, A., Liguori, S., & Iulianelli, A. (2015). Membrane reactors for methane steam reforming (MSR). In Angelo Basile, Luisa Di Paola, Faisal l. Hai, and Vincenzo Piemonte (Eds.), *Membrane reactors for energy applications and basic chemical production* (pp. 31–59). Elsevier.

Beghi, G. E. (1986). A decade of research on thermochemical hydrogen at the Joint Research Centre-Ispra. In T. Nejat Veziroglu, Zhu Yajie, and Bao Deyou (Eds.), *Hydrogen systems* (pp. 153–171).

Bell, D. A., Towler, B. F., & Fan, M. (2010). *Coal gasification and its applications.* William Andrew.

Budama, V. K., Duarte, J. P. R., Roeb, M., & Sattler, C. (2023). Potential of solar thermochemical water-splitting cycles: A review. *Solar Energy*, *249*, 353–366.

Choi, H., Kim, S. H., Bae, J., Katikaneni, S. P. R., Jamal, A., Harale, A., Paglieri, S. N., & Lee, J. H. (2022). CFD analysis and scale up of a baffled membrane reactor for hydrogen production by steam methane reforming. *Computers & Chemical Engineering*, *165*, 107912.

Diver, R. B., Miller, J. E., Allendorf, M. D., Siegel, N. P., & Hogan, R. E. (2008). Solar thermochemical water-splitting ferrite-cycle heat engines. *Journal of Solar Energy Engineering*, *130*(4), 041001: American Society of Mechanical Engineers (ASME).

Fazlikeshteli, S., Vendrell, X., & Llorca, J. (2023). Low-temperature partial oxidation of methane over Pd–Ni bimetallic catalysts supported on CeO_2. *International Journal of Hydrogen Energy*, *48*(32), 12024–12035.

Fleys, M., Simon, Y., Swierczynski, D., Kiennemann, A., & Marquaire, P.-M. (2006). Investigation of the reaction of partial oxidation of methane over Ni/La_2O_3 catalyst. *Energy & Fuels*, *20*(6), 2321–2329.

Funk, J. E. (2001). Thermochemical hydrogen production: past and present. *International Journal of Hydrogen Energy*, *26*(3), 185–190.

Gao, S., Zhang, Y., He, J., Zhang, X., Jiao, F., Liu, T., Li, H., Wu, C., & Ma, M. (2023). Coal gasification fine slag residual carbon decorated with hollow-spherical Fe_3O_4 nanoparticles for microwave absorption. *Ceramics International*, *49*(11), 17554–17565.

Hangun, B., Eyecioglu, O., & Beken, M. (2022). *Investigating the energy production trends of countries and its relationship between economic development.* 2022 11th International Conference on Renewable Energy Research and Application (ICRERA), 597–600.

Ibrahim, I. A., Ötvös, T., Ghanem, C., Rocca, E., & Gilmanova, A. (2021). International Energy Agency (IEA).

Izanloo, M., Mehrpooya, M., & Delpisheh, M. (2021). Integrated thermochemical Mg-Cl-Na hydrogen production cycle, carbon dioxide capture, ammonia production, and methanation. *International Journal of Energy Research*, *45*(7), 10719–10737.

Jiang, L., Chen, S., Chen, Y., Chen, Z., Sun, F., Dong, X., & Wu, K. (2023). Underground coal gasification modelling in deep coal seams and its implications to carbon storage in a climate-conscious world. *Fuel*, *332*, 126016.

Kapusta, K., Wiatowski, M., Thomas, H. R., Zagorščak, R., Sadasivam, S., Masum, S., Kempka, T., Otto, C., Basa, W., & Szyja, M. (2023). Experimental simulations of methane-oriented underground coal gasification using hydrogen-The effect of coal rank and gasification pressure on the hydrogasification process. *International Journal of Hydrogen Energy*, *48*(3), 921–932.

Karaca, A. E., Qureshy, A. M. M. I., & Dincer, I. (2023). An overview and critical assessment of thermochemical hydrogen production methods. *Journal of Cleaner Production*, *385*, 135706.

Khaleel, A., Pillantakath, A. R., & Adamson, A. (2023). Significant role of well-dispersed Fe^{2+} ions in the support of Ni catalysts in enhancing coking resistance during partial oxidation of methane. *International Journal of Hydrogen Energy*, *48*(87), 33913–33926.

Khan, Z. U., Alsaiari, M., Ahmed, M. A., Muhammad, N., Tariq, M., Rahim, A., & Niaz, A. (2023). Catalysts for electrochemical water splitting for hydrogen production. In I. Inamuddin, T. Altalhi, S. M. Adnan, & M. A. Amin (Eds.), *Materials for hydrogen production, conversion, and storage* (pp. 225–247), Wiley.

Kodama, T., Kondoh, Y., Yamamoto, R., Andou, H., & Satou, N. (2005). Thermochemical hydrogen production by a redox system of ZrO_2-supported Co (II)-ferrite. *Solar Energy*, *78*(5), 623–631.

Lee, J. E., Shafiq, I., Hussain, M., Lam, S. S., Rhee, G. H., & Park, Y.-K. (2022). A review on integrated thermochemical hydrogen production from water. *International Journal of Hydrogen Energy*, *47*(7), 4346–4356.

Likkasit, C., Maroufmashat, A., Elkamel, A., Ku, H., & Fowler, M. (2018). Solar-aided hydrogen production methods for the integration of renewable energies into oil & gas industries. *Energy Conversion and Management*, *168*, 395–406.

Lim, J., Joo, C., Lee, J., Cho, H., & Kim, J. (2023). Novel carbon-neutral hydrogen production process of steam methane reforming integrated with desalination wastewater-based CO_2 utilization. *Desalination*, *548*, 116284.

Liu, K., Song, C., & Subramani, V. (2010). *Hydrogen and syngas production and purification technologies*. John Wiley & Sons.

Ma, L., Ding, C., Ma, Z., Wang, J., Xu, H., Zhang, K., & Li, G. (2023). Defect engineering by steam treatment over Pt/CeO_2 catalyst promoting C–H activation in partial oxidation of methane. *International Journal of Hydrogen Energy*, *48*(45), 17055–17064.

Ma, Z., Davenport, P., & Saur, G. (2022). System and technoeconomic analysis of solar thermochemical hydrogen production. *Renewable Energy*, *190*, 294–308.

Manna, J., Jha, P., Sarkhel, R., Banerjee, C., Tripathi, A. K., & Nouni, M. R. (2021). Opportunities for green hydrogen production in petroleum refining and ammonia synthesis industries in India. *International Journal of Hydrogen Energy*, *46*(77), 38212–38231.

Onuki, K., Nakajima, H., Ioka, I., Futakawa, M., & Shimizu, S. (1994). *IS process for thermochemical hydrogen production.* Japan Atomic Energy Research Institute, Tokyo, Japan.

Pregger, T., Graf, D., Krewitt, W., Sattler, C., Roeb, M., & Möller, S. (2009). Prospects of solar thermal hydrogen production processes. *International Journal of Hydrogen Energy, 34*(10), 4256–4267.

Razi, F., Hewage, K., & Sadiq, R. (2023). A comparative exergoenvironmental evaluation of chlorine-based thermochemical processes for hydrogen production. *International Journal of Hydrogen Energy, 48*(95), 37108–37123.

Rosen, M. A. (2010). Advances in hydrogen production by thermochemical water decomposition: A review. *Energy, 35*(2), 1068–1076.

Salahi, F., Zarei-Jelyani, F., Farsi, M., & Rahimpour, M. R. (2023). Optimization of hydrogen production by steam methane reforming over Y-promoted Ni/Al_2O_3 catalyst using response surface methodology. *Journal of the Energy Institute, 108*, 101208.

Sato, S. (1979). Thermochemical hydrogen production. In *Solar-hydrogen energy systems* (pp. 81–114). Oxford and New York: Pergamon Press.

Siddik, M., Islam, M., Zaman, A., & Hasan, M. (2021). Current status and correlation of fossil fuels consumption and greenhouse gas emissions. *International Journal of Energy, Environment, and Economics, 28*, 103–119.

Yang, J., Wang, C., Xing, J., & Tang, J. (2023). Palladium decorated anatase for photocatalytic partial oxidation of methane to ethane. *Surfaces and Interfaces, 40*, 103108.

Yin, Z., Xu, H., Chen, Y., Zhao, T., & Wu, J. (2023). Experimental simulate on hydrogen production of different coals in underground coal gasification. *International Journal of Hydrogen Energy, 48*(19), 6975–6985.

York, A. P. E., Xiao, T., & Green, M. L. H. (2003). Brief overview of the partial oxidation of methane to synthesis gas. *Topics in Catalysis, 22*, 345–358.

Zhang, X., Zhang, Y., He, J., Li, H., Bai, Y., & Gao, S. (2023). $ZnFe_2O_4$ nanospheres decorated residual carbon from coal gasification fine slag as an ultrathin microwave absorber. *Fuel, 331*, 125811.

Chapter 9

Recent technology and challenges in cost-effective hydrogen production from renewable energy sources

Sukanta Roga, Swarali Kulkarni, Hrushikesh Jadhav, Mrunali Ukey, Sanket Bahadure, and Swapan Bhaumik

9.1 INTRODUCTION

Due to its chemical properties, hydrogen can easily mix with other elements like water, hydrocarbons, and alcohol, which makes it a fundamental and common element on Earth. For energy security, hydrogen is an important energy carrier due to the wide range of domestic energy sources available. The global energy demand is primarily driven by economic development and population growth, with power production significantly influencing industrial development. Fossil fuels fulfill this demand, but their rapid depletion puts them in a precarious position, leading to increased CO_2 emissions and global warming (Kalamaras & Efstathiou, 2013). The use of renewable energy sources will likely become more common as environmental issues worsen. Environmental concerns, global warming, and greenhouse gas emissions require the world to transition to renewable energy. The most viable option for replacing fossil fuels is renewable energy, which includes biomass, ocean thermal energy conversion (OTEC), geothermal, wind, and solar power. Since certain ideal renewable energy sources, such as the sun and wind, are unpredictable, hydrogen can best use these resources as both a fuel and an energy carrier. Decarbonizing the main sectors might also be necessary to reach net-zero CO_2 emissions by 2050 (Kalamaras & Efstathiou, 2013). Because of its carbon-free solutions, hydrogen is becoming recognized as a unique energy source and potential fuel. Utilizing the fuel distribution and storage infrastructure now in place for other chemical fuels provides an additional means of storing and providing hydrogen.

The primary sources of energy in the globe are coal (35%), gas (23%), hydropower (16%), and nuclear energy (10%). About 13% of the total energy comes from renewable sources, with biomass and biogas providing the remaining amounts. Of this, solar energy accounts for 4%, wind energy for 7%, and hydrogen for the remaining 2%. The majority of emissions

 DOI: 10.1201/9781003537816-9

from burning fossil fuels come from the combustion of coal, oil, and gas. More than any other fossil fuel, coal will be responsible for 40% of all world emissions in 2022. Cement production comes in fourth. The production of gasoline follows in second place with a 32% contribution from oil and comes in first with a 4% contribution. These days, hydrogen accounts for very little of the energy sector, and when it is generated using renewable energy, its net carbon dioxide emissions are far lower than those of other fuels (Dash et al., 2023). Hydrogen burns the most quickly compared to all other fuels, whether they be gases or liquids. Because they are exempt from the thermal Carnot cycle's thermal efficiency limitations, hydrogen fuel cells can operate at high levels of efficiency. It is projected that a significant number of hydrogen refueling stations will be constructed in 2022 to support fuel cells that run on hydrogen and hybrid vehicles that use fuel cells as their foundation.

Different processes for generating hydrogen from renewable energy have been compared in recent research. Hydrogen generation using steam methane reforming (SMR) has been reviewed in a study to evaluate its economic and environmental benefits using various feedstocks. The process of solar thermal methane reforming, which produces hydrogen and syngas, was also reviewed. There was a suggestion to use solar power to reform steam methane as a potential method to help with the decarbonization of fossil fuels and the execution of a transition path toward a hydrogen economy powered by the sun. The versatility of hydrogen makes it an ideal alternative to solar and wind energy since it can be used as fuel, storage, methanol production, and energy transport. A thorough analysis of hydrogen generation is needed given the growing interest in hydrogen applications. The essential characteristics of renewable energy-powered hydrogen-generating devices must also be considered in this evaluation (Dash et al., 2023).

9.2 LITERATURE REVIEW

In addition to fossil fuel reforming, hydrogen can be produced in a variety of ways. Hydrogen can be produced in water using electrolysis, photoelectrolysis, and thermochemical water splitting. Also discussed is the production of hydrogen from biomass-based processes. In addition, gasification, pyrolysis, and aqueous phase reforming are other possible processes.

9.2.1 Gasification of biomass

Among organic alternatives to petroleum, biomass will likely be the most sustainable. Many materials can be used to produce biomass, such as waste paper, maize, animal feces, crop leftovers, agricultural waste, sawdust, and a variety of other materials. The process of producing biodiesel from organic, renewable feedstock is referred to as biomass gasification. After being dried at 150°C, the biomass is heated to 800–900°C in a gasifier with an oxidizing substance (Kalamaras & Efstathiou, 2013). Due to the rising heat, the complex solid hydrocarbons in the dry refuse residue disintegrate and release combustible gases like hydrogen and syngas. These gases can be used as fuel after being isolated and purified.

9.2.2 Pyrolysis

Another current and promising technique of hydrogen synthesis is pyrolysis, also known as co-pyrolysis. Raw organic material is cooked to temperatures between 500°C and 900°C and gasified at pressures between 0.1 and 0.5 MPa. The lack of oxygen and air during the process virtually eliminates the chance of dioxin production. Because no water or air is present, carbon dioxide cannot develop, so secondary reactors (WGS, PrOx, etc.) are not required (e.g., CO or CO_2) (Kalamaras & Efstathiou, 2013).

The following happens as a result:

$$C_nH_m + \text{Heat} \rightarrow nC + 0.5mH_2 \tag{9.1}$$

9.2.3 Electrolysis

Hydrogen production techniques like water electrolysis could be promising in the future. This process produces only 4% of the hydrogen generated globally today. The molecular bonds that hold water molecules together disintegrate when a direct current is run between two conductors in a water solution, releasing hydrogen and oxygen (Kalamaras & Efstathiou, 2013):

$$\text{Cathode}: 2H_2O(1) + 2_e \rightarrow H_2(g) + 2OH \tag{9.2}$$

$$\text{Anode}:\ 4OH - (aq) \rightarrow O_2(g) + 2H_2O(1) + 4e - \text{Overall}: 2H_2O \rightarrow 2H_2 + O_2 \tag{9.3}$$

During electrolysis, room temperature is used as the temperature of the electrolysis process. Platinum (Pt), the material used for the electrodes, does not combine with sulfuric acid, a common electrolyte in water electrolysis. Due to the absence of greenhouse gases during the process, it is environmentally friendly. In addition, the generated oxygen is used for industrial purposes as well.

9.2.4 Photoelectrolysis

Water undergoes electrolysis with the use of photoelectrochemical (PEC) light-collecting devices. In an electrolyte with an aqueous semiconductor photoelectrode, sunlight can accelerate the evolution of hydrogen and oxygen, known as oxygen evolution reaction (OER) and hydrogen evolution reaction (HER). OER needs free electrons, which is why the HER releases electrons into the electrolyte (Kalamaras & Efstathiou, 2013). Photoelectrolysis is currently the least costly and most efficient method of generating renewable hydrogen. It combines water electrolysis and solar energy gathering to produce a semiconductor photoelectrode. Despite being in the testing stage, this technology already has encouraging costs and rates of efficiency for producing hydrogen.

As shown in Figure 9.1, hydrogen can be used to produce solar energy in several ways. It includes thermal methods, photovoltaic, and photochemical methods.

9.2.5 Thermochemical water splitting

Water is divided into hydrogen and oxygen using only heat in a process known as thermolysis, also known as thermochemical water splitting. Theoretically, these techniques could result in close to 50% total efficiency. The following justifies how water thermally separates into a single phase.

$$H_2O + \text{Heat} \rightarrow= H_2 + 0.5O_2 \quad (4)$$

Water electrolysis is one of the easiest ways to create hydrogen without byproducts. Figure 9.2 helps us to understand this. It shows the water in the container getting reduced to oxygen at the anode and the hydrogen atom through the membrane gets oxidized at the cathode to form a hydrogen molecule. To create hydrogen (natural gas), steam reforming of methane is currently the most common and cost-effective method. This process produces more than half of the hydrogen generated globally (Kalamaras & Efstathiou, 2013). There are still thermochemical processes in use today, such as gasification and pyrolysis of vegetation. When hydrogen is

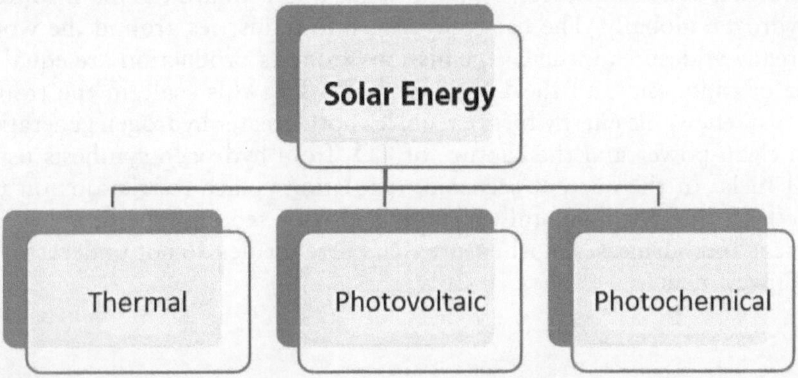

Figure 9.1 H_2 production using solar energy.

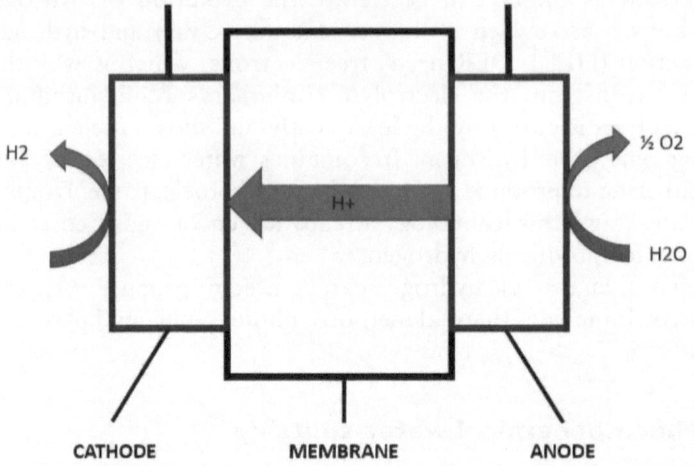

Figure 9.2 Water electrolysis.

obtained in this manner, it costs about three times as much as when hydrogen is obtained through the SMR process. Consequently, it is not generally thought of as being cost-competitive with steam reforming. One of the easiest ways to produce hydrogen without polluting the environment is through water electrolysis.

9.2.6 Hydrogen production challenges

Currently, hydrogen production is costly. However, due to declining renewable energy costs, hydrogen production may increase by 30% by 2030 due to increased renewable electricity production. Electrolyzers, fuel cells, and refueling devices all benefit from mass manufacturing (Dash et al., 2023). The slow development of hydrogen infrastructure precludes widespread use. The amount hydrogen users pay depends largely on the availability of refueling stations. Currently, natural gas and coal are the main sources of hydrogen globally. The use of hydrogen in industries around the world is already widespread; the CO_2 emissions from its production are equal to those of Indonesia and the UK combined. Using this scale in the transition to a renewable energy future requires both greater hydrogen generation from clean power and the capture of CO_2 from hydrogen synthesis using fossil fuels. In the current scenario, regulations place restrictions on the growth of the environmentally friendly hydrogen sector. Together, the government and businesses must ensure that current rules do not unnecessarily limit investment.

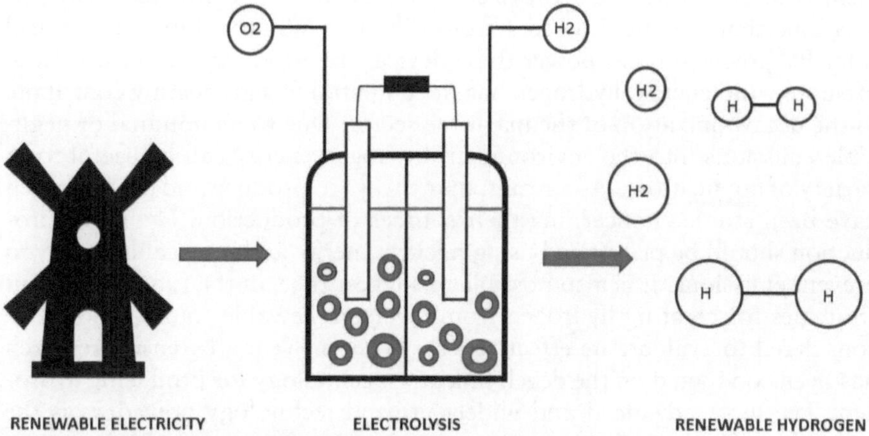

Figure 9.3 Conversion of renewable electricity to hydrogen through electrolysis.

9.2.7 Methodology

Specific keywords helped narrow down the analyzed material's selected sample. Using a strict methodology, the most reliable and significant papers for the research were examined and compiled. Literature reviews provide a foundation for the development of new ideas and theories.

9.2.8 Validation and comparative study with existing work

When it comes to the creation of hydrogen-powered energy systems, Europe has been at the vanguard. Additionally, there is a sizable market for growing a successful company.

As mentioned previously, the cost of the electrolyzer and electricity greatly affects the price of green hydrogen in India. Currently, the price of hydrogen generated in a country like India by electrolysis ranges from around $7/kg to $4.10/kg, contingent upon the chosen technique and related indirect expenses. This makes it challenging to compete with the current pricing of brown or gray hydrogen. Despite being a net importer of natural gas, India has among the lowest levelized costs of electricity (LCOE) for wind and solar power in the world. The conversion of renewable electricity to hydrogen through electrolysis is shown in Figure 9.3.

9.3 RESULTS AND DISCUSSION

Further research is needed to commercialize hydrogen generation technologies that use renewable energy sources, as well as to build markets and infrastructure. Water electrolysis equipment can also be utilized in isolated locations that are disconnected from the grid. Wind turbines or off-grid solar PV power sources power these devices. By producing clean fuel from offshore wind energy, hydrogen has the potential to significantly contribute to the decarbonization of the maritime sector. Due to its minimal or negligible emissions into the environment, hydrogen energy can be useful for a variety of applications. As a result, many ways to produce and use hydrogen have been studied concerning their sources of production. Hydrogen production should be prioritized using nuclear energy and renewable energy to prevent emissions. It is also possible to suggest that, in the future, different strategies for creating hydrogen from diverse renewable sources should be considered to evaluate its effectiveness. In recent years, extensive research has been conducted on the development of technology for producing hydrogen. The most advanced and widely utilized technology nowadays is the reforming of hydrocarbons. Hydrogen generation from renewable resources like biomass and water is being considered to reduce fossil fuel dependence. The "in-house" H_2 production could improve the country's economic and energy security. Every location in the world may be able to manufacture a large portion of its energy due to the potential of H_2 to be produced from a wide range of feedstocks and using a wide range of procedures. It seems obvious that H_2 may end up becoming the most widely used fuel as technologies advance and mature. There are a few small-scale outliers, even though pure hydrogen is rarely used as a fuel in today's power plants. For example, a hydrogen-fired gas turbine in Kobe, Japan, produces 1.1 MW of electricity and 2.8 watts of heat energy for a local neighborhood, and for a 12 MW combined-cycle gas turbine in Italy, hydrogen is provided by a nearby petrochemical complex. Utilizing hydrogen-rich gases from refineries, petrochemical plants, and steel mills is more common. Hydrogen is currently used in 70% of reciprocating gas engines, but it will soon be used in 100% of them. Currently, natural gas provides 76% of the dedicated hydrogen produced, with coal accounting for nearly all of the remaining 23%. There is a lot of potential for electrolysis to create additional low-carbon hydrogen, even though it only accounts for 2% of hydrogen produced globally (IEA report, 2019). Modern technology and energy sources are used to produce hydrogen. The majority of hydrogen produced worldwide now comes from the use of fossil fuels. A small quantity of hydrogen is created electrolytically, using energy and water; yet, it was a significant industrial hydrogen source from the 1920s through the 1960s utilizing hydroelectric electricity, until natural gas replaced it. As the cost of renewable energy (especially solar PV and wind) continues to drop, more attention is being paid to the production of hydrogen through water electrolysis and the possibility of further converting that hydrogen into hydrogen-based fuels or feedstocks; the current infrastructure is more compatible with synthetic hydrocarbons and ammonia than hydrogen. Methane splitting is a newly suggested technique for producing hydrogen from natural gas. Since the 1990s, many technologies have been developed. The main method uses electricity as a fuel and a feedstock, and it is based on three-phase alternating current plasma. It produces hydrogen and solid carbon, but not any CO_2 emissions (Fulcheri, 2023). For the same amount of hydrogen generated, methane splitting takes three to five times less power than electrolysis; nevertheless, its efficiency advantage is limited by the need for high-temperature plasma and significant thermal losses. It produces extremely little CO_2 and converts carbon to carbon black, a solid form of carbon (Nnabuife et al., 2022). As part of their study, it is examining the current state of blue and green hydrogen production, including the source of energy, storage, transportation, and distribution, as well as commercialization opportunities and challenges. Due to its ability to produce hydrogen from renewable sources, hydrogen is an excellent candidate for combining fuel energy carrier and energy storage roles. Hydrogen can be produced from solar and wind power; however, they are intermittent. In addition, this study compares non-renewable and renewable hydrogen production systems based on their economics, efficiency, and sustainability (Ishaq et al., 2022). Cleaner energy sources such as hydrogen are one of the promising alternatives for reducing global warming at the same time. Currently, hydrogen is mostly produced by steam reforming fossil fuels. In this study, however, renewable feedstocks and alternative technologies are examined (Yusuf et al., 2021). The study by Fragiacomo et al. (2022) discusses the different applications of electrolyzers for hydrogen production and the versatility of fuels. Fuel cell systems and water electrolysis units can facilitate the transition to a hydrogen economy in the energy sector, enhancing performance and flexibility. As part of the energy transition, hydrogen plays a crucial role—it enables all major sectors of the economy to be electrified and decarbonized, as well as facilitates sectoral integration. In addition, it poses a great opportunity to make Europe, including Poland, a leader in energy innovation technology, improving urban air quality and creating new high-skilled jobs (Drela, 2021). The most cost-effective thermochemical process is biomass supercritical water gasification (SCWG). This has been highly moisturized and can be used directly in SCWG without the need for expensive drying processes (Hosseini & Wahid, 2016). Both industrial and commercial applications of this process are anticipated to be highly promising for the production of bio-hydrogen (Hosseini, 2015). The article reviews recent industrial techniques for hydrogen production, focusing on thermochemical water splitting using sunlight and renewable sources like plasma reformation and biological processes. It highlights emerging non-conventional techniques for biomass production (Chaubey et al., 2013). Especially hydrogen is a promising alternative to fossil fuels, and this review discusses the growing demand for renewable energy resources to replace fossil fuels. It highlights recent advancements in renewable energy and the prospects of hydrogen production, focusing on various energy systems and technologies (Pareek et al., 2020).

The review discusses limitations and future perspectives in H_2 production, operational challenges, and the role of nanotechnology in achieving a steady state and stable H_2 (Wu et al., 2023). Future research should focus on cost reduction, decarbonization, and designing robust H_2 storage systems compatible with long-distance distribution networks and fueling stations (Ahmed et al., 2022). For decentralized/off-grid hydrogen generation from sustainable, renewable sources, the study suggests solar integrated plasmolysis, which offers great energy efficiency, affordable costs, and an easy-to-build reactor (Younas et al., 2022). The storage of hydrogen underground in porous media offers a large, affordable storage capacity, but safety concerns and research gaps must be addressed (Muhammed et al., 2023). The purpose of this review is to evaluate the options, prospects, and challenges of using renewable energy sources to produce hydrogen, an issue of critical importance for a sustainable world (Abbasi & Abbasi, 2011). Three electrolysis systems are available for generating solar hydrogen. It highlights challenges, pros, cons, and commercialization processes, underscoring the need for further maturation (Hosseini et al., 2020; Qureshi et al., 2022). H_2 storage advances and challenges, risks associated with production, handling, and transportation, safety recommendations, and challenges related to techno-economic feasibility are discussed (Qureshi et al., 2023; Monga et al., 2023). Research on hydrogen production methods, storage, distribution, and end-use discusses competitive hydrogen cost values, with ongoing research promoting clean, carbon-free energization in the near future (Seelam et al., 2020; El-Emam et al., 2019). This study reviews hydrogen production and storage, identifying future research trends and providing suggestions for a hydrogen economy. It emphasizes the need for alternative methods to meet global energy demand and highlights the shortcomings and opportunities of clean energy and hydrogen. The review also provides insights into policy-making and technology development (Razi et al., 2020; Rasul et al., 2022)

Avoiding the usage of hydrogen gas would be a more effective strategy than using it directly. As a result of current technology-related CO_2 emissions, carbon black demand is expected to rise from 12 Mt to 16 Mt globally over the next five years. With methane splitting, less than 5 Mt of H_2/year of hydrogen could be produced, meeting all of these needs and preventing these emissions. Even though graphene, carbon nanotubes, and carbon fibers are among the rarest forms of solid carbon and are far smaller than those of carbon black, they might develop quickly as more batteries and carbon-reinforced concrete are used. There may be alternatives available in other solid carbon markets.

Meanwhile, other SMR process innovations are being explored. Despite the possibility of producing steam with alternative energy sources, natural

Table 9.1 Capital cost of different methods to produce hydrogen

Methods	*Capital cost ($)*
Direct biophotolysis	50
Indirect biophotolysis	135
Dark fermentation	50
Photofermentation	24
Solar PV electrolysis	54
Solar thermal electrolysis	421
Wind electrolysis	499
Nuclear electrolysis	500
Nuclear thermolysis	2107
Solar thermolysis	16
Photoelectrolysis	2

gas would still be required as feedstock. As a result, the concentrated "process" CO_2 stream may be simpler to capture. The comparison between various hydrogen production processes with their respective capital cost in (M$) is represented in Figure 9.4.

It is possible to produce the required high-temperature steam using electricity, and in locations with the right solar resources, concentrating solar heat may be used (IEA report, 2019).

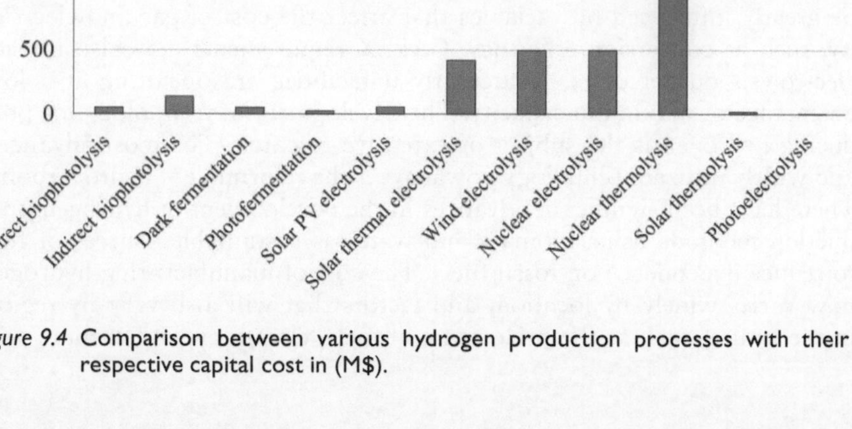

Figure 9.4 Comparison between various hydrogen production processes with their respective capital cost in (M$).

Table 9.2 Statistics including production and retailing cost of hydrogen

Methods	*Energy source*	*Feedstock*	*Capital cost ($)*	*Hydrogen cost ($/kg)*
Direct bio-photolysis	Solar	H_2O + algae	50	2.13
Indirect bio-photolysis	Solar	H_2O + algae	135	1.42
Dark fermentation	—	Organic biomass	50	2.57
Photofermentation	Solar	Organic biomass	24	2.83
Solar PV electrolysis	Solar	H_2O	54	20
Solar thermal electrolysis	Solar	H_2O	421	10.4
Wind electrolysis	Wind	H_2O	499	6.03
Nuclear electrolysis	Nuclear	H_2O	500	7
Nuclear	Nuclear	H_2O	2107	2.63
Solar thermolysis	Solar	H_2O	16	8.4
Photoelectrolysis	Solar	H_2O	2	10.36

Figure 9.5 shows the comparison between various hydrogen production processes with their respective H_2 cost in $/kg. The discovery of less expensive materials for electrodes and membranes, for example, and economies of scale in the manufacturing processes will have an impact on future cost reductions (Nnabuife et al., 2022). There are numerous low-impact methods now being developed. The advancement of these technologies might reduce the world's reliance on fuels, many of which originate in unstable places.

Fuel costs account for the majority of hydrogen production costs, except for hydrogen derived from coal. Hydrogen pricing in the future will thus be greatly influenced by variables that affect the cost of gas and electricity, such as conversion efficiency. CAPEX requirements may also impact electrolysis output costs, particularly if facilities are operating at a low percentage of maximum capacity. The development of technology for producing hydrogen is the subject of extensive research. The most advanced and widely utilized technology nowadays is the reforming of hydrocarbons. There have been significant advances in the development of hydrogen production methods using biomass and water as sustainable sources of fuel to reduce dependence on fossil fuels. The cost of manufacturing hydrogen now varies widely by location, and factors that will also vary by region in the future, such as the price of fossil fuels, electricity, and carbon, will

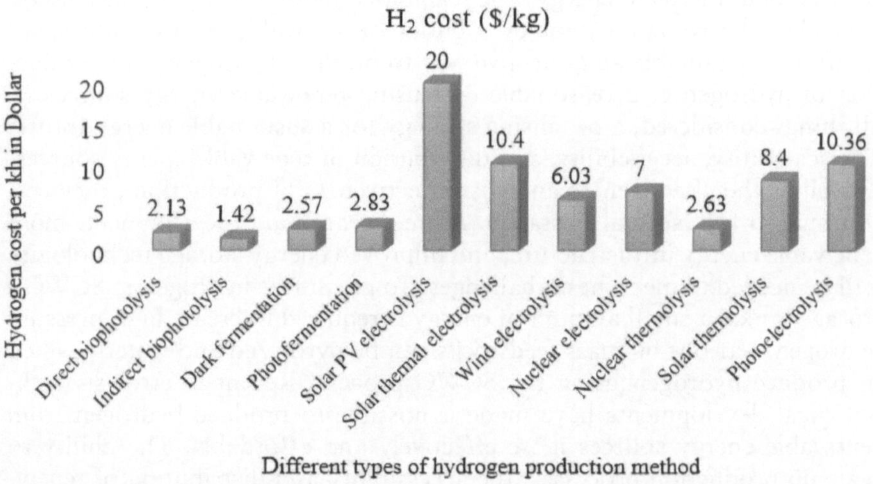

Figure 9.5 H_2 cost against different production methods.

determine whether hydrogen will remain economically viable. When few other low-carbon options are accessible but not cost-effective at the current price, converting hydrogen into other hydrogen-based fuels might be intriguing.

9.4 CONCLUSIONS

In conclusion, recent technology has made it possible to produce hydrogen from renewable energy sources more sustainably and cost-effectively. Each method has benefits and drawbacks, but electrolysis is currently the most promising method for producing hydrogen at a reasonable price. Hydrogen can be produced sustainably and effectively through electrolysis with no emissions, but several issues need to be resolved regarding the scalability, accessibility, and distribution of renewable energy sources. Another method that has gained popularity recently for producing hydrogen from renewable energy sources is biomass gasification. In the process of "biomass gasification," biomass is transformed into a gas that can be used to create hydrogen. The production of hydrogen using this method is sustainable, but some important issues need to be resolved, including the method's scalability and the accessibility of biomass. Two promising processes for generating hydrogen from renewable energy sources, utilizing photosynthesis and solar energy, respectively, are photobiological hydrogen production and photoelectrochemical hydrogen production. Another technique for generating hydrogen from renewable energy sources is known as integration, which entails using a variety of renewable energy sources to power electrolysis cells, such as solar and wind energy. The scalability and reliability of this method depend on the availability and distribution of renewable energy sources, but it offers a sustainable and effective way to produce hydrogen. The production of hydrogen at a reasonable cost using renewable energy sources is, all things considered, a promising strategy for a sustainable energy future. The scalability, accessibility, and distribution of renewable energy sources, as well as the effectiveness and cost-effectiveness of production processes, continue to face several obstacles. More research and development, more renewable energy infrastructure, and improved energy storage technologies will be needed to meet these challenges. To pressurize hydrogen in SCWG's storage tanks, a small amount of energy is required to create high-pressure hydrogen. Various biomass feedstocks can be pyrolyzed and water-gasified to produce hydrogen using the SCWG process. Recent electrolysis technological developments have made it possible to produce hydrogen from renewable energy sources more effectively and affordably. The ability to scale up production processes, the accessibility and distribution of renewable energy sources, and the need for further technological development all continue to be issues. More research, development, infrastructure for renewable energy, and improved energy storage technologies will be needed to address these issues. If these issues can be resolved, hydrogen could play a significant role in the development of a sustainable energy future by offering a reliable and efficient energy source for many different industries.

REFERENCES

Abbasi, T., & Abbasi, S. A. (2011). Renewable hydrogen: Prospects and challenges. *Renewable and Sustainable Energy Reviews*, *15*(6), 3034–3040.

Ahmed, S. F., Mofijur, M., Nuzhat, S., Rafa, N., Musharrat, A., Lam, S. S., & Boretti, A. (2022). Sustainable hydrogen production: Technological advancements and economic analysis. *International Journal of Hydrogen Energy*, *47*(88), 37227–37255.

Chaubey, R., Sahu, S., James, O. O., & Maity, S. (2013). A review on development of industrial processes and emerging techniques for production of hydrogen from renewable and sustainable sources. *Renewable and Sustainable Energy Reviews*, *23*, 443–462.

Dash, S. K., Chakraborty, S., & Elangovan, D. (2023). A brief review of hydrogen production methods and their challenges. *Energies*, *16*(3), 1141.

Drela, K. (2021). Harnessing solar energy and green hydrogen—the energy transition. *Procedia Computer Science*, *192*, 4942–4951.

El-Emam, R. S. & Özcan, H. (2019). Comprehensive review on the techno-economics of sustainable large-scale clean hydrogen production. *Journal of Cleaner Production*, *220*, 593–609. https://doi.org/10.1016/j.jclepro.2019.01.309

Fragiacomo, P., Piraino, F., Genovese, M., Corigliano, O., & De Lorenzo, G. (2022). Strategic overview on fuel cell-based systems for mobility and electrolytic cells for hydrogen production. *Procedia Computer Science*, *200*, 1254–1263.

Fulcheri, L., Rohani, V. J., Wyse, E., Hardman, N., & Dames, E. (2023). An energy-efficient plasma methane pyrolysis process for high yields of carbon black and hydrogen. *International Journal of Hydrogen Energy, 48*(8), 2920–2928.

Hosseini, S. E., Wahid, M. A., Jamil, M. M., Azli, A. A. M., & Misbah, M. F. (2015). A review on biomass-based hydrogen production for renewable energy supply. *International Journal of Energy Research, 39*, 1597–1615. doi: 10.1002/er.3381

Hosseini, S. E., & Wahid, M. A. (2016). Hydrogen production from renewable and sustainable energy resources: Promising green energy carrier for clean development. *Renewable and Sustainable Energy Reviews*, *57*, 850–866.

Hosseini, S. E. & Wahid, M. A. (2020). Hydrogen from solar energy, a clean energy carrier from a sustainable source of energy. International *Journal of Energy Research, 44*(6): 4110–4131. https://doi.org/10.1002/er.493.

Ishaq, H., Dincer, I., & Crawford, C. (2022). A review on hydrogen production and utilization: Challenges and opportunities. *International Journal of Hydrogen Energy*, *47*(62), 26238–26264.

Kalamaras, C. M., & Efstathiou, A. M. (2013). Hydrogen production technologies: Current state and future developments. In Yousef Al-Assaf & Andreas Poullikkas (Eds.), *Conference papers in science* (Vol. 2013). Hindawi.

Monga, D., Shetti, N. P., Basu, S., & Kakarla, R. R. (2023). Recent advances in various processes for clean and sustainable hydrogen production. *Nano-Structures & Nano-Objects*, *33*, 100948.

Muhammed, N. S., Gbadamosi, A. O., Epelle, E. I., Abdulrasheed, A. A., Haq, B., Patil, S., & Kamal, M. S. (2023). Hydrogen production, transportation, utilization, and storage: Recent advances towards sustainable energy. *Journal of Energy Storage*, *73*, 109207.

Nnabuife, S. G., Ugbeh-Johnson, J., Okeke, N. E., & Ogbonnaya, C. (2022). Present and projected developments in hydrogen production: A technological review. *Carbon Capture Science & Technology*, *3*, 100042.

Pareek, A., Dom, R., Gupta, J., Chandran, J., Adepu, V., & Borse, P. H. (2020). Insights into renewable hydrogen energy: Recent advances and prospects. *Materials Science for Energy Technologies*, *3*, 319–327.

Qureshi, F., Yusuf, M., Kamyab, H., Vo, D. V. N., Chelliapan, S., Joo, S. W., & Vasseghian, Y. (2022). Latest eco-friendly avenues on hydrogen production towards a circular bioeconomy: Currents challenges, innovative insights, and future perspectives. *Renewable and Sustainable Energy Reviews*, *168*, 112916.

Qureshi, F., Yusuf, M., Khan, M. A., Ibrahim, H., Ekeoma, B. C., Kamyab, H., & Chelliapan, S. (2023). A state-of-the-art review on the latest trends in hydrogen production, storage, and transportation techniques. *Fuel*, *340*, 127574.

Rasul, M. G., Hazrat, M. A., Sattar, M. A., Jahirul, M. I., & Shearer, M. J. (2022). The future of hydrogen: Challenges on production, storage and applications. *Energy Conversion and Management*, 272, 116326.

Razi, F. & Dincer, I. (2020). A critical evaluation of potential routes of solar hydrogen production for sustainable development. *Journal of Cleaner Production, 264*, 121582,10.1016/j.jclepro.2020.121582.

Seelam, P. K., Rathnayake, B., Pitkäaho, S., Turpeinen, E., & Keiski, R. L. (2020). Overview on recent developments on hydrogen energy: Production, catalysis, and sustainability. In *Current trends and future developments on (bio-) membranes* (pp. 3–32). Elsevier.

The future of hydrogen, report prepared by the IEA for the G20 (2019), Japan. seizing today's opportunities. (n.d.).

Wu, H., Alkhatami, A. G., Farhan, Z. A., AbdalSalam, A. G., Hamadan, R., Aldarrji, M. Q., & Cao, Y. (2023). Recent developments in the production of hydrogen: Efficiency comparison of different techniques, economic dimensions, challenges and environmental impacts. *Fuel Processing Technology, 248*, 107819.

Younas, M., Shafique, S., Hafeez, A., Javed, F., & Rehman, F. (2022). An overview of hydrogen production: current status, potential, and challenges. *Fuel, 316*, 123317.

Yusuf, M., Alnarabiji, M. S., & Abdullah, B. (2021). *Clean hydrogen production technologies. Advances in sustainable energy: Policy, materials and devices* (pp. 159–170). Springer International Publishing.

Chapter 10

Design and simulation of industrial automation system for hydrogen production

Harsh Raval, Aliyu Aliyu, and Faiz Iqbal

10.1 INTRODUCTION

Sustainable energy solutions is one of the keys to the effective fulfilment of sustainable development goals (SDGs). Governments around the globe are promoting the transition towards a green economy to achieve the target of SDGs. The hydrogen economy (HE) is an integral part of the green economy. However, before transitioning to the hydrogen economy, it is essential to make a robust supply chain of hydrogen. Hence, research on the production, storage, and supply system of hydrogen is required.

There are multiple processes to produce hydrogen such as steam methane reforming (SMR), partial oxidation (PO), autothermal reforming (ATR), and coal gasification. While SMR, PO, and ATR use natural gas, the coal gasification process involves the use of coal resulting in carbon emissions. On the other hand, water electrolysis is a process that does not emit carbon at all, providing clean hydrogen. However, electricity is required for the electrolysis process, and the production of electricity may emit carbon dioxide. In solution to that, the use of electricity produced by renewable sources eliminates carbon emissions. This will eventually enable the production of green hydrogen (Simbolotti, 2009).

Electrolysis is a prominent process due to its capability of producing green hydrogen when an electrolyser coupled with renewable energy sources. The electrolysis process can further be categorised as alkaline water electrolysis, polymer electrolyte membrane electrolysis, and high-temperature electrolysis. This categorisation is based on electrolyte or electrode material and working temperature.

10.1.1 Alkaline water electrolysis

Alkaline water electrolysers (AWEs) use an aqueous solution such as KOH or NaOH as an electrolyte. The electrolyte is circulated through the electrolysis cell through which DC current is circulated. Water splits into oxygen and hydrogen due to current circulation. The technology is prominent and mature, and it is available on a commercial scale. Developing AWE

DOI: 10.1201/9781003537816-10

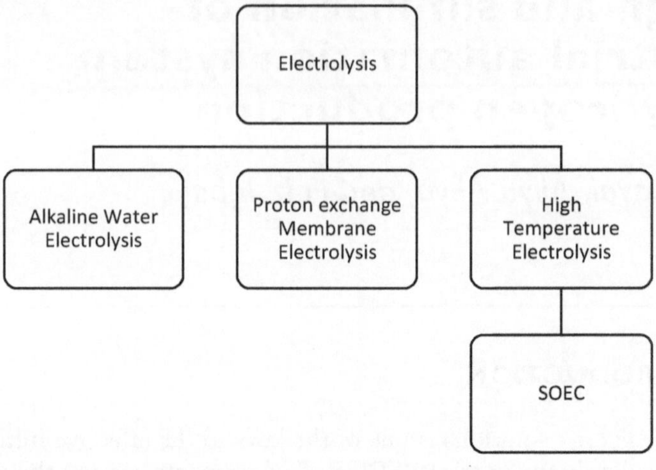

Figure 10.1 Types of electrolysis.

plants requires less CAPEX compared to other electrolysis plants. However, it has a limitation of low-pressure hydrogen delivery such as 25 bar, and hence postproduction compression of the hydrogen produced is required for high-pressure storage (Bodner et al., 2015). Moreover, gas crossover and electrode corrosion due to the aqueous solution is also a limitation of the plant, which increases the maintenance of the AWE plant. However, this problem can be solved with improvements in the material of electrode and separator designs.

10.1.2 Polymer electrolyte membrane (PEM) electrolysis

The PEM electrolysis process uses a solid polymer electrolytic membrane instead of the circulation of the electrolytes into the cell. The operating principle of the PEM electrolysis process can be described by Equations (10.1) and (10.2) (Bodner et al., 2015). The PEM electrolysis overcomes the drawbacks of AWE regarding corrosion and postproduction of high-pressure compression due to high operating pressure of PEM electrolysis. However, PEM electrolysis incorporates high-value electrodes, increasing CAPEX. Moreover, it has a low hydrogen production capacity in comparison to AWE and hence lower efficiency (Bodner et al., 2015). The technology is not yet ready for commercial use. However, some lab-scale and demonstration plants are established.

$$\text{Anode: } H_2O \rightarrow {}^1\!/_2\, O_2 + 2H^+ + 2e^- \tag{10.1}$$

$$\text{Cathode: } 2H^+ + 2e^- \rightarrow H_2 \tag{10.2}$$

10.1.3 High-temperature electrolysis

High-temperature electrolysis is an electrolysis process with operating temperatures of 700°C to 1000°C in the solid-oxide electrolyser (SOEC). Typical operating temperatures in AWEs are 100°C. The benefit of SOEC is that this type of electrolyser can operate at high efficiencies. At temperatures of 700°C to 1000°C, electrode reactions become more reversible in nature. However, SOEC is still in the development phase. Some of the problems include thermal stresses developed due to high operating temperatures. The development of ceramic-based materials is necessary for a solution to that. SOEC technology is not yet ready for commercialisation.

10.2 PROCESS FLOW CHART AND CONTROL SYSTEM

A typical process flow chart of an AWE is presented by Bodner et al. (2015), in their article, which gives some good insight into material flow with some of the BoP (Balance of Plant) components such as the separator, drier, and purifier. Material flow in the electrolyser can be well understood based on their presentation. Additionally, their work generates some of the requirements for components such as a storage tank for distilled water where purified water can be stored. Moreover, a formulation tank is required to prepare the KOH solution of a specified concentration. In addition, arrangements for the compression and storage of produced gas need to be made. Also, while designing the AWE, one must consider safety issues in handling hydrogen and oxygen. Gas crossover, a phenomenon in which oxygen partially gets mixed with hydrogen or vice versa, is a serious issue at higher current densities that needs to be handled well. Additionally, a heat exchanger should be employed to maintain the process temperature (Sanchez et al., 2020). They have given an Aspen Plus model of a 15 kW electrolyser setup in which they have provided a simplified process diagram considering these factors. In addition, they have utilised heat loss from the process to maintain the process temperature. The Aspen Plus model shows purging requirements and equipment through which purging is done. A detailed layout based on these researchers' work is combined and presented in Figure 10.2.

10.3 INTRODUCTION OF INDUSTRIAL AUTOMATION SYSTEM

Industrial automation systems play a pivotal role in enhancing efficiency, reliability, and safety across various industries, and their potential use in hydrogen production is particularly promising. In the context of hydrogen production, automation systems can streamline the entire process, from raw

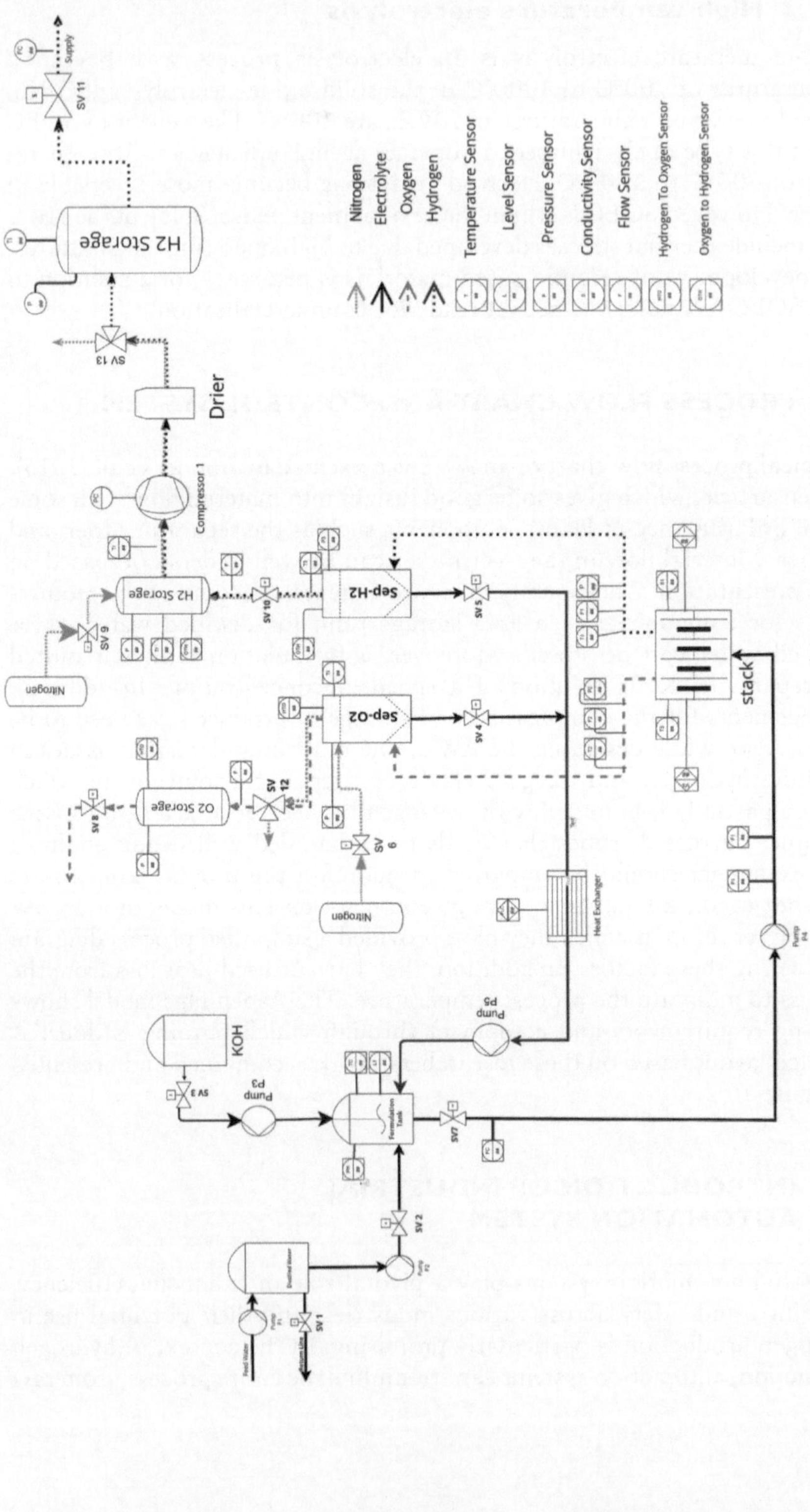

Figure 10.2 Proposed process flow diagram and control system.

material handling to the final product distribution. Automated control systems can optimise the operation of electrolysers, which are commonly used in water electrolysis to generate hydrogen, by adjusting parameters such as voltage, current, and temperature for maximum output. Additionally, automation can facilitate the monitoring and maintenance of equipment, ensuring optimal performance and minimising downtime. Integration with advanced sensors and data analytics enables real-time process monitoring, allowing operators to respond swiftly to fluctuations in demand or variations in input conditions. There have been many attempts reported by researchers wherein they have developed automated systems and utilised them for research in fields related to mechanical engineering as a wider category. Some of the research studies reported are from references (Iqbal et al., 2016, 2022; Reddy et al., 2023; Borra et al., 2023; Vuppugalla et al., 2023).

10.3.1 Operating parameters and control systems

It is necessary to control the AWE process when working on an industrial scale to meet various safety and load requirements. The main control parameters involved in the hydrogen production process through AWE are operating current density, operating temperature, solution concentration, and operating pressure. Apart from this, several monitoring parameters are involved such as storage pressure and temperature, tank water levels, state of charge of gas storages, and hydrogen to oxygen (HTO) and oxygen to hydrogen (OTH) contents. Additionally, monitoring parameters are affected when there is a change in operating parameters. For instance, research work has shown that stack voltage increases when processes are performed at higher current densities. In addition to this, gas crossover also increases when operating at higher current densities, which can be controlled by maintaining the operating temperature. There is a complex relationship among such variables. However, there is a possibility to optimise the process parameters using data analytics. Some of the research establishing relationships between operating parameters can be found in references (Dobó & Palotás, 2016, 2017; Tijani et al., 2014; De Groot et al., 2022; Abdelghani-Idrissi et al., 2021; Li et al., 2022; Rocha et al., 2022). The control system instrument can be employed on the layout based on these research works, which is displayed in Figure 10.2.

10.4 DESCRIPTION OF PROCESS DIAGRAM AND CONTROL SYSTEM

10.4.1 Storage system for distilled water tank

A separate storage system is proposed for distilled water. The input water from the purification plant can be pumped to the distilled water storage tank. The level sensor in the storage tank continuously monitors the level of water in the storage tank and provides information on the water level to the controller, based on which Pump P1 can be turned on or off. A return line is provided in case of emptying the tank required for the maintenance of the tank. The size of the storage tank can be calculated based on the stoichiometric consumption of water during the electrolysis process and the length and size of pipes in the plant.

10.4.2 Storage for KOH

A separate KOH storage tank is proposed. The KOH will be stored in a liquid solution with a higher concentration form in the tank, which will be then fed to the formulation tank with the help of a pump. The size of the tank can be calculated based on the consumption of electrolyte solution and feeding frequency to the formulation tank. The size of the plant in terms of hydrogen production capacity, length and size of pipe, and losses are other factors that should be considered while calculating the size of the KOH storage tank. The tank is mounted with a level sensor that monitors the solution level. The level reading can then be transmitted to the operator.

10.4.3 Formulation tank

The formulation tank is an important part of the proposed scheme. The tank plays a key role in managing the concentration of electrolyte solution. The tank is mounted with several sensors, such as level sensors, which continuously monitor and maintain the level of the solution in the tank. Additionally, the tank is equipped with a pH sensor, which detects the pH level of the electrolyte solution. Pump P2 feeds distilled water to the tank, while Pump P3 feeds a high-concentration electrolyte solution to the tank. Flowmeter FM1 tracks the flow rate of electrolyte solution while Flowmeter FM2 tracks the flow rate of distilled water fed to the formulation tank. Moreover, the tank is equipped with a stirrer that ensures uniformity of the solution. Flowmeter FM3 measures the flow rate of the outlet of electrolyte to the electrolysis stack. Flowmeter FM4 tracks the flow rate of the inlet quantity of electrolyte from the stack, completing the circulation of the electrolyte. SV2 and SV3 are proportional solenoid valves that can control the flow rate of distilled water and ensure the feed of respective materials to the formulation tank. Solenoid valve SV3 and Flowmeter FM1 form a feedback-driven control system to ensure process accuracy. Similarly, Solenoid valve SV2 and Flowmeter FM 2 form a feedback-driven control system for the distilled water inlet to the formulation tank. Pump P5 pumps return electrolyte from the stack to the formulation tank, completing the circulation of the electrolyte.

10.4.4 Electrolyser stack

The electrolyser stack is the critical component where the actual electrolysis process takes place and, hence, close monitoring of the electrolysis process is required. Pump P4 feeds the electrolyte solution to the stack at the required pressure. The flow of electrolytes can be controlled by the proportional solenoid valve SV7. Pressure sensor PM1 records the operating pressure and forms a closed-loop control system with a pump, ensuring the required operating pressure. Temperature sensor T1 records and provides information on electrolyte temperature to the controller. The stack is mounted with Temperature sensor TS2, which records electrode temperature. Measurement of electrode temperature is required to prevent overheating of electrodes. The process input to Solenoid valve SV7, power supply, and heat exchanger can be given based on the temperature readings from TS2. The input of TS2 also helps prevent a dry run of the electrolyser cell stack in case of failure of any component of the electrolyte circulation circuit of the plant. Additionally, the stack is mounted with a voltmeter that records cell voltage. The input of cell voltage is required to ensure the safety of the electrode, as the system should not run below the designed cell voltage. Furthermore, the power management system of the power supply controls the operating current density of the stack. The ammeter attached to the stack ensures the operating current density to the electrolyser. The hydraulic streamline is divided into two parts to the respective separators.

10.4.5 Separators

Both separators are mounted with a pressure sensor and a temperature sensor. Pressure sensor PM4 tracks the pressure of O_2 in the O_2 separator, while PM5 tracks the pressure of the H_2 separator. The O_2 separator is mounted with an HTO sensor, which records the hydrogen content present in oxygen. Moreover, the OTH sensor is mounted in the hydrogen sensor for the measurement of oxygen content in hydrogen. SV4 and SV5 are solenoid valves that control the outlet of electrolyte from the separator to the formulation tank.

10.4.6 Nitrogen purging system

A nitrogen purging system is required to ensure the safety of a plant. It is necessary to discard oxygen when the presence of hydrogen is more than 2%. Hydrogen purging is used in that case to dilute the gas mixture and release it to the environment. Nitrogen purging systems at separators and nitrogen purging systems at hydrogen buffer storage are proposed as shown in Figure 10.2.

SV6 and SV13 are solenoid valves that control the flow of nitrogen to the separator. Similarly, SV9 controls the flow of nitrogen to the H_2 storage.

SV6, SV13, and SV9 form a closed-loop system with an HTO sensor mounted on the O_2 separator, H_2 separator, and H_2 buffer storage. SV10, SV12, and SV14 are three-way solenoid valves that control the direction of flow. In the case of purging, the flow of gas is diverted towards the exhaust in an open environment.

10.4.7 Oxygen storage system

Oxygen storage is mounted with a pressure sensor to monitor the pressure of the stored oxygen. Moreover, Flowmeter FM7 records the flow of oxygen, based on which the total produced amount of oxygen can be calculated. Leakage in the system can be detected by mass balance.

10.4.8 Hydrogen storage system

The hydrogen storage system consists of a buffer storage tank mounted with an OTH sensor, temperature sensor, and pressure sensor. A three-way solenoid valve, SV14, connects the buffer storage tank with the drier. Dried hydrogen is then compressed to storage pressure via Compressor 1 that is mounted with a pressure sensor. The outlet of the compressor is connected to the hydrogen storage tank.

The volume of all storage tanks can be calculated based on the plant's production capacity, pipe diameter and size, ambient temperature, and storage pressure.

Automation system architecture can be prepared based on the process flowchart and control system layout.

10.5 AUTOMATION SYSTEM ARCHITECTURE

Automation system architecture is designed based on the operational requirements of the electrolysis system and the process flowchart mentioned in Figure 10.2. The process flowchart can be configured between systems such as a fluid flow control system, fluid monitoring system, pressure control system, pressure monitoring system, temperature monitoring system, temperature control system, power management system, and operation safety system.

The proposed architecture consists of PLC. PLC is a short form of Programmable Logic Control. Typical PLC units incorporate a CPU, power supply circuit, input/output circuit, communication circuit, and I/O interface. Additionally, in case of shortcomings in several of the I/O interfaces, an expansion module can be attached to increase the number of I/O interfaces (Zhao, 2021). The architecture consists of an HMI to provide interaction with the system. Additionally, various sensors mentioned in the process flow chart are included in the architecture. The architecture is mentioned in Figure 10.3.

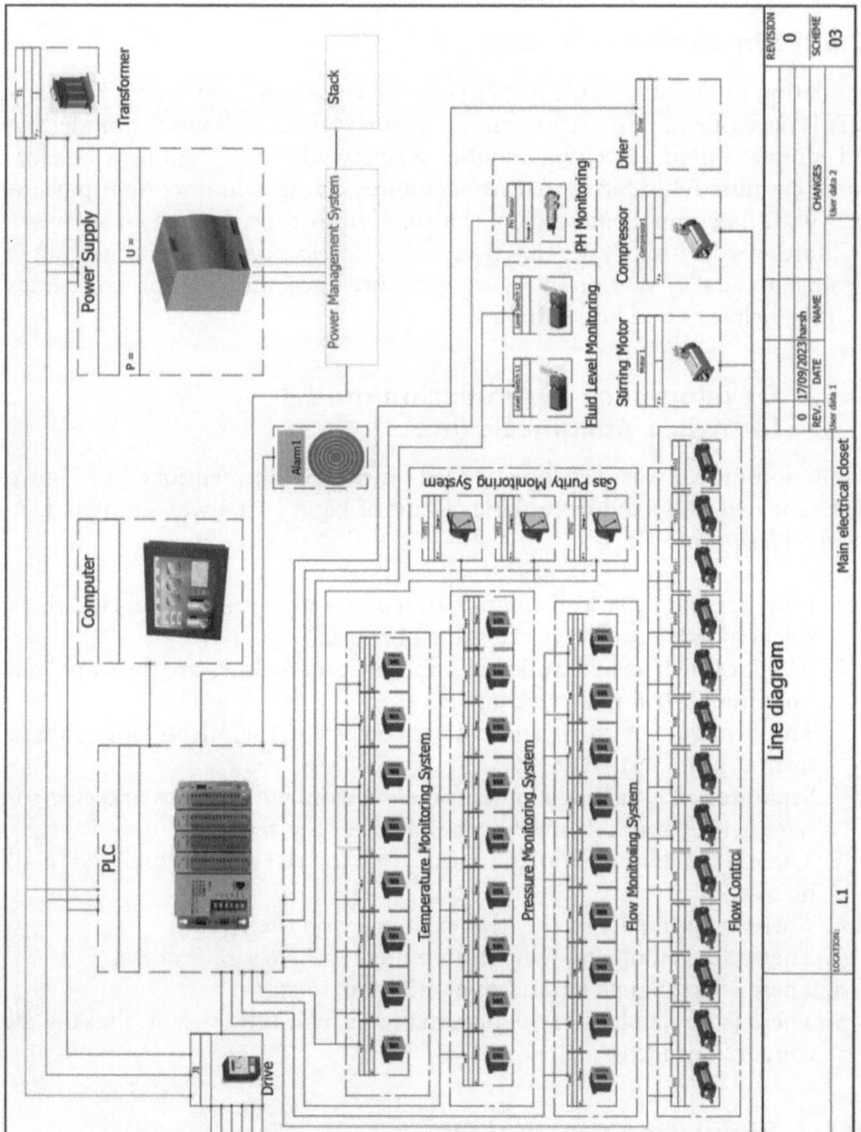

Figure 10.3 Automation system architecture.

10.6 MODELLING AND SIMULATION

10.6.1 Simulation software

Simulation model was made in "Fluid Sim" version 6.1 software. The software is developed by an art system. Fluid Sim software allows the modelling and simulation of electropneumatic, electrohydraulic, digital, and electronic circuits. Fluid Sim software facilitates changes in operation parameters with its components during the simulation time. Fluid Sim software facilitates verification of the function of an electric circuit. Additionally, it is possible to identify suction lines, low-pressure lines, and high-pressure lines with the help of Fluid Sim software.

10.6.2 Development of simulation model for hydro-pneumatic line

Simulation model was developed based on the layout mentioned in Figure 10.4. The hydraulic and pneumatic circuit of Figure 10.4 was simulated.

Considerations:

1. Proportional valve and combination of variable venturi and two-third solenoid valve can serve the same purpose.
2. All pipes such as hydraulic lines and pneumatic lines are the same and function with similar mechanics.
3. The function of the system stays consistent for all pressure ranges from 0.1 MPa to 70 MPa.
4. Separators have a single outlet pipeline from which gases and electrolytes get separated from the pipeline after the separator.
5. Theories of fluid dynamics are the same for all types of fluid and in all phases.
6. There is no impact of the heat exchanger on the circuit.
7. There is no change in temperature during the entire process.
8. There is no change in fluid dynamics due to stack.
9. There is no line loss in the pneumatic or hydraulic circuit, making all connections ideal.

10.6.3 Simulation parameters

Distilled water tank T1 is supplying distilled water with a maximum flow rate of 100 l/min, which is then pumped through Pump P1 having a maximum permissible flow rate of 100 l/min. Proportional Solenoid valve SV1 allows a flow of 70% nominal flow rate. Tank T2 supplies KOH with a nominal flow rate of 100 l/min, which is pumped by Pump P2 having a maximum permissible flow rate of 100 l/min to the Formulation tank T3. Proportional Solenoid valve SV2 controls the amount of KOH blending to

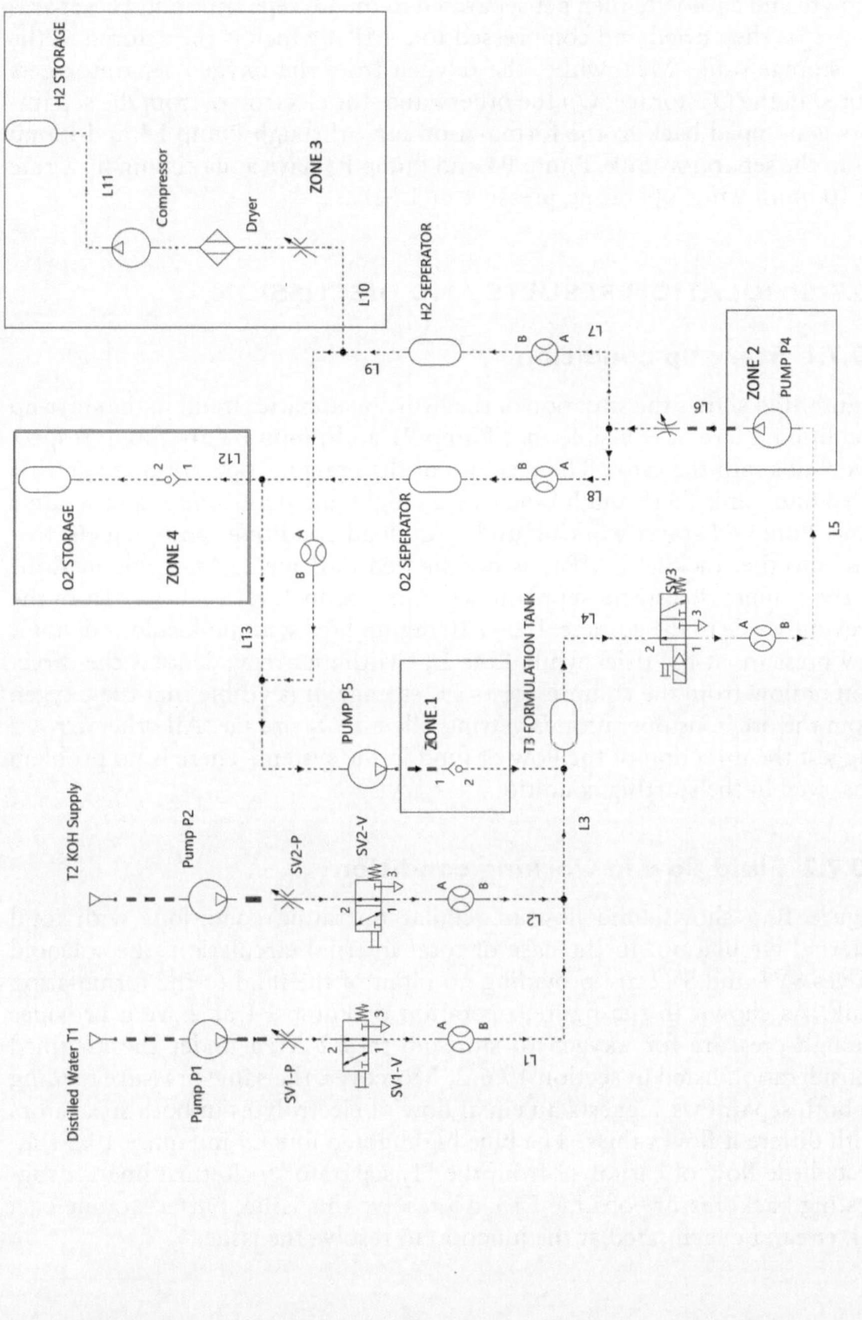

Figure 10.4 Simulation for the start-up condition of electrolyser

30%. The system makes a 30% KOH solution in Formulation tank T3. The electrolyte solution is supplied from the formulation tank to the stack through Pump P4 having a maximum flow rate of 10 l/min. However, the flow rate is controlled by proportional Solenoid valve SV3 to 6 l/min. The anolyte and catholyte then get separated in the O_2 separator and H_2 separator. H_2 is then dried and compressed to 2 MPa, which is then stored in the H_2 storage tank. Meanwhile, the oxygen from the oxygen separator gets stored in the O_2 storage. On the other hand, the electrolyte from the separators is pumped back to the formulation tank through Pump P4 and Pump P5 to the separator tank. Pump P4 and Pump P3 have a maximum flow rate of 10 l/min while operating pressure at 1 MPa.

10.7 SIMULATION RESULTS AND DISCUSSION

10.7.1 Start-up condition

Figure 10.4 shows the situation of the hydropneumatic circuit in the start-up condition. Here, it is visible that Pump P1 and Pump P2 are filling respective fluids into the tank. The residues of the previous operation are getting filled into Tank T3 through Line L14, as highlighted in Zone 1. At the same time, Pump P4 starts working under full load condition, pumping electrolyte into the stack at 2 MPa, as highlighted in Zone 2. As visible in Zone 3, the compressor starts supplying compressed hydrogen leftover from the previous cycle to the storage. Line L10 highlighted with pink colour denotes low pressure at the drier while Line L11 with an arrow denotes the direction of flow from the compressor to H_2 storage. It is visible that the oxygen from the previous operation is getting filled in O_2 storage. All other arrows suggest the direction of the flow of fluid in the system. There is no problem observed in the starting condition.

10.7.2 Fluid flow in working condition

Figure 10.5 shows fluid flow in regular operating conditions with total internal circulation. In the case of total internal circulation, the solenoid valves SV1 and SV2 are providing no input of the fluid to the formulation tank. As shown in the figure, operation of Pump P4 at 2 MPa provides enough pressure for oxygen to store up to 1.2 MPa under the assumed consideration listed in section 10.6.2. Moreover, the same pressure reading at both separators suggests an equal flow of electrolytes in both separators with different flow values. The blue highlighted line L9 in Figure 10.5 suggests little flow of catholyte from the H_2 separator to Return line L9 suggesting backpressure of Line L13. To resolve this issue, buffer storage or a mixer can be facilitated at the junction to resolve the issue.

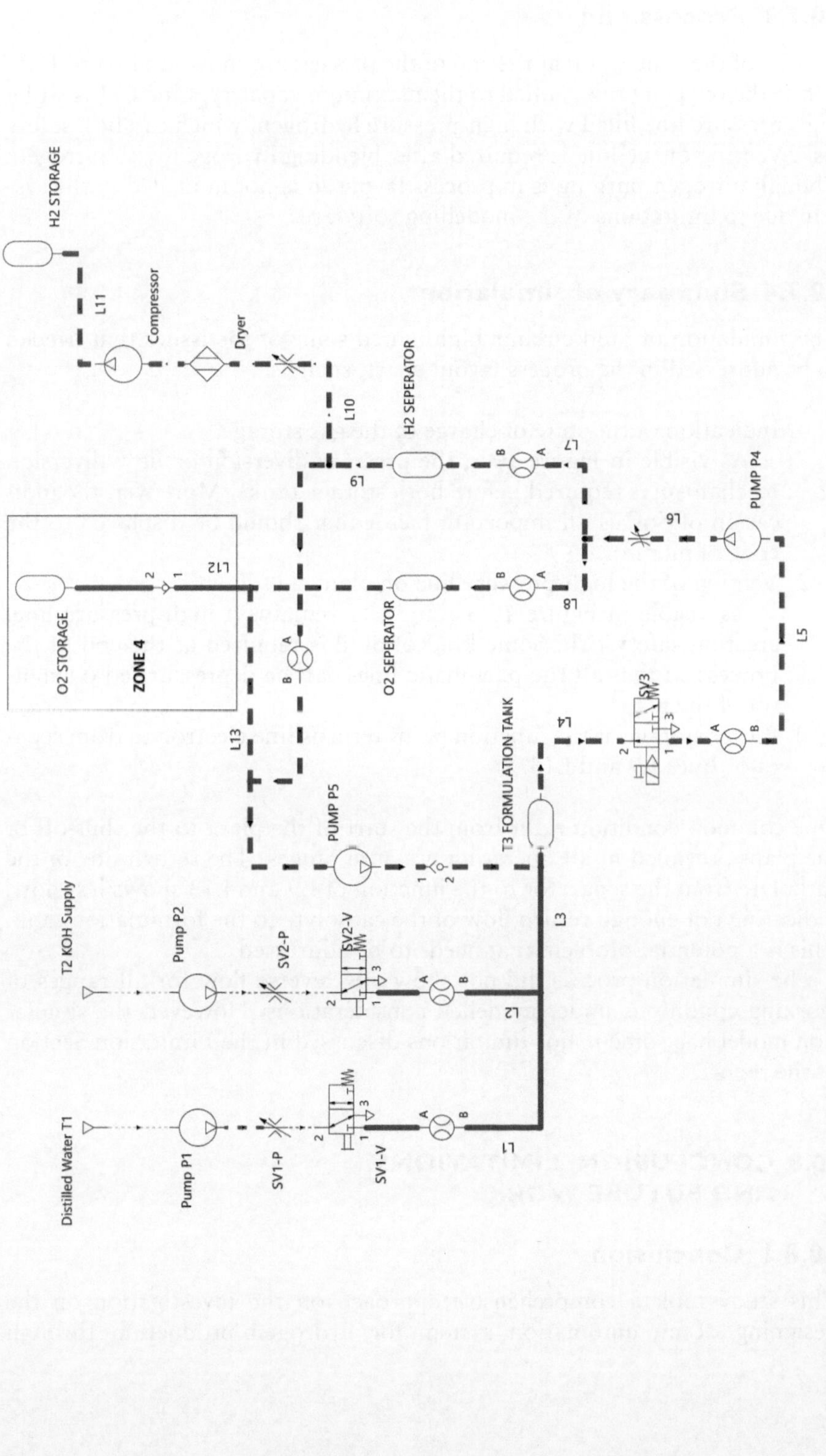

Figure 10.5 Flow of fluid in regular working condition.

10.7.3 Process end

Results of the simulation at the end of the process are shown in Figure 10.6. While the oxygen tank is filled to the maximum capacity, Line L11 is still a high-pressure line filled with high-pressure hydrogen, which can be a safety risk. Venting of the line is required after blending hydrogen with nitrogen. Though nitrogen purging is in process layout, it is not modelled in the system due to limitations of the modelling software.

10.7.4 Summary of simulation

The simulation of fluid circuits highlighted some of the issues that needed to be addressed in the process layout chart, such as:

1. Indication of the state of charge of the gas storage.
   As visible in Figure 10.5, the pressure diversion or flow diversion mechanism is required before both storage tanks. Moreover, the indication of SoC is an important factor that should be displayed to the control monitor.
2. Venting of the high-pressure line on plant shutdown
   As visible in Figure 10.6, Line L11 remains a high-pressure line, creating safety risk. Some kind of SOP is required at the end of the process so that all the pneumatic lines can be depressurised on shut-off of the plant.
3. Buffer storage at the junction point returns line electrolyte from separator lines L9 and L13.

One common condition right from the start of the plant to the shut-off of the plant prevailed in all operating pressure ranges. The return line of the catholyte from the separator to the junction of L9 and L13 shows less flow, indicating not enough return flow of the catholyte to the formulation tank. This is a potential problem that needs to be addressed.

The simulation process did not show any reverse flow for all ranges of working conditions under modelled considerations. However, the simulation model has some major limitations discussed in the Limitation Section of the report.

10.8 CONCLUSION, LIMITATIONS, AND FUTURE WORK

10.8.1 Conclusion

This study took a comprehensive approach on the investigation on the designing of an automation system for hydrogen production through

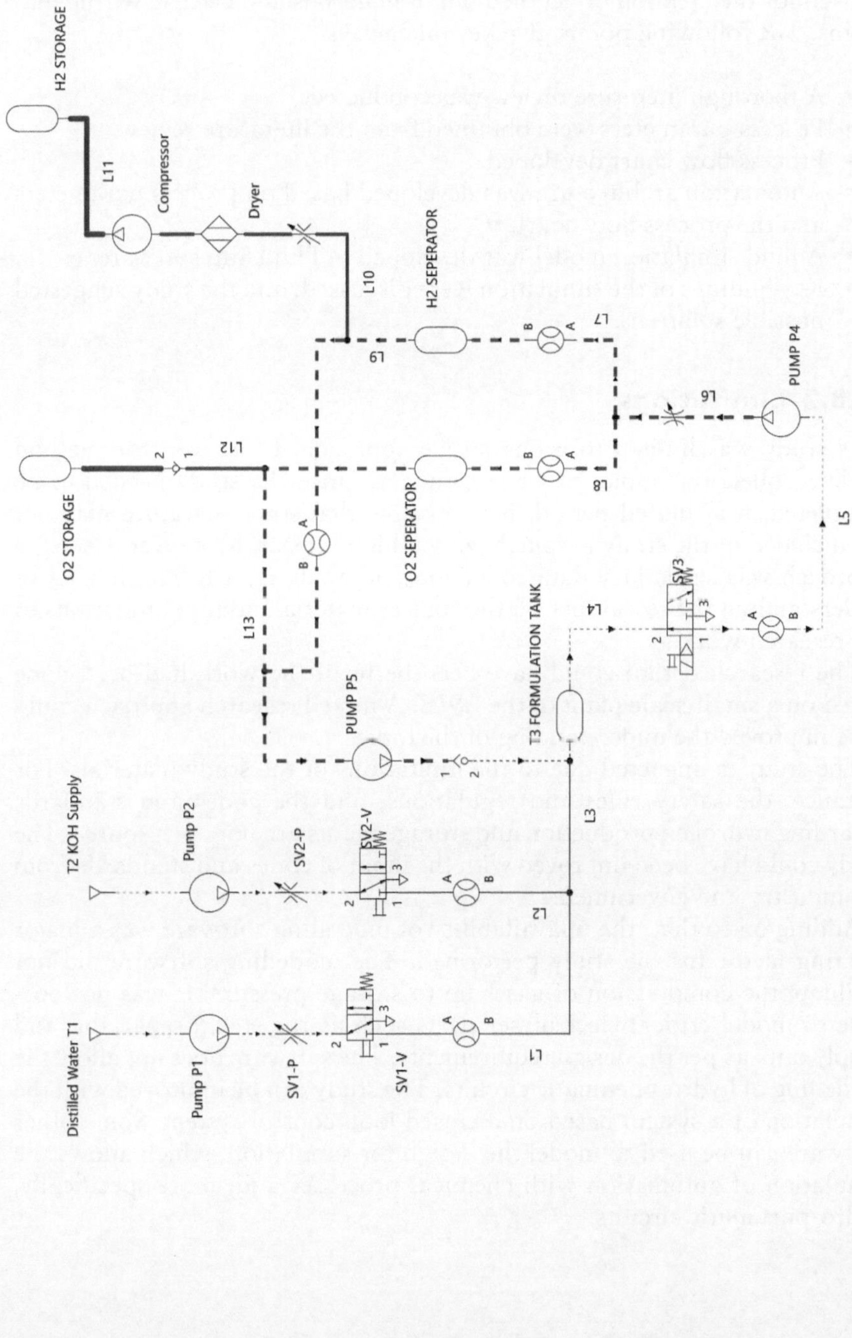

Figure 10.6 Fluid flow before shutting off the electrolyser.

alkaline water electrolyser (AWE). It offered a thorough understanding of the integration of automation systems and hydrogen generation systems using alkaline water electrolysers. The study proposed a fundamental automation architecture from a system engineering perspective, which can be utilised for the creation of detailed bill of materials and electric wiring diagrams. The following points give key takeaways.

- A thorough literature review was conducted.
- Process parameters were obtained from the literature review.
- Process flow chart developed.
- Automation architecture was developed based on process parameters and the process flow chart.
- A fluid simulation model was developed in Fluid Sim software.
- Key findings of the simulation were discussed, and the study suggested possible solutions.

10.8.2 Limitations

This study was limited to a qualitative approach. Moreover, the method of data collection employed was secondary. Since the study needed to be completed in a limited period, it is possible that some literature may not get included in the study despite having a high impact. Moreover, since the approach was secondary data collection and analysis, it has an impact of understanding the researcher on the subject material, adding limitations to the research work.

The research design could have been better if the work had been done based on a small-scale plant of the AWE. A mixed research approach could have improved the understanding of the topic.

The study is impacted due to the limitations of the study materials. For instance, the safety rules and regulations, and the codes and standards regarding hydrogen production and storage plants are not open source. The study could have been improved with the input of codes and standards from the industry and government.

Adding on to that, the unavailability of modelling software was a major limiting factor for the study performed. The modelling software did not facilitate the compression of gases up to storage pressure. It was not possible to model critical electrolyser parts such as the stack, separator, and supply tank as per the design requirements. The software does not allow the modelling of hydro-pneumatic circuits. The study can be improved with the simulation of a system based on a closed-loop control system. Some other software can be used to model the design for simulation, which allows the simulation of automation with chemical processes and, more specifically, hydro-pneumatic circuits.

REFERENCES

Abdelghani-Idrissi, S., Dubouis, N., Grimaud, A., Stevens, P., Toussaint, G., & Colin, A. (2021). Effect of electrolyte flow on a gas evolution electrode. *Scientific Reports, 11*(1), 4677.

Bodner, M., Hofer, A., & Hacker, V. (2015). H2 generation from alkaline electrolyzer. *Wiley Interdisciplinary Reviews: Energy and Environment*, *4*(4), 365–381.

Borra, S. V., Guntipally, H., Vuppugalla, S., Iqbal, F., & Vistapalli, J. (2023). Developing digital twin for transforming CNC machine to industry 4.0 for remote monitoring. *In 2023 IEEE international conference on mechatronics and automation (ICMA)* (pp. 100–105). IEEE.

De Groot, M. T., Kraakman, J., & Barros, R. L. G. (2022). Optimal operating parameters for advanced alkaline water electrolysis. *International journal of hydrogen energy, 47*(82), 34773–34783.

Dobó, Z., & Palotás, Á. B. (2016). Impact of the voltage fluctuation of the power supply on the efficiency of alkaline water electrolysis. *International Journal of Hydrogen Energy, 41*(28), 11849–11856.

Dobó, Z., & Palotás, Á. B. (2017). Impact of the current fluctuation on the efficiency of Alkaline Water Electrolysis. *International Journal of Hydrogen Energy, 42*(9), 5649–5656.

Iqbal, F., Rammohan, R., Patel, H. A., & Jha, S. (2016). Design and development of automated workpiece cleaning system for ball end magneto-rheological finishing process. *In International conference on advances in materials & manufacturing* (Vol. 19, No. 5, pp. 289–295). IOP Publishing.

Iqbal, F., Semwal, T., & Stokes, A. A. (2022) Transforming a standalone machine tool to a cyber-physical system: A use case of BEMRF machine tool to tackle the COVID-19 restrictions. In F. Iqbal & T. Semwal (Eds.), *Cyber-physical systems* (pp. 330–344). CRC Press.

Li, J. Q., Li, J. C., Kwon, J. T., & Shang, C. (2022). The effect of internal pressure change on the temperature rise and the amount of filling hydrogen of high pressure storage tank. *Advances in Mechanical Engineering, 14*(8), 16878132221121030.

Reddy, V. A., Krishna, V. V., Charishma, A., Iqbal, F., & Vistapalli, J. (2023). Transformation of industry 3.0 CNC machines to industry 4.0 machines: Sensor selection and integration. In *2023 IEEE international conference on mechatronics and automation (ICMA)* (pp. 285–290). IEEE.

Rocha, F., Delmelle, R., Georgiadis, C., & Proost, J. (2022). Effect of pore size and electrolyte flow rate on the bubble removal efficiency of 3D pure Ni foam electrodes during alkaline water electrolysis. *Journal of Environmental Chemical Engineering, 10*(3), 107648.

Sánchez, M., Amores, E., Abad, D., Rodríguez, L., & Clemente-Jul, C. (2020). Aspen Plus model of an alkaline electrolysis system for hydrogen production. *International journal of hydrogen energy, 45*(7), 3916–3929. doi: 10.1016/j.ijhydene.2019.12.027

Simbolotti, G. (2009). *R&D priorities and gaps hydrogen production and storage hydrogen implementing agreement hydrogen implementing agreement.*

Tijani, A. S., Yusup, N. A. B., & Rahim, A. A. (2014). Mathematical modelling and simulation analysis of advanced alkaline electrolyzer system for hydrogen production. *Procedia Technology, 15*, 798–806.

Vuppugalla, S., Guntipally, H., Borra, S. V., Iqbal, F., & Vistapalli, J. (2023). Development of a smart interfacing dashboard between management and shop-floor for industry 3.0 machines and small and medium enterprises (SME). In *2023 IEEE international conference on mechatronics and automation (ICMA)* (pp. 403–408). IEEE.

Zhao, B. (2021). The design and research of electric automation control system based on PLC. *Intelligence, 3*, 4.

Chapter 11

Storage of hydrogen in solid-state materials

Yogesh Kumar Yadav, Mohammad Abu Shaz and Thakur Prasad Yadav

11.1 INTRODUCTION

Although the idea of using H_2 as fuel is not new, H_2 is both one of the elements that is most common in the universe and the most potential alternative fuel for the future (Züttel, 2004; Güneş, 2019). But you cannot just harvest it; you must extract it from another material. Energy is needed to extract H_2 for use in fuel cells or burning because H_2 atoms are bonded together in molecules with other elements. H_2 has a lower environmental impact than fossil fuels because it produces less gas that contributes to global warming, such as sulfur, oxidized nitrogen, and extremely small particles (Watson & McMichael, 2001; Huston, 1984; Steinfeld & Palumbo, 2001). The energy sector uses H_2 because of its extremely high energy density/unit mass, i.e., the total amount of energy in a system per unit volume (142 MJ/kg) .

Because H_2 is bound to other elements in molecules, it must be removed before it can be used in fuel cells or combustion. The detailed physical characteristics of H_2 are compiled in Table 11.1, while a selection of those characteristics is listed in Table 11.2 along with a comparison to other conventional hydrocarbon fuels. A clean, sustainable energy source for both fixed and mobile uses is H_2.

In space, H_2 has other uses (Jena, 2011; Ramachandran & Menon, 1998; Züttel, 2003). It might reduce our dependency on hydrocarbon-based energy. Although it always mixes with other elements, such as oxygen and carbon, once it is separated, clean energy can be produced from it. H_2 and oxygen (O) can be separated from water (H_2O) to generate H_2 (Dillon et al., 2000; Hirscher et al., 2001). H_2 can be produced using a variety of materials, including coal, lignite, natural gas, fossil fuels, and renewable energy sources (Bargigli et al., 2004). H_2 comes in a variety of hues from different sources. For instance, H_2 comes in the following colors: black, brown, gray, blue, and green (Incer-Valverde et al., 2023). It can also come from coal, lignite, natural gas, fossil fuels, and renewable sources. If H_2 is derived from sustainable sources, it can be produced and utilized in a clean, natural cycle (Hosseini & Wahid, 2016, 2020). One consequence of consuming H_2 as fuel is water vapor (Dönitz & Erdle, 1985; Veziro & Barbir, 1992). Burning

DOI: 10.1201/9781003537816-11

Table 11.1 Physical properties of hydrogen

Molecular formula	*Hydrogen (H_2)*
Phase at STP	Gas
Melting point	14.025 K
Boiling point	20.268 K
Molar volume	11.42×10^{-3} m^3mol
Enthalpy of fusion	0.05868 kJ/mol
Enthalpy of vaporization	0.44936 kJ/mol
Density	0.0899 kg/m^3
Speed of sound	1270 m/s at 298.15 K
Electronegativity	2.2 (Pauling scale)
Specific heat capacity	14304 J/kgK
Electric conductivity	N/A
Thermal conductivity	0.1815 W/mK
Ionization energy	1312.06 kJ/mol
Lower calorific value	110.9–10.1 (MJ/kg-MJ/Nm^3)
Minimum ignition energy	0.02 mJ
Stoichiometric flame speed	2.337 m/s
Density	0.084 kg/Nm^3
Boilng point	20.4 K
Critical point	32.9 K
Specific heat	14.9kJ/(kg-K)
Flammability limits (by volume percentage)	4–75%

H_2 in air that contains nitrogen can result in the production of nitrogen oxides, which are also generated in gasoline engines. These oxides can be eliminated from H_2 engines by reducing the temperature of the combustion engine. Acid rain, ozone depletion, and carbon dioxide buildup can all be reduced by H_2 energy (Kern, 2021). Normal conditions result in H_2 existing as a gas. Among all known compounds, it has the lowest melting and boiling points, only exceeded by He. H_2 exists in two states: as a liquid at −253°C, and as a solid at −259°C, both below the melting point and atmospheric pressure. Figure 11.1 displays the H_2 phase diagram. At 0°C and 1 bar, H_2 has a density of 0.08988 g/l and a molecular weight of 2.015 g. For comparison, Table 11.3 shows the densities of fuels in liquid and gaseous forms.

In its gaseous, liquid, and solid phases, H_2 has a significantly lower density than methane under normal circumstances (Abbas & Daud, 2010). In particular, H_2 has a density that is about 8 times lower than methane when it is gaseous, 6 times lower when it is liquid, and 55 times lower when it is solid. Notably, a given volume of gasoline or water has a higher concentration of H_2 than pure liquid H_2, demonstrating its lower volumetric density (111 kg/m^3, 84 kg/m^3, and 71 kg/m^3, respectively), despite having a lower

Table 11.2 A few physical properties of propane, methane, H_2, and gasoline

Property	*Hydrogen*	*Methane*	*Propane*	*Gasoline*
Molecular weight (u^a)	2.02	16.04	44.06	~107
Density (kg/m^3) at normal conditions	0.089	0.651	1.87	4.4
Buoyancy (density with respect to air)	0.07	0.55	1.52	3.4–4.0
Diffusion coefficient (cm^2/s)	0.61	0.16	0.12	0.05
Lean flammability limit in air (% by volume)	4.1	5.3	2.1	1.0
Rich flammability limit in air (% by volume)	75	15	10	7.8
Minimum ignition energy (m-J)	0.02	0.29	0.26	0.24
Minimum self-ignition energy (K)	858	813	760	501–744
Lean detonability limit in air (% by volume)	18	6.3	3.1	1.1
Rich detonability limit in air (% by volume)	59	13.5	7.0	3.3
Explosion energy (kg equivalent TNT per m^3 of vapor)	2.02	7.02	20.2	44.2
Molecular weight (u^a)	2.02	16.04	44.06	~107
Density (kg/m^3) at normal conditions	0.084	0.651	1.87	4.4
Buoyancy (density with respect to air)	0.07	0.55	1.52	3.4–4.0
Diffusion coefficient (cm^2/s)	0.61	0.16	0.12	0.05
Lean flammability limit in air (% by volume)	4.1	5.3	2.1	1.0

[a]Unified atomic mass unit or atomic mass unit (u) = 1.660538782(83) × 10^{-24} g.

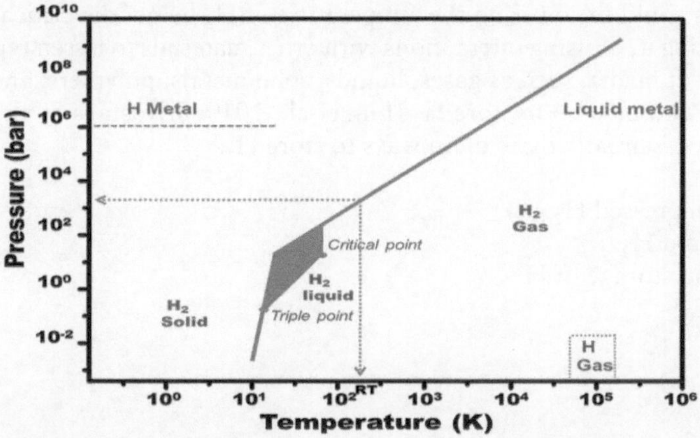

Figure 11.1 Basic H_2 phase diagram

Table 11.3 Fuel densities in gaseous and liquid forms

Substance	*Gas at 20°C, 1 bar (kg/m³)*	*Liquid at boiling point, 1 bar (kg/m³)*
Hydrogen	0.089	70.8
Methane	0.65	422
Gasoline	4.4	700

density. The atomic structure of H_2 is particularly intriguing since it has the largest fraction of valence electrons to protons of any element in the periodic table. Every H_2 atom contains one proton and one electron, meaning that each electron has a significant amount of energy gained. When related to other chemical fuels such as liquid hydrocarbons (which have a chemical energy/unit mass of about 13.1 $kWh{\cdot}kg^{-1}$), H_2 has about three times that amount, or 39.4 $kWh{\cdot}kg^{-1}$.

11.2 HYDROGEN STORAGE

The energy storage capacity of H_2 is 142 MJ/kg, which is far larger than that of liquid hydrocarbons, which have a capacity of just 47 MJ/kg (Gregg et al., 1967; Muratov & Gromov, 2017; Rzepka et al., 1998; Williams & Eklund, 2000). The significantly low energy density of H_2 at ambient temperature (AT) and atmospheric pressure (AP) presents a hurdle, though (Beebe et al., 1947; Leung et al., 1976; Ma et al., 2001; Srivastava et al., 2015; Züttel et al., 2002). Surmounting this obstacle is essential to improving the dependability of renewable energy sources, with storage being a key component. H_2 of 1 kg takes up a sizable volume of 11 m^3 under typical conditions (Olabi et al., 2021; Srivastava et al., 2015; Ye et al., 1999). Several techniques can be used to expand the density of H_2 in a storage system in order to address this. This could be lowering the temperature of H_2 below the critical point, compressing it, or using interactions with other materials to lessen repulsion. A variety of media, such as gases, liquids, solid metals, polymers, and liquid hydrides, can be used to store H_2 (Lim et al., 2010; Srivastava et al., 2015). There are essentially three main ways to store H_2.

1. Compressed H_2 gas
2. Liquid H_2
3. Solid storage of H_2

11.3 METHODOLOGICAL GOALS FOR ON-BOARD H_2 STORAGE SYSTEMS

Compact, lightweight, and reasonably priced containment solutions are required for vehicles that use on-board energy storage. During a 400 km journey, a modern, commercially available car with optimal fuel economy uses about 24 kg of gasoline. On the other hand, just 4 kg of H_2 is needed for an electric vehicle with a fuel cell, whereas 8 kg of H_2 is needed for a combustion engine (Cunanan et al., 2021). It is crucial to remember that 1 kg of H_2 takes up a sizable 11 m^3 at room temperature and atmospheric pressure, making it unsuitable for on-board storage in automobiles (Srivastava et al., 2015; Ye et al., 1999). To put this into perspective, this capacity is equivalent to a 5 m diameter balloon, which is obviously not a practical choice for vehicle storage. The necessary technical specifications for on-board H_2 storage systems are listed in Table 11.4. These goals were defined by the US DOE, the US Council for Automotive Research (USCAR), and energy companies. They function as crucial performance standards for H_2 storage systems installed in light-duty vehicles. These objectives are part of the coordinated efforts under the US DRIVE Partnership. The storage of H_2 gas at room temp in a high-pressure compressed form is known as high-pressure compressed H_2 gas storage (Makridis, 2017; Zheng et al., 2012). High-pressure cylinders (HPC) of the gas that can endure pressures of up to 200 bar are needed for this storage technique. However, new lightweight composite cylinders that can withstand pressures of up to 800 bar have been developed, thanks to technical advancements (Klell, 2010; Züttel, 2004). As a result, H_2 can now have a volumetric density of 36 kg/m³ or about half that of liquid H_2 at the standard boiling point (Chen et al., 1999; Elam, 1996; Fan et al., 1999; Liu et al., 1999; Nijkamp et al., 2001; Ströbel et al., 1999; Yang et al., 2023). Because the walls of the pressure cylinder thicken as the pressure increases, the gravimetric H_2 density drops. The volumetric density of H_2 reaches 20 kg/m³ at 300 bar; it does not increase linearly with pressure. Although it is not practically viable to compress the gas to pressures of 2000 bar, one can increase the volumetric density to approximately 40–70 kg/m³ by further compressing the gas to pressures between 700 and 2000 bar. Figure 11.2 displays the volumetric density of H_2 as a function of pressure at three different temperatures: 273 K, 298 K, and 373 K. The storage of liquid H_2 requires cryogenic temperatures. The study of extremely low temperatures, or cryogenics, is crucial for managing such circumstances. At 20.3 K, H_2 boils, and at 32.9 K, it reaches its critical temperature and turns into a gas.

A closed storage system usually keeps its pressure at about 104 bar at room temperature. The volumetric density of liquid H_2 is 70.8 $kg \cdot m^{-3}$, which is marginally greater than the density of solid H_2, which is 70.6 $kg \cdot m^{-3}$. Mitigating H_2 boil-off in liquid H_2 storage systems requires careful attention to two important factors: the thermal padding of cryogenic storage tanks and the efficient control of energy-efficient liquefaction

Table 11.4 Technical goals for systems that store H_2 on board

Storage parameter	*Units*	*2020*	*2025*	*Ultimate*
System gravimetric capacity				
Usable, specific-energy from H_2 (net useful energy/max system mass)	kWh/kg (kg H_2/kg system)	1.5 (0.045)	1.8 (0.055)	2.2 (0.065)
System volumetric capacity				
Usable energy density from H_2 (net useful energy/max system volume)	kWh/L (kg H_2/L system)	1.0 (0.030)	1.3 (0.040)	1.7 (0.050)
Storage system cost				
Storage system cost	$/kWh net ($/kg H_2)	10 (333)	9 (300)	8 (266)
Fuel cost	$/gge at pump	4	4	4
Durability/operability				
Operating ambient temperature	°C	−40/60 (sun)	−40/60 (sun)	−40/60 (sun)
Min/max delivery temperature	°C	−40/85	−40/85	−40/85
Operational cycle life (1/4 tank to full)	cycles	1500	1500	1500
Min delivery pressure from storage system	bar (abs)	5	5	5
Max delivery pressure from storage system	bar (abs)	12	12	12
On-board efficiency	%	90	90	90
"Well" to power plant efficiency	%	60	60	60
Charging/discharging rates				
System fill time	min	3–5	3–5	3–5
Minimum full flow rate (e.g., 1.6 g/s target for 80 kW rated fuel cell power)	(g/s)/kW	0.02	0.02	0.02
Average flow rate	(g/s)/kW	0.004	0.004	0.004
Start time to full flow (20°C)	s	5	5	5
Start time to full flow (−20°C)	s	15	15	15
Transient response at operating temperature 10%–90% and 90%–0% (based on full flow rate)	s	0.75	0.75	0.75
Fuel quality				
Fuel quality (H_2 from storage)	% H_2	Meet or exceed SAE J2719		

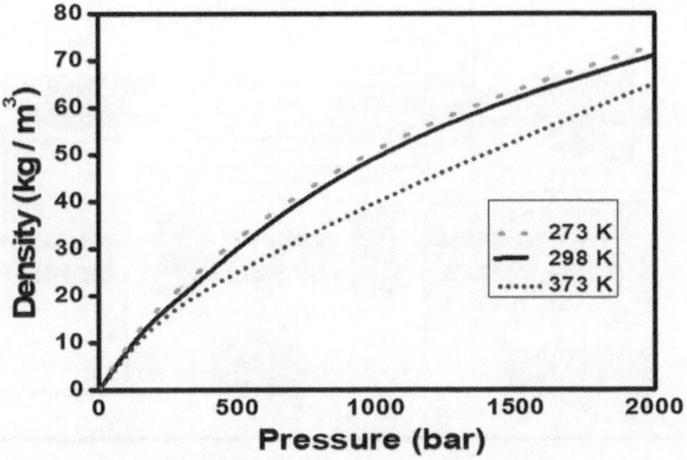

Figure 11.2 The volumetric density of H_2 at three distinct temperatures is shown to evolve with pressure.

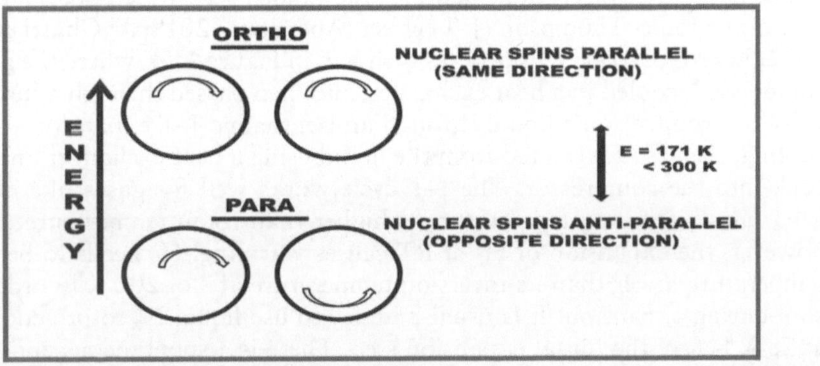

Figure 11.3 There are two types of H_2 molecules.

operations. As shown in Figure 11.3, H_2 occurs at room temperature as a mixture of 25% normal H_2 (with nuclear spin I = 0) and 75% ortho form (with nuclear spin I = 1). Ortho H_2 cannot be isolated in its purest state. All H_2 molecules assume the para form at 0 K or absolute zero (Woolley et al., 1948; Zarur & Rabitz, 1974). The energy difference between ortho and para H_2 is the cause of their distinct physical properties. The melting and boiling temperatures of para H_2 are approximately 0.1 K lower than those of conventional H_2.

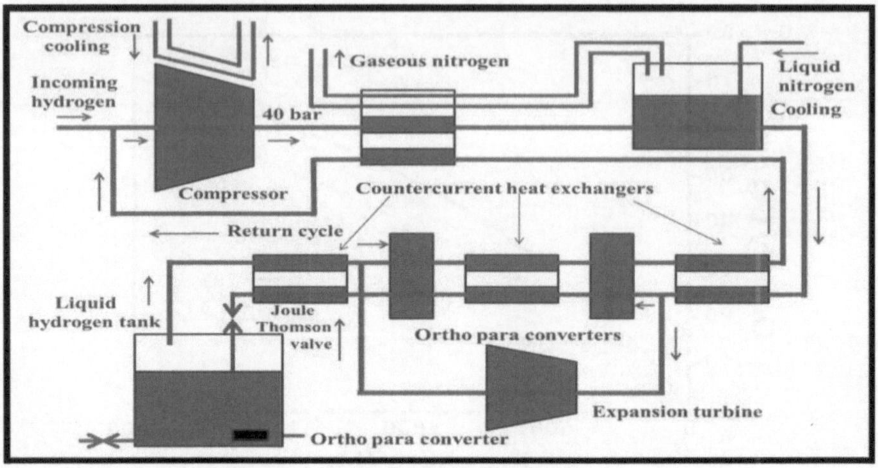

Figure 11.4 The Linde cycle or Joule–Thompson cycles.

The basic cycle of H_2 liquefaction is the Linde cycle, which takes advantage of the Joule–Thompson (J–T) effect (Abdi et al., 2019; Al Ghafri et al., 2022; Eckroll, 2017). This process is shown in Figure 11.4, where the gas is compressed, cooled in a heat exchanger, and then passed through a throttle valve to produce some liquid through an isenthalpic J–T expansion. After cooling, the gas is extracted from the liquid using a heat exchanger and fed back into the compressor. The J–T cycle works well for gases like nitrogen, whose inversion temperature is higher than room temperature (RT); however, the expansion of H_2 at RT causes warming. H_2 needs to be at a temperature lower than its inversion temperature (Ti) of 202 K in order to cool during expansion. It is usual practice to use liquid N_2 to pre-cool H_2 at 78 K before the initial expansion step. There is a spontaneous enthalpy change of 11640 kJ/kg when H_2 transitions from gas at 300 K to liquid at 20 K. About 15.2 kWh/kg of technical work is needed to liquefy H_2 at RT, which is almost half of the lower heating value of H_2 combustion. The energy requirement, calculated theoretically, is W^{th} = 3.23 kWh/kg. Certain characteristics of a liquid storage tank, like its size, shape, and thermal insulation, affect how quickly H_2 boils off of it. Since controlled proportional decreases in heat leakages due to the surface-to-volume ratio (SA:V), larger storage tanks have lower boil-off rates. For example, boil-off losses in double-walled, vacuum-insulated spherical dewars with a 50 m^3 capacity are usually 0.4% per day, but losses in 100 m^3 and 20,000 m^3 capacities are 0.2% and 0.06%, respectively. Systems for storing liquid H_2 are most useful in situations where the expense of H_2 is not as important and when the gas will be used quickly, like in space applications. This is because H_2 continues to boil out and liquefaction requires a significant amount of energy.

11.4 SOLID-STATE STORAGE OF HYDROGEN

Some substances have the capacity to store H_2 on their surface (adsorption) or inside their structure (absorption) (Blagojević et al., 2012; Kumar et al., 2023; Kumar et al., 2024; Langmi et al., 2003; Lee et al., 2000; Pandey et al., 2022; Rosi et al., 2003; Shukla et al., 2024; Verma et al., 2023, 2024; Yadav & Awasthi, 2022). By taking advantage of the bonds that exist between H atoms and molecules in a compound, it is possible to store larger amounts of H_2 in smaller volumes at lower pressures and temperatures that are close to ambient. Absorption is the process by which H_2 is chemically incorporated into the metals' and alloys' storage media (Kumar et al., 2022; Shukla et al., 2022; Verma et al., 2023; Yadav et al., 2022). On the other hand, adsorption takes place when H_2 sticks to the surface of porous materials (Shukla & Yadav, 2022; Verma et al., 2022). During absorption, H atoms attach themselves chemically to metal atoms and produce chemical hydrides. Based on the storage phenomenon, H_2 storage materials may be roughly divided into four types:

- Intermetallic or metal hydrides
- Complex hydrides
- Chemical hydrides
- High-surface-area sorbents

The process of retaining H_2 on a material's surface as either H_2 atoms (H) or H_2 molecules (H_2) is known as adsorption (Figure 11.5a). During absorption, H_2 molecules break apart to become hydrogen atoms, which are then incorporated into the solid lattice structure (Figure 11.5b, c). It could be possible to store more H in smaller amounts at room temperature and low pressure using this method. Tight bonds between H atoms can occur within molecular structures, as seen in the chemical compounds with H atoms in Figure 11.5d. The volumetric and gravimetric H_2 densities of several H_2 storage devices are presented in Figure 11.6.

11.5 INTERACTION OF H_2 WITH SURFACE OF SOLID

Physisorption is the term used to describe the adsorption of H_2 on a material's surface (Kumar et al., 2024; Shukla & Yadav, 2022).

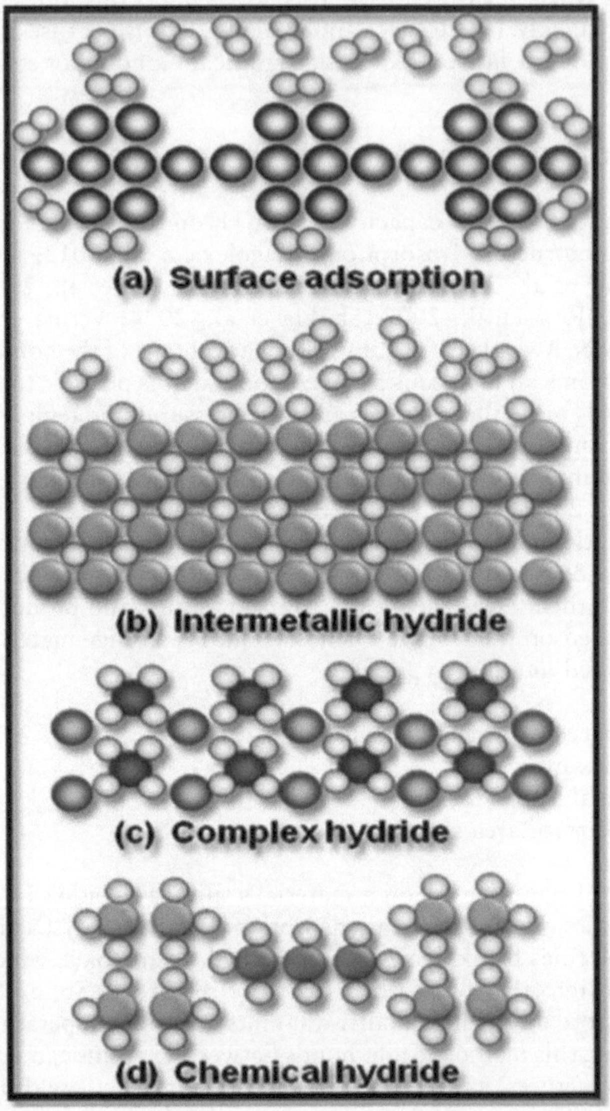

Figure 11.5 Materials for storing H_2 depending on the purpose of the storage.

On the surface of the metal, the process of creating solid solutions from molecular H_2 begins. Chemisorption is the process by which H_2 molecules dissociate at a metal surface and diffuse into the bulk subsequent to

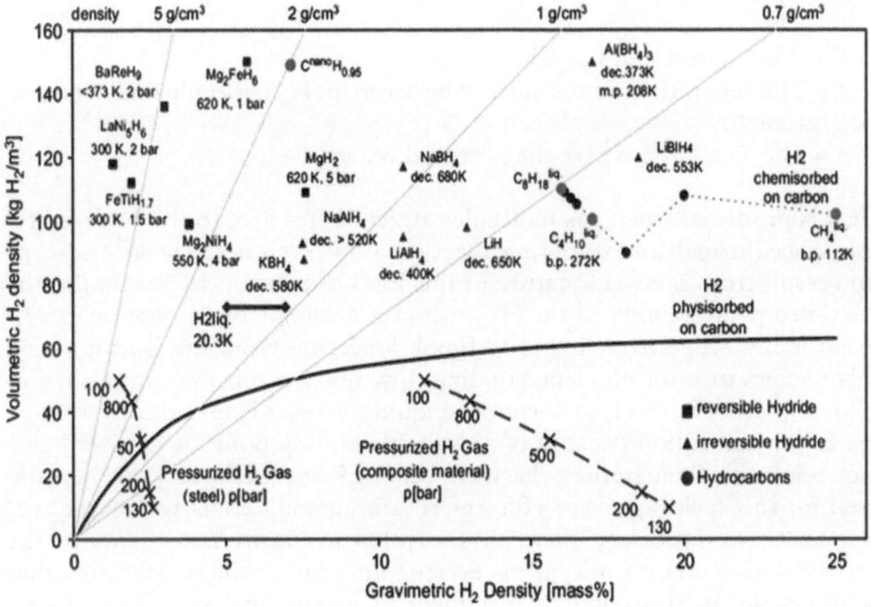

Figure 11.6 Densities of several H_2 storage techniques, both volumetric and gravimetric.

adsorption. This action at the metal surface transforms the H_2 (chemically) into atomic H prior to its absorption into the bulk.

11.6 PHYSISORPTION OF H_2

H_2 physically adsorbs onto the solid surface as a result of the diatomic molecule's retention there due to weak van der Waals' interactions among the gas molecules and the atoms (Dixit et al., 2017; Pandey et al., 2021; Patel et al., 2018; Verma et al., 2020). These forces between molecules are generated by the interaction of transitory dipoles. The combination of short-range repulsive interactions and attractive van der Waals' forces between a gas molecule and an atom on the adsorbent surface outcomes in a potential energy curve that may be described by the Lennard–Jones equation, which is shown below.

$$V(r_i) = 4\mathcal{E}\left[\left(\frac{\sigma}{r_i}\right)^{12} - \left(\frac{\sigma}{r_i}\right)^{6}\right]$$

where

$\mathscr{E}$ = The energy well's depth.
r_i = The separation on the surface between an H_2 molecule and a generic atom "i."
σ = The value of r_i where the potential is zero.

Heat is produced when gas molecules attach themselves to the surface of a substance through an exothermic reaction. Low levels of the heat of adsorption result from the weak nature of this kind of contact. In particular, the restricted polarizability of the H_2 causes the enthalpy of adsorption (ΔH_a) to normally range from 1 to 10 kJ/mol. Since the weak interactions need this temperature for efficient bonding, low temperatures are necessary to achieve considerable H_2 adsorption. Liquid nitrogen is useful for monitoring H_2 physisorption because of its suitable boiling point of –196.1°C at 1 atm pressure. Temperatures between –203.15 and –193.15°C are usually used for this technique. For efficient H_2 storage, materials with large surface areas are necessary since physisorption primarily takes place at the surface. Increasing porosity is necessary to maximize the surface area that is accessible. In this context, a number of porous materials have drawn attention, including graphene and, more recently, polymers with intrinsic microporosity (PIMs), high-surface-area (HSA) carbons, carbon nanotubes (CNTs), zeolites, and metal-organic frameworks (MOFs). These materials' varied architectures and qualities allow for customized methods for maximizing H_2 storage. A comparison of several characteristics for several kinds of physisorption materials is shown in Table 11.5. Porous materials exhibit good cyclability because the material structure usually does not change much during the adsorption/desorption cycle. As H_2 physisorption on HSA

Table 11.5 Features and attributes of common porous materials examples for storing H_2

Materials	*Typical framework elements*	*Surface area* (m^2g^{-1})	*Porosity* (cm^3g^{-1})	*Excess hydrogen capacity at 77 K and 20 bar* (wt%)	(kgH_2l^{-1})
Porous carbons	C	3150	1.95	6.9	—
CNTs	C	1160	—	3.8	—
Zeolites	O, Al, Si	670	—	2.2	0.031
MOFs	H, O, C, TM	2200	0.89	6.1	0.039
PIMs	H, O, C	1050	0.40	2.7	—
Graphene	C	925	—	1.38	—

materials is exothermic, liquid nitrogen must be used to keep the temperature down during rehydrogenation procedures.

Porous H_2 storage materials are less desirable for on-board vehicle storage when taken as a whole. However, their use as a more affordable substitute for liquid H_2 storage in stationary applications is gaining traction. If H_2 bound to the substrate more firmly, room temperature H_2 storage with an ideal enthalpy of contact of about 30 kJ/mol would be possible, eliminating the need for cryogenic temperatures.

11.7 CHEMISORPTION OF H_2 IN METALS

Among the several types of chemical storage systems, interstitial hydrides offer the safest way to store H_2. This section will describe the principles of the metal–H_2 interaction that occurs during chemical H_2 storage (Mishra et al., 2020; Pandy et al.,, 2017; Shahi et al., 2015, 2017; Shukla et al., 2015, 2019). A schematic representation of the interstitial hydride production is shown in Figure 11.7.

The weak van der Waals' interaction causes the H_2 molecules to be adsorbed when they first approach the surface of the alloy or intermetallic compound, as shown in Figure 11.7(a). Subsequently, as illustrated in Figure 11.7c, the physisorbed H_2 molecules dissociate into individual H_2 atoms and are chemically absorbed at the surface. An interstitial solid solution is then created as a result of the H_2 atoms diffusing into the bulk and moving to the interstitial positions of the alloy or intermetallic compound (Figure 11.7d). The alpha (α) phase is the name given to the solid state. H_2 atoms in the interstitial region may generate a hydride phase by trapping more H_2 atoms, which could change the metal's crystal structure. This formation is known as the beta (β) phase. Dehydrogenation releases H_2 atoms, which rise to the surface from the bulk. At the surface, the H atoms recombine to form the H_2 molecule, which then desorbs into H_2 gas.

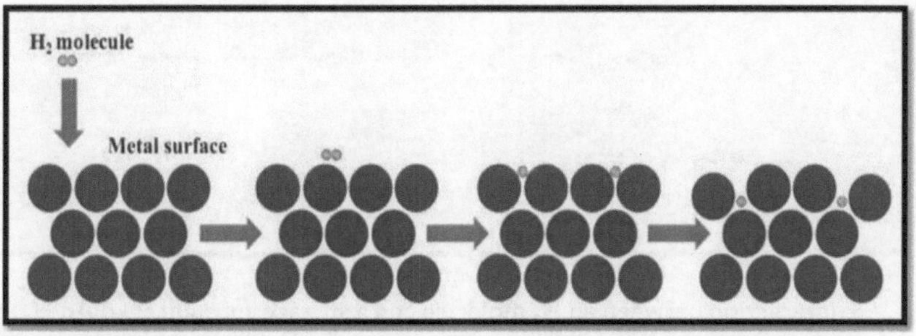

Figure 11.7 The processes through which interstitial hydride is formed.

11.8 H_2 STORAGE IN METALS AND INTERMETALLIC

Many metals' interstitial regions have the capacity to hold large amounts of H_2 (Yadav et al., 2012). These metals can be charged with H_2 using molecular H_2. Molecular H_2 splits into H atoms close to the metal surface before being stored in the interstitial sites. During the desorption process, two H atoms recombine to produce the H_2 molecule H_2. H_2 reacts with metal via a series of stages to generate metal hydride. Figure 11.8 shows a schematic representation of the synthesis of metal hydride.

Once the metal surface has undergone chemisorption-induced dissociation, the H atoms usually diffuse rapidly through the bulk metal to form an M-H solid solution, which is also referred to as the α-phase. H_2 atoms will occupy interstitial regions following hydrogenation. Figure 11.9 displays the tetrahedral (T) and octahedral (O) interstitial gaps of the three major hydride crystal forms. The fourfold symmetry axis is used to separate the "O" sites into Ox, Oy, and Oz sites. Similarly, a BCC lattice's "T" sites are separated into their symmetry axes, Ty, Tz, and Tx. Every H_2 metal atom has six tetrahedral and three octahedral positions. In BCC metals, H_2 will preferably occupy the tetrahedral positions. The sites will fill up from the lowest to the greatest energies during hydrogenation. However, due to the H atoms' mutual repulsion, some sites will continue to be vacant. Saturation happens as the H_2 pressure rises, and the metal hydride MHx begins to develop. Absorption is the term for the reaction of H_2 gas with a metal.

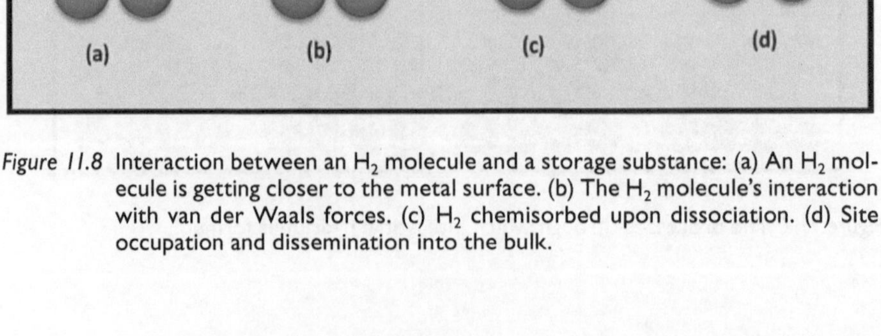

Figure 11.8 Interaction between an H_2 molecule and a storage substance: (a) An H_2 molecule is getting closer to the metal surface. (b) The H_2 molecule's interaction with van der Waals forces. (c) H_2 chemisorbed upon dissociation. (d) Site occupation and dissemination into the bulk.

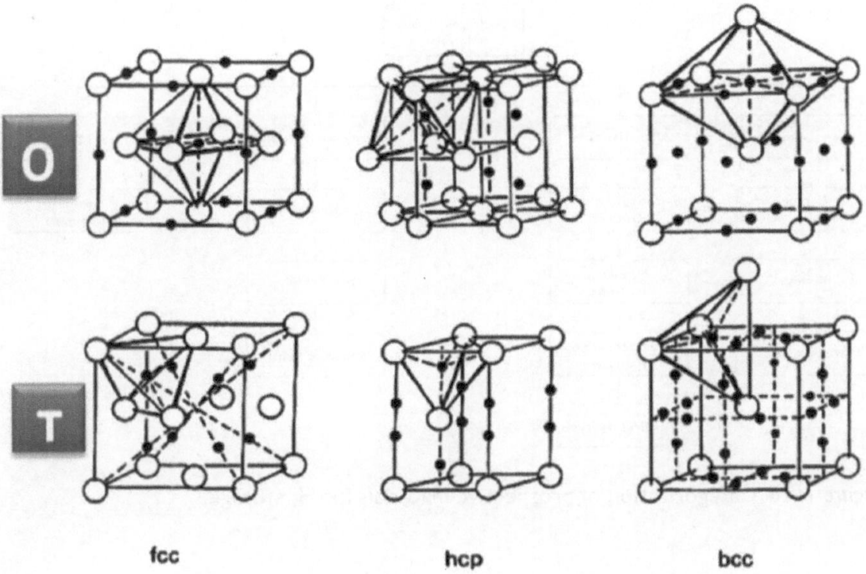

Figure 11.9 Metals of the BCC, FCC, and HCP types have interstitial sites that are octahedral (O) and tetrahedral (T).

11.9 MATERIALS OF CHOICES FOR H_2 STORAGE

The classification of H_2 storage materials is based on the type of material used to store H_2. Figure 11.10 shows the various types of materials for storing H_2.

11.10 H_2 STORAGE IN POROUS MATERIALS

There are different allotropic forms of carbon, each with a unique structure. The most well known are non-porous materials, such graphite and diamond. But there are also many porous carbon nanostructures, most of which are made of benzene-like hexagonal carbon units. These structures are distinguished from one other by the configuration of these hexagons inside the material. Using a well-ordered inorganic porous prototype, like zeolites or mesoporous silica, on which the carbon precursor is placed and then carbonized, is one technique that produces porous carbons with a distribution of pore sizes. The carbon sample that remains after the inorganic template is removed has a regular pore structure. The diversity of carbon nanostructures is best illustrated by Iijima's groundbreaking finding of carbon nanotubes in 1991. Beyond gravimetric or volumetric uptake, two metrics stand out in evaluating the quantity of H_2 stored in absorbent

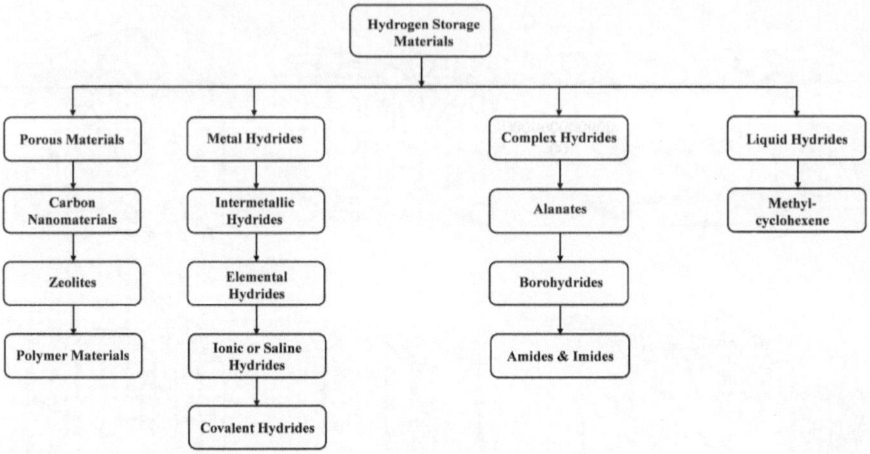

Figure 11.10 Categorization of prospective materials for H_2 storage.

materials: (a) absolute storage capacity and (b) surplus storage capacity. The relationship between absolute adsorption (N_{ads}) and excess adsorption (N_{ex}) can be easily determined using a conventional adsorption experiment. The quantity of gas adsorbed on the sample (N_{ads}) is calculated by deducting the free molecules in the gas phase from the total gas supplied in the sample cell (N_i).

$$N_{ads} = N_i - V_g \rho_g$$

where V_g is the free volume engaged by the gas together with the pore volume and ρ_g is the gas density.

11.10.1 Zeolites

Zeolites are specifically defined as crystalline microporous minerals that fall under the class of alumo silicates; however, the name also encompasses other compositions, such as alumina phosphates. Natural zeolites are found in more than 50 different varieties. Their microporous structure appears to be perfect for H_2 adsorption in principle. Since the pore size of zeolites is almost the same as that of a H_2 molecule, heat cycling can be used to either adsorb or desorb H_2 (Muduli & Kale, 2023). With a projected pore system of 0.37 cm^3g^{-1} for the pure silica polymorph, ITQ-33 is the zeolite material with the maximum micropore volume. If H_2 were injected at liquid H_2 density, the system's storage capacity would be restricted to about 2–3 wt%. This performance is much below the benchmark set by the DOE.

11.10.2 Metal-organic frameworks (MOFs)

The category of porous materials that include materials with an organic and metal structure. Together with organic molecules—most frequently polyvalent carboxylic acids—which serve as organic vertices linking the metal ions, they are composed of nitrogen donor linkers. They are notable because of the massive pore volumes and very large surface areas of metal-organic frameworks. MOFs offer greater design freedom than zeolites since they can change the ligand linkages. For this reason, using metal-organic framework in gas storage applications is appealing. Numerous investigations have been carried out to evaluate metal-organic framework's capacity to store H_2 throughout the last few decades. For the first time, MOF-5 was shown to be able to store up to 4.5 wt% of H_2 at 78 K and 20 bar H_2 pressure by Rosi et al. in their investigation into the H_2 storage properties of metal-organic framework materials (Rosi et al., 2003). Subsequently, MOF-5 became the focus of other research studies concerning the storage of H_2. Moreover, these materials may degrade if they come into contact with water or air. High surface area MOF materials need to be more stable under typical operating circumstances in order to be used in automobiles. Zeolitic imidazolate frameworks (ZIFs) perform better than MOFs in terms of chemical and thermal stabilities, according to current research. Zeolitic imidazolate framework-8 (ZIF-8) has been reported to exhibit a maximum H_2 uptake of 3.3% at 77 K and 30 bar. To be a promising material for H_2 storage, exceptionally stable MOFs in bulk scale (very porous) with very large surface area require accessible production processes.

11.10.3 Polymers with intrinsic microporosity

Potential exists for H_2 storage materials with intrinsic microporosity in polymers (McKeown et al., 2006; Ramimoghadam et al., 2016). The polymer based on a triptycene monomer had the highest H_2 storage capacity, reaching up to 2.7 wt% H_2 at 77 K under 1 MPa. It is possible to manufacture a wide variety of polymers with BET surface areas ranging from 400 to 1100 m^2/g. Wide-surface-area polymers have a substantial H_2 storage capability. One of the primary problems with these kinds of storage systems is the low temperature required for H_2 storage.

11.10.4 Metal hydrides for H_2 storage

Between 0.5 and 2 wt% of H_2 molecules can be retained by metal hydrides, making them suitable for usage in automobiles (Sakintuna et al., 2007). Hydrides exhibit an exothermic reaction when charged with compressed H_2; conversely, an endothermic process occurs when heat is absorbed and H_2 is released. There are a ton of possibilities for hydride compounds because almost any metallic element in the periodic table that is greater than 50 can absorb H_2 in considerable amounts. But only a few are suitable for holding H_2 at the required temperature and pressure. The materials can store H_2 at normal temperature and release it below 100°C, respectively, in a realistic manner. In order to improve kinetics and alter thermodynamics, alloying is added to all intermetallic compounds used in practical compositions, which are mixes of elements "A" (hydride forming element) and elements "B" (no hydride forming element). Intermetallic compounds come in four varieties: AB, A_2B, AB_2, and AB_5. Fe, Ni, Mn, Al, and Cu are alloying elements; La, Ti, Mg Ca, Ce, Sm, and Zr are hydride formers. The first material was discovered in the past by Philips Laboratories (Holland) as the AB_5 compound $LaNi_5$. The two essential components of the AB and A_2B kinds have been determined to be TiFe and Mg_2Ni by Brookhaven National Laboratories (USA). There's a lot to think about when selecting an alloy for a specific use. The plateau slope, plateau pressure at various temperatures, and pressure difference between the absorption and desorption isotherms (hysteresis) are all determined by the thermodynamic properties. The heat of reaction determines how much energy is required to bring about these changes. The following characteristics of metal hydrides are crucial for applications involving the storage of H_2:

1. Easy of activation
2. Plateau pressures at different temperature
3. Plateau slope
4. Hydriding and dehydriding kinetics
5. Absorption desorption hysteresis
6. H_2 capacity
7. Ductility
8. Volume change on hydriding
9. Mechanical strength
10. Stability in cycling
11. Tolerance to gaseous impurities

Common intermetallic alloys that can store H_2 at pressures ranging from 0.1 to 1.5 MPa are shown in Table 11.6. The special group includes materials with x being a variable from the AB_5, AB_2, and AB types, such as $LaNi_5$_xAlx, $FeTi_1$_xMnx, TiV_2_xMnx, etc.

Table 11.6 The amount of energy in intermetallic hydrides

Media type	*Material*	*Hydride form*	*H_2 capacity (wt%)*	*Energy density (kJ/kg hydride)*	*Enthalpy of hydrogenation (kJ/mol H_2)*
AB	TiFe	$TiFeH_{1.95}$	1.7	2560	28.1
AB_2	$ZrMn_2$	$ZrMn_2H_4$	1.7	2419	53.2
A_2B	Mg_2Ni	Mg_2NiH_4	7.0	10,0000	64.5
AB_5	$LaNi_5$	$LaNi_5H_6$	1.4	1850	30.1
Metal	Mg	MgH_2	7.7	11,0000	74.2

11.10.5 Alkali metal hydrides

Group I of the periodic table contains alkali metals, which produce hydrides with a high degree of ionic bonding property. High breakdown temperatures are caused by the metal's strong bond with H_2. Compared to the alkaline metals, which are light elements that crystallize in a BCC lattice arrangement, the hydrides have an even lower density. This is partly explained by the way the hydrides are arranged in an FCC lattice, which has a higher packing density (74%), compared to BCC (68%). Because the alkali metal hydrides melt at low temperatures above the melting point of the metal, they dissolve into pure metal and H_2. One potential material for storing H_2 is magnesium hydroxide (MgH_2), which has an H_2 content of 7.6 wt% and a more complicated bonding structure. A significant negative enthalpy (–75 kJ mol^{-1} H_2) in the synthesis of MgH_2 results in a desorption temperature of roughly 300°C at 1 bar of H pressure. While beryllium and magnesium crystallize in the HCP structure, their equivalent hydrides take on tetragonal and orthorhombic structures, respectively. The hydrides of the alkaline earth metals Be, Mg, and Ca have band gaps ranging from 3 to 6 eV, making them insulators. The alkaline earth metal hydrides must be heated to high temperatures in order to separate into metal and H_2.

11.10.6 Transition metal hydrides

The physical characteristics of transition metal hydrides are significantly more diverse than those of Group I and II elements in the periodic table. Early transition metal hydrides are frequently rather stable; however, the hydrides lose stability when the metals shift to the right. For instance, TiH_2 requires a temperature greater than 450°C to break down. Another important difference between the hydrides of transition metals and those of major groups I and II is their metallic bonding behavior. Because magnesium's H_2 is readily dissociable, the impact of low concentrations of transition metals on the kinetics of magnesium's hydrogenation and dehydrogenation has been well investigated. Less stable hydrides are produced when magnesium hydroxide (MgH_2) is destabilized by alloying with transition metals; nonetheless, no magnesium-based metal hydride that permits H_2 desorption below 250°C has been found.

11.10.7 Complex hydrides for H_2 storage

Compounds such as alanates, imides, amides, and boro-hydrides are examples of complex hydrides that have large volumetric and gravimetric H_2 capacities (Bogdanović et al., 2009; Jepsen et al., 2012; Orimo et al., 2007). Complex hydrides, which were once thought to be irreversible, are now seen as attractive materials for storing H_2 because of their remarkable storage capabilities. The current focus of study is on doping these complex hydrides with appropriate catalysts in order to improve their H_2 storage capabilities by lowering the activation energy barrier for H_2 absorption/desorption. A range of catalysts, such as compounds based on transition metals, rare-earth metals, and carbon nanostructures, have demonstrated potential in enhancing the sorption characteristics of complicated hydrides. Catalyst doping can reduce the high activation energy (E_a) of complex hydrides without changing the enthalpy (ΔH). In desorption reactions, lowering the reaction enthalpy may result in an increase in the desorbed H_2 flux. Optimizing the enthalpies of reactions through scaffolding, nanocondensation, and nanocomposites creation could lead to better kinetics of H_2 desorption at the appropriate pressure and temperature ranges. It is imperative to lower the H_2 desorption activation energy (E_a) and improve the thermodynamics of H_2 processes in order to employ complex hydrides as reversible H_2 storage materials. Both goals can be accomplished with nanotechnology, thanks to its ability to precisely control particle size and modify structural elements. The complicated dynamics of hydrides are a result of their complexity. The slowest step in the sequence determines the kinetics of the subsequent elementary reaction steps that lead to desorption of H_2 from complex hydrides. H_2 extraction from single-phase complex hydrides requires a number of prerequisites, such as the dissociation of the ionic bond between cationic metallic ions and anionic ligands, atomic rearrangement resulting in the growth and nucleation of the product phase, and the production and detachment of H_2 molecules from particle surfaces. H_2 desorption from complex hydrides is further complicated by the thermodynamic potential outcomes and interaction processes.

In the last ten years, many methods have been used to solve H_2 desorption kinetics, including scaffolding-based nanoconfinement, in-situ product seeding, catalyst admixing, and adjusting reaction enthalpy through interactions with other materials. Merely $NaAlH_4$ demonstrates thermodynamic reversibility among alanates within reasonable operating temperatures and pressure ranges. $TiCl_3$ doping of $NaAlH_4$ allows for effective recharging at 120°C and 100 bar H_2. Furthermore, at about 110°C, nanocluster Ti compound-doped $NaAlH_4$ shows better H_2 desorption kinetics due to its smaller particle size. $LiAlH_4$ hasn't been as researched as $NaAlH_4$, which has drawn a lot of attention for optimizing reversible capacity through increased doping and catalyst admixing. Nevertheless, theoretical and experimental research has provided important new insights into its thermal breakdown behavior. Furthermore, as compared to $LiAlH_4$, $NaAlH_4$ exhibits higher thermodynamic stability. Generally, complex hydrides go through a multi-step process of heat decomposition that can be stated as:

$$MAlH_4 \rightarrow Al + \frac{3}{2}H_2 \rightarrow M + \frac{1}{2}H_2$$

Complex hydrides are suitable for a variety of applications due to their huge H_2 storage capacity and wide range of thermodynamic stabilities, which vary depending on the metal combination. Borohydrides are not only more stable than alanates, but they also show greater stability throughout the Li, Na, and K series. Figure 11.11 shows the theoretical H_2 concentration of a few selected complex hydrides.

The finding of reversibility in the complex hydrides proved revolutionary for solid-state H_2 storage, despite the slow kinetics. About twice as

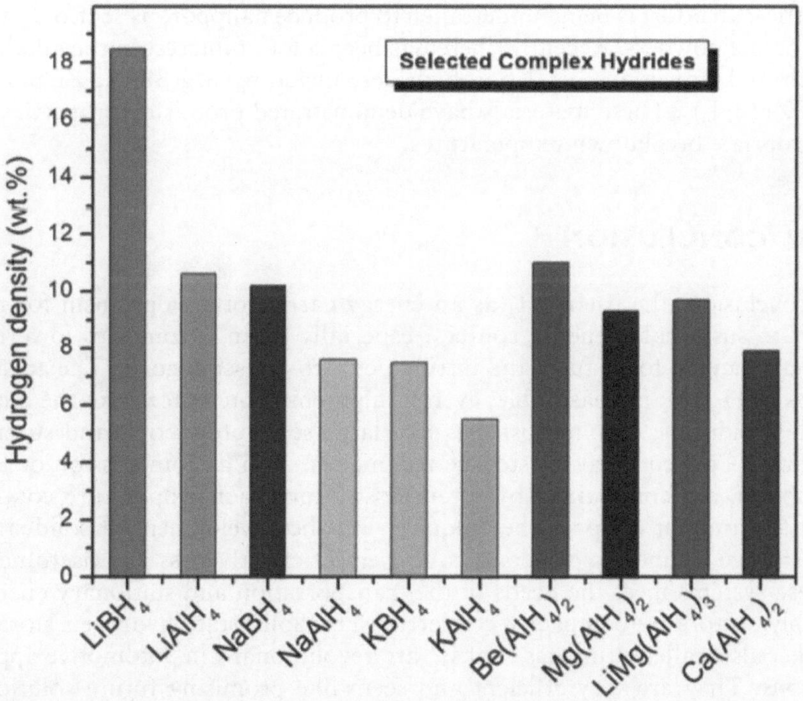

Figure 11.11 Selective complex hydrides with theoretical H_2 density.

much reversible energy can be stored in the catalyzed Na alanate system as in a typical metal hydride. Together with $NaAlH_4$ and Na_3AlH_6, other alanates that can undergo rehydrogenation at the required temperatures and pressures are Na_2LiAlH_6, $KAlH_4$, K_3AlH_6, K_2NaAlH_6, and K_2LiAlH_6. Nevertheless, the gravimetric H_2 storage densities of these materials are all quite low. It is known that many alanates have larger gravimetric H_2 densities than sodium alanate and are less stable than sodium alanate. These alanates include $LiAlH_4$, Mg $(AlH_4)_2$, Li_3AlH_6, Ca $(AlH_4)_2$, and Ti $(AlH_4)_4$.

11.10.8 Borohydrides

The search for a high-capacity reversible/cyclic hydride has shifted its attention to borohydrides. However, high temperatures are required for the release of H_2 from borohydrides (Kojima, 2005; Li et al., 2011). The well-known borohydride $LiBH_4$ undergoes the following reaction, which yields 13.5 wt% H_2 at 438°C after dissolving into LiH and B:

$$2LiBH_4 \rightarrow 2LiH + 2B + H_2$$

The outstanding thermal stability of the borohydrides is driving current research attempts to minimize the breakdown. To improve the reaction kinetics, an effort is being undertaken to produce nanoporous scaffolds and nanoscale powders. Recently, there has been a lot of interest in the alkaline earth and transition metal borohydrides, including $Mg(BH_4)_2$, $Zr(BH_4)_4$, and $Zn(BH_4)_2$. These materials have demonstrated promising properties at appropriate breakdown temperatures.

11.11 CONCLUSION

In conclusion, the use of H_2 as an energy carrier offers optimism for the shift to sustainable energy sources, especially when it comes to lowering dependency on fossil fuels and hitting net zero-emission goals. The advantages of H_2 are immeasurable, even in high-emission sectors like the automobile industry, despite obstacles like large-scale production and storage security. Conventional H_2 storage techniques, such as compression or liquefaction, are constrained by the necessity for exact temperature control and the amount of space they require. But the development of solid-state H_2 storage technology ushers in a new era of effectiveness and usefulness. These systems meet the needs of the transportation and stationary energy supply sectors by compactly converting H_2. Solid-state hydrogen storage tanks, also called "tiny gas tanks," are revolutionary in automotive applications. They are very efficient and seem like promising future solutions because they can function at reasonable temperatures and allow for quick

H_2 absorption and release. This book chapter presents a thorough analysis that makes it clear that solid-state H_2 devices have the ability to completely change the energy landscape. They are essential to achieving our shared goal of a sustainable future free of emissions, in addition to providing a competitive substitute for fossil fuels. A cleaner and more robust energy ecosystem for future generations is within reach as we advance the development and application of these technologies.

REFERENCES

Abbas, H. F., & Daud, W. W. (2010). Hydrogen production by methane decomposition: A review. *International Journal of Hydrogen Energy*, *35*(3), 1160–1190.

Abdi, A., Chiu, J., & Martin, V. (2019). State of the art in hydrogen liquefaction. In Abdi, AmirChiu, Justin NingWeiMartin, Viktoria (Eds.), *Proceedings of the ISES solar world congress 2019* (pp. 1–10).

Al Ghafri, S. Z., Munro, S., Cardella, U., Funke, T., Notardonato, W., Trusler, J. M., ... May, E. F. (2022). Hydrogen liquefaction: A review of the fundamental physics, engineering practice and future opportunities. *Energy & Environmental Science*, *15*(7), 2690–2731.

Bargigli, S., Raugei, M., & Ulgiati, S. (2004). Comparison of thermodynamic and environmental indexes of natural gas, syngas and hydrogen production processes. *Energy*, *29*(12–15), 2145–2159.

Beebe, R. A., Biscoe, J., Smith, W. R., & Wendell, C. B. (1947). Heats of adsorption on carbon black. I. *Journal of the American Chemical Society*, *69*(1), 95–101.

Blagojević, V. A., Minić Dejan, G., Grbović Novaković, J., & Minic Dragica, M. (2012). Hydrogen economy: Modern concepts, challenges and perspectives. In Dragica Minic (Ed.), *Hydrogen energy–challenges and perspectives* (pp. 3–28).

Bogdanović, B., Felderhoff, M., & Streukens, G. (2009). Hydrogen storage in complex metal hydrides. *Journal of the Serbian Chemical Society*, *74*(2), 183–196.

Chen, P., Wu, X., Lin, J., & Tan, K. L. (1999). High H2 uptake by alkali-doped carbon nanotubes under ambient pressure and moderate temperatures. *Science*, *285*(5424), 91–93.

Cunanan, C., Tran, M. K., Lee, Y., Kwok, S., Leung, V., & Fowler, M. (2021). A review of heavy-duty vehicle powertrain technologies: Diesel engine vehicles, battery electric vehicles, and hydrogen fuel cell electric vehicles. *Clean Technologies*, *3*(2), 474–489.

Dillon, A. C., Gennett, T., Alleman, J. L., Jones, K. M., Parilla, P. A., & Heben, M. J. (2000). Carbon nanotube materials for hydrogen storage. In *Proc. 2000 DOE/NREL Hydrogen program review* (pp. 8–10) NREL/CP-570-28990 Sponsored by the Office of Power Delivery Systems, Office of Power Technologies, U.S. Department of Energy Presented at the 2000 U.S. DOE Hydrogen Program Review San Ramon, California May 9–11, 2000. National Renewable Energy Laboratory 1617 Cole Boulevard Golden, Colorado 80401-3393 NREL is a U.S. Department of Energy Laboratory Operated by Midwest Research Institute • Battelle • Bechte.

Dixit, V., Bhatnagar, A., Pandey, S., Yadav, T. P., Shaz, M. A., Sinha, A. S. K., & Srivastava, O. N. (2017). Carbon derived from solid kernel of coconut: Its structural, microstructural characteristics and use as new Hydrogen storage material. *Materials Focus*, *6*(5), 502–507.

Dönitz, W., & Erdle, E. (1985). High-temperature electrolysis of water vapor—status of development and perspectives for application. *International Journal of Hydrogen Energy*, *10*(5), 291–295.

Eckroll, J. (2017). *Concepts for large scale hydrogen liquefaction plants* (Master's thesis), NTNU.

Elam, C. C. (1996). IEA agreement on the production and utilization of Hydrogen. *USA Annual Report.*

Fan, Y. Y., Liao, B., Liu, M., Wei, Y. L., Lu, M. Q., & Cheng, H. M. (1999). Hydrogen uptake in vapor-grown carbon nanofibers. *Carbon*, *37*, 1649–1652.

Gregg, S. J., Sing, K. S. W., & Salzberg, H. W. (1967). Adsorption surface area and porosity. *Journal of the Electrochemical Society*, *114*(11), 279Ca.

Güneş, İ. (2019). Hydrogen production from renewable energy sources and its storage. *Research & Reviews in Engineering – Summer*, *2019*, 41.

Hirscher, M., Becher, M., Haluska, M., Dettlaff-Weglikowska, U., Quintel, A., Duesberg, G. S., ... Bernier, P. (2001). Hydrogen storage in sonicated carbon materials. *Applied Physics A*, *72*, 129–132.

Hosseini, S. E., & Wahid, M. A. (2016). Hydrogen production from renewable and sustainable energy resources: Promising green energy carrier for clean development. *Renewable and Sustainable Energy Reviews*, *57*, 850–866.

Hosseini, S. E., & Wahid, M. A. (2020). Hydrogen from solar energy, a clean energy carrier from a sustainable source of energy. *International Journal of Energy Research*, *44*(6), 4110–4131.

Huston, E. L. (1984, July). *Liquid and solid storage of hydrogen.* Hydrogen Energy Progress V, Proceedings of the 5th World Hydrogen Energy Conference, Toronto, Canada.

Incer-Valverde, J., Korayem, A., Tsatsaronis, G., & Morosuk, T. (2023). "Colors" of hydrogen: Definitions and carbon intensity. *Energy Conversion and Management*, *291*, 117294.

Jena, P. (2011). Materials for hydrogen storage: Past, present, and future. *The Journal of Physical Chemistry Letters*, *2*(3), 206–211.

Jepsen, J., von Colbe, J. M. B., Klassen, T., & Dornheim, M. (2012). Economic potential of complex hydrides compared to conventional hydrogen storage systems. *International Journal of Hydrogen Energy*, *37*(5), 4204–4214.

Kern, D. J. (2021). Atmospheric chemistry: An overview–ozone, acid rain, and greenhouse gases. In S. Schroth & J. Daniels (Eds.), *Building STEM skills through environmental education* (pp. 172–218). IGI Global. https://doi.org/10.4018/978-1-7998-2711-5.ch00.

Klell, M. (2010). Storage of hydrogen in the pure form. In Dr. Michael Hirscher (Ed.), *Handbook of hydrogen storage* (pp. 1–37). Wiley.

Kojima, Y. (2005). Hydrogen storage and generation using sodium borohydride. *r&d Review of Toyota CRDL*, *40*(2), 31–36.

Kumar, A., Yadav, T. P., & Mukhopadhyay, N. K. (2022). Notable hydrogen storage in Ti–Zr–V–Cr–Ni high entropy alloy. *International Journal of Hydrogen Energy*, *47*(54), 22893–22900.

Kumar, A., Yadav, T. P., Shaz, M. A., & Mukhopadhyay, N. K. (2023). Hydrogen storage performance of C14 type Ti0. 24V0. 17Zr0. 17Mn0. 17Co0. 17Fe0. 08 high entropy intermetallics. *Transactions of the Indian National Academy of Engineering*, 1–9 .

Kumar, A., Yadav, T. P., Shaz, M. A., & Mukhopadhyay, N. K. (2024). Hydrogen storage properties in rapidly solidified TiZrVCrNi high-entropy alloys. *Energy Storage*, *6*(1), e532.

Langmi, H. W., Walton, A., Al-Mamouri, M. M., Johnson, S. R., Book, D., Speight, J. D., ... Harris, I. R. (2003). Hydrogen adsorption in zeolites A, X, Y and RHO. *Journal of Alloys and Compounds*, *356*, 710–715.

Lee, S. M., Park, K. S., Choi, Y. C., Park, Y. S., Bok, J. M., Bae, D. J., ... Lee, Y. H. (2000). Hydrogen adsorption and storage in carbon nanotubes. *Synthetic Metals*, *113*(3), 209–216.

Leung, W. B., March, N. H., & Motz, H. (1976). Primitive phase diagram for hydrogen. *Physics Letters A*, *56*(6), 425–426.

Li, H. W., Yan, Y., Orimo, S. I., Züttel, A., & Jensen, C. M. (2011). Recent progress in metal borohydrides for hydrogen storage. *Energies*, *4*(1), 185–214.

Lim, K. L., Kazemian, H., Yaakob, Z., & Daud, W. W. (2010). Solid-state materials and methods for hydrogen storage: A critical review. *Chemical Engineering & Technology: Industrial Chemistry-Plant Equipment-Process Engineering-Biotechnology*, *33*(2), 213–226.

Liu, C., Fan, Y. Y., Liu, M., Cong, H. T., Cheng, H. M., & Dresselhaus, M. S. (1999). Hydrogen storage in single-walled carbon nanotubes at room temperature. *Science*, *286*(5442), 1127–1129.

Ma, Y., Xia, Y., Zhao, M., Wang, R., & Mei, L. (2001). Effective hydrogen storage in single-wall carbon nanotubes. *Physical Review B*, *63*(11), 115422.

Makridis, S. (2017). Hydrogen storage and compression. *arXiv preprint arXiv:1702.06015*.

McKeown, N. B., Gahnem, B., Msayib, K. J., Budd, P. M., Tattershall, C. E., Mahmood, K., ... Walton, A. (2006). Towards polymer-based hydrogen storage materials: engineering ultramicroporous cavities within polymers of intrinsic microporosity. *Angewandte Chemie*, *118*(11), 1836–1839.

Mishra, S. S., Yadav, T. P., Mukhopadhyay, N. K., & Srivastava, O. N. (2020). Synthesis of fine skeletal structure on Al–Cu–Co decagonal quasicrystals for hydrogen production through steam reforming of methanol. *International Journal of Hydrogen Energy*, *45*(46), 24491–24501.

Muduli, R. C., & Kale, P. (2023). Synergetic effect of porous silicon–Nickel composite on its solid-state hydrogen energy storage properties. *International Journal of Hydrogen Energy*, *48*(90), 35185–35196.

Muratov, D. S., & Gromov, S. (2017). Evaluating hydrogen uptake for two types of multi-wall carbon nanotubes from nitrogen adsorption/desorption data. *Nano Hybrids and Composites*, *13*, 341–347.

Nijkamp, M. G., Raaymakers, J. E. M. J., Van Dillen, A. J., & De Jong, K. P. (2001). Hydrogen storage using physisorption–materials demands. *Applied Physics A*, *72*, 619–623.

Olabi, A. G., Abdelghafar, A. A., Baroutaji, A., Sayed, E. T., Alami, A. H., Rezk, H., & Abdelkareem, M. A. (2021). Large-scale hydrogen production and storage technologies: Current status and future directions. *International Journal of Hydrogen Energy*, *46*(45), 23498–23528.

Orimo, S. I., Nakamori, Y., Eliseo, J. R., Züttel, A., & Jensen, C. M. (2007). Complex hydrides for hydrogen storage. *Chemical Reviews*, *107*(10), 4111–4132.

Pandey, S. K., Bhatnagar, A., Mishra, S. S., Yadav, T. P., Shaz, M. A., & Srivastava, O. N. (2017). Curious catalytic characteristics of Al–Cu–Fe quasicrystal for de/rehydrogenation of MgH2. *The Journal of Physical Chemistry C*, *121*(45), 24936–24944.

Pandey, S. K., Bhatnagar, A., Shahi, R. R., Yadav, T. P., & Srivastava, O. N. (2014). Co-catalytic effect of carbon based nanostructures and TiO2 on sorption behavior of nanocrystalline MgH2. *Advanced Science Letters*, *20*(5–6), 1120–1123.

Pandey, S. K., Bhatnagar, A., Shukla, V., Kesarwani, R., Deshpandey, U., & Yadav, T. P. (2021). Catalytic mechanism of TiO2 quantum dots on the de/re-hydrogenation characteristics of magnesium hydride. *International Journal of Hydrogen Energy*, *46*(75), 37340–37350.

Pandey, S. K., Verma, S. K., Bhatnagar, A., & Yadav, T. P. (2022). Catalytic characteristics of titanium-(IV)-isopropoxide (TTIP) on de/re-hydrogenation of wet ball-milled MgH2/Mg. *International Journal of Energy Research*, *46*(12), 17602–17615.

Patel, C. R. P., Tripathi, P., Vishwakarma, A. K., Talat, M., Soni, P. K., Yadav, T. P., & Srivastava, O. N. (2018). Enhanced hydrogen generation by water electrolysis employing carbon nano-structure composites. *International Journal of Hydrogen Energy*, *43*(6), 3180–3189.

Ramachandran, R., & Menon, R. K. (1998). An overview of industrial uses of hydrogen. *International Journal of Hydrogen Energy*, *23*(7), 593–598.

Ramimoghadam, D., Gray, E. M., & Webb, C. J. (2016). Review of polymers of intrinsic microporosity for hydrogen storage applications. *International Journal of Hydrogen Energy*, *41*(38), 16944–16965.

Rosi, N. L., Eckert, J., Eddaoudi, M., Vodak, D. T., Kim, J., O'Keeffe, M., & Yaghi, O. M. (2003). Hydrogen storage in microporous metal-organic frameworks. *Science*, *300*(5622), 1127–1129.

Rzepka, M., Lamp, P., & De la Casa-Lillo, M. A. (1998). Physisorption of hydrogen on microporous carbon and carbon nanotubes. *The Journal of Physical Chemistry B*, *102*(52), 10894–10898.

Sakintuna, B., Lamari-Darkrim, F., & Hirscher, M. (2007). Metal hydride materials for solid hydrogen storage: A review. *International Journal of Hydrogen Energy*, *32*(9), 1121–1140.

Shahi, R. R., Bhatanagar, A., Pandey, S. K., Shukla, V., Yadav, T. P., Shaz, M. A., & Srivastava, O. N. (2015). MgH2–ZrFe2Hx nanocomposites for improved hydrogen storage characteristics of MgH2. *International Journal of Hydrogen Energy*, *40*(35), 11506–11513.

Shahi, R. R., Yadav, T. P., Shaz, M. A., & Srivastava, O. N. (2017, February). Synthesis characterization and hydrogenation behaviour of as quenched Ti41. 5+ XZr41. 5-XNi17 (x= 0, 3.5, 11.5 and 13.5) nano quasicrystalline ribbons. In An Pang Tsai, Hem Raj Sharma, Ron Lifshitz, & Ryuji Tamura (Eds.), *Journal of physics: Conference series* (Vol. 809, No. 1, p. 012011). IOP Publishing.

Shukla, V., Bhatnagar, A., Pandey, S. K., Shahi, R. R., Yadav, T. P., Shaz, M. A., & Srivastava, O. N. (2015). On the synthesis, characterization and hydrogen storage behavior of ZrFe2 catalyzed Li–Mg–N–H hydrogen storage material. *International Journal of Hydrogen Energy*, *40*(36), 12294–12302.

Shukla, V., Bhatnagar, A., Singh, S., Soni, P. K., Verma, S. K., Yadav, T. P., ... Srivastava, O. N. (2019). A dual borohydride (Li and Na borohydride) catalyst/additive together with intermetallic FeTi for the optimization of the hydrogen sorption characteristics of Mg (NH 2) 2/2LiH. *Dalton Transactions*, *48*(30), 11391–11403.

Shukla, V., Bhatnagar, A., Soni, P. K., Vishwakarma, A. K., Shaz, M. A., Yadav, T. P., & Srivastava, O. N. (2017). Enhanced hydrogen sorption in a Li–Mg–N–H system by the synergistic role of Li 4 (NH 2) 3 BH 4 and ZrFe 2. *Physical Chemistry Chemical Physics*, *19*(14), 9444–9456.

Shukla, V., Bhatnagar, A., Verma, S. K., Pandey, A. P., Vishwakarma, A. K., Srivastava, P., ... Srivastava, O. N. (2021). Simultaneous improvement of kinetics and thermodynamics based on SrF 2 and SrF 2@ Gr additives on hydrogen sorption in MgH 2. *Materials Advances*, *2*(13), 4277–4290.

Shukla, V., Shaz, M. A., & Yadav, T. P. (2024). Improved hydrogen sorption characteristics of Mg (NH2) 2–2 LiH/Li4BH4 (NH2) 3 using SrH2 and MgF2 auto-catalyst. *International Journal of Hydrogen Energy*, *57*, 1244–1252.

Shukla, V., & Yadav, T. P. (2022). Notable catalytic activity of CuO nanoparticles derived from metal-organic frameworks for improving the hydrogen sorption properties of MgH2. *International Journal of Energy Research*, *46*(9), 12804–12819.

Shukla, V., Yadav, T. P., & Shaz, M. A. (2022). Achievement of excellent hydrogen sorption through swift hydrogen transport in 1: 2 Mg (NH2) 2–LiH catalyzed by Li4BH4 (NH2) 3and carbon nanostructures. *International Journal of Hydrogen Energy*, *47*(56), 23679–23693.

Srivastava, O. N., Yadav, T. P., Shahi, R. R., Pandey, S. K., Shaz, M. A., & Bhatnagar, A. S. H. I. S. H. (2015, September). Hydrogen energy in India: storage to application. In *Proceedings of the Indian national science academy* (Vol. 81, No. 4, pp. 915–937).

Steinfeld, A., & Palumbo, R. (2001). Solar thermochemical process technology. In R. A. Meyers (Ed.), *Encyclopedia of physical science and technology* (3rd ed., Vol. 15, pp. 237–256). Academic Press.

öbel, R., Jörissen, L., Schliermann, T., Trapp, V., Schütz, W., Bohmhammel, K., ... Garche, J. (1999). Hydrogen adsorption on carbon materials. *Journal of Power Sources*, *84*(2), 221–224.

Verma, S. K., Bhatnagar, A., Shukla, V., Soni, P. K., Pandey, A. P., Yadav, T. P., & Srivastava, O. N. (2020). Multiple improvements of hydrogen sorption and their mechanism for MgH_2 catalyzed through TiH2@ Gr. *International Journal of Hydrogen Energy*, *45*(38), 19516–19530.

Verma, S. K., Mishra, S. S., Mukhopadhyay, N. K., & Yadav, T. P. (2024). Superior catalytic action of high-entropy alloy on hydrogen sorption properties of MgH2. *International Journal of Hydrogen Energy*, *50*, 749–762.

Verma, S. K., Shaz, M. A., & Yadav, T. P. (2022). Facile synthesis of M2 (m-dobdc)(M= Fe and Mn) metal-organic frameworks for remarkable hydrogen storage. *Energy Storage*, *4*(4), e333.

Veziro, T. N., & Barbir, F. (1992). Hydrogen: The wonder fuel. *International Journal of Hydrogen Energy*, *17*(6), 391–404.

Watson, R. T., & McMichael, A. J. (2001). Global climate change—the latest assessment: Does global warming warrant a health warning?. *Global Change and Human Health*, *2*(1), 64–75.

Williams, K. A., & Eklund, P. C. (2000). Monte Carlo simulations of H2 physisorption in finite-diameter carbon nanotube ropes. *Chemical Physics Letters*, *320*(3–4), 352–358.

Woolley, H. W., Scott, R. B., & Brickwedde, F. G. (1948). Compilation of thermal properties of hydrogen in its various isotopic and ortho-para modifications. *Journal of Research of the National Bureau of Standards*, *41*, 379–475.

Yadav, T. P., & Awasthi, K. (2022). Carbon nanomaterials: fullerene to graphene. *Transactions of the Indian National Academy of Engineering*, *7*(3), 715–737.

Yadav, T. P., Kumar, A., Verma, S. K., & Mukhopadhyay, N. K. (2022). High-entropy alloys for solid hydrogen storage: Potentials and prospects. *Transactions of the Indian National Academy of Engineering*, *7*(1), 147–156.

Yadav, T. P., Shahi, R. R., & Srivastava, O. N. (2012). Synthesis, characterization and hydrogen storage behaviour of AB2 (ZrFe2, Zr (Fe0. 75V0. 25) 2, Zr (Fe0. 5V0. 5) 2 type materials. *International Journal of Hydrogen Energy*, *37*(4), 3689–3696.

Yang, M., Hunger, R., Berrettoni, S., Sprecher, B., & Wang, B. (2023). A review of hydrogen storage and transport technologies. *Clean Energy*, *7*(1), 190–216.

Ye, Y., Ahn, C. C., Witham, C., Fultz, B., Liu, J., Rinzler, A. G., ... Smalley, R. E. (1999). Hydrogen adsorption and cohesive energy of single-walled carbon nanotubes. *Applied Physics Letters*, *74*(16), 2307–2309.

Zarur, G., & Rabitz, H. (1974). Effective potential formulation of molecule-molecule collisions with application to H2–H2. *The Journal of Chemical Physics*, *60*(5), 2057–2078.

Zheng, J., Liu, X., Xu, P., Liu, P., Zhao, Y., & Yang, J. (2012). Development of high pressure gaseous hydrogen storage technologies. *International Journal of Hydrogen Energy*, *37*(1), 1048–1057.

Zhou, L. (2005). Progress and problems in hydrogen storage methods. *Renewable and Sustainable Energy Reviews*, *9*(4), 395–408.

Züttel, A. (2003). Materials for hydrogen storage. *Materials Today*, *6*(9), 24–33.

Züttel, A. (2004). Hydrogen storage methods. *Naturwissenschaften*, *91*, 157–172.

Züttel, A., Sudan, P., Mauron, P., Kiyobayashi, T., Emmenegger, C., & Schlapbach, L. (2002). Hydrogen storage in carbon nanostructures. *International Journal of Hydrogen Energy*, *27*(2), 203–212.

Chapter 12

Absorption and adsorption-based storage of hydrogen as a fuel

Jashveer Singh and Rajesh Kumar

12.1 INTRODUCTION

Hydrogen is a widely available substance in nature. It is the smallest element available in the atmosphere. Its density and concentration are very small compared to atmospheric air, while a very large amount of hydrogen is available in water, which is chemically bonded by the oxygen atom (Schlapbach & Züttel, 2001; Züttel et al., n.d.). Hydrogen has very favourable qualities as a renewable fuel due to its exceptional combustion properties. It offers a considerable amount of specific energy and does not generate any carbon or hazardous emissions into the environment. It is an abundantly available renewable fuel found in the natural environment. However, the low volumetric density of hydrogen gas poses a challenge in terms of achieving efficient and cost-effective storage (Ahmed et al., 2017; Siegel et al., 2015; Yang et al., 2010). Hydrogen, being a very volatile gas, is held at extremely low temperatures by liquefaction and cryogenic storage. However, this method may be rather costly owing to the significant amount of power needed to sustain the cryogenic storage chambers (Aceves et al., 2000). In addressing this issue, the use of solid adsorbents for hydrogen storage has attracted significant interest among academics, presenting a promising alternative to high-pressure storage tanks (Allendorf et al., 2018). Adsorbents have the capability to store hydrogen at atmospheric pressure via the formation of weak van der Waals (vdW) forces. They may then release the stored hydrogen when required by applying sufficient heat and pressure (Shet et al., 2021). Metal-organic frameworks (MOFs) are the adsorbents used to store hydrogen by the physisorption process (Rowsell & Yaghi, 2005). According to recent studies (Ahmed et al., 2019; Collins & Zhou, 2007; Ghoufi et al., 2012; Suh et al., 2012), MOFs such as PCN-610/NU-100, UMCM-9, IRMOF-20, and SNU-70 have impressive hydrogen storage potential. Physisorption of hydrogen in MOFs is shown to be the most stable and cost-effective method of storage when compared to other conventional approaches. The porous crystalline structure of the MOFs has a very large surface area and the ability to adsorb the hydrogen atoms in their voids and to desorb the hydrogen atoms upon the application of a

DOI: 10.1201/9781003537816-12

small amount of heat and pressure. This way, the re-production of hydrogen gas is done at atmospheric pressure and temperature without too much expense on high-pressure storage chambers (Singh et al., 2021). However, the thermal conductivity of the MOFs is very poor, which may affect the process of adsorption and desorption of hydrogen (Purewal et al., 2012). If the reinforcement of the MOFs is done with some nanofillers that have high thermal conductivity, then the overall performance of the MOF nanocomposites to absorb or desorb the hydrogen atom will be enhanced. Graphene and boron nitride (hBN)-based nanofillers with excellent thermal properties are an attractive engineering proposition for designing nanocomposites with excellent thermal and mechanical properties (Kumar & Parashar, 2016; Rajasekaran et al., 2016). These nanofillers are potential candidates for solid-state adsorption-based hydrogen storage because storing hydrogen in the solid state is much safer than at high pressures (Ströbel et al., 2006). Improving MOF's thermal conductivity and optimizing the heat transfer for faster uptake and release is one of the key challenges for MOFs. Hence, the use of MOF-based nanocomposites with better thermal and heat transfer characteristics can be highly efficient for hydrogen storage (D. Liu et al., 2012; Purewal et al., 2012). In order to comprehend the atomistic behaviour of these MOF nanocomposites, molecular dynamics simulations appear to be a favourable option among several traditional methodologies. This is primarily due to their somewhat less extensive character compared to experimental and density functional theory (DFT) simulations. This book chapter will explore the improved thermal and mechanical strength capabilities of MOF nanocomposites for hydrogen storage.

12.2 MATERIALS CHARACTERIZATION TECHNIQUES

Numerous methods have been utilized by researchers to anticipate the mechanical and thermal characteristics of materials, with experimental techniques being the primary driver for characterization techniques. Despite the associated higher expenses and time requirements, experimental techniques are still being most utilized for the characterization of nanofillers. Transmission electron microscopy (TEM) and atomic force microscopy (AFM) (Arenal et al., 2011; Golberg et al., 2007) based techniques dominate experimental approaches. Since traditional methods are time-consuming and expensive, an alternative is developing in the form of computer methods based on the finite element method (FEM), density functional theory (DFT), and molecular dynamics (MD). Researchers often use finite element method (FEM)-based atomic truss frame models for the modelling of carbon–carbon bonding in carbon nanotubes (CNTs) and graphene (Li & Chou, 2003; Parashar & Mertiny, 2012). Researchers have also developed methods to predict the mechanical characteristics of hexagonal boron nitride (hBN) nanofillers using molecular mechanics (MM) (Boldrin et al., 2011a). It is believed that these numerical methods are adequate for simulating global characteristics. For evaluating the atomic level characteristics of nanofillers, researchers often use higher fidelity simulations like density functional theory (DFT) (Andrew et al., 2012; Kudin et al., 2001) in addition to the above numerical approaches. To evaluate the elastic characteristics of nanofillers, DFT-based techniques use a quantum mechanical framework (Peng et al., 2012). DFT simulations are computationally expensive; hence, they can only be done on a small number of atoms at a time.

In contrast to the more sophisticated numerical simulations and experimental approaches with higher accuracy, there is significant potential in the employment of molecular dynamics (MD)-based simulations. Notably, advancements in interatomic potentials have played a vital role in the prediction of nanofiller behaviour (Tersoff, 1986, 1988a, 1988b). Among various material properties, MD has gained excellent ground in its use for material behaviour prediction. Out of various properties, a simple example of calculating Young's modulus (as below) stands out as a well-established and dependable parameter. The stress–strain behaviour and elastic modulus of the considered nanofiller can be anticipated using the strain energy (*U*), a common factor in MD-based simulations. The connection between strain energy (*U*) and strain (ε) is elucidated by the subsequent equations to derive Young's modulus (*Y*, measured in N/m^2) and 2D Young's modulus (*YS*, measured in N/m).

$$Y = (\partial^2 U / \partial \varepsilon^2) / V \tag{12.1}$$

$$Y = (\partial^2 U / \partial \varepsilon^2) / S \tag{12.2}$$

Here, *V* is the volume and *S* is the surface area of the atomic system.

Similarly, to calculate the thermal behaviour of nanomaterials and nanocomposites, the thermal conductivity calculation can be done using Equation (12.3):

$$q_x = -k\, A_c \frac{dT}{dx} \tag{12.3}$$

Here, qx is the heat applied to the material, k is the thermal conductivity, A_c is the cross-sectional area of the structure, and dT/dx is the temperature gradient.

12.2.1 Interatomic potentials

MD-based simulations often use interatomic potentials that have been either established with experiments or higher fidelity numerical simulations.

Mathematical formulations representing the potential energy (*E*) of a system of atoms are called interatomic potentials. The interatomic potentials may be categorized into three main groups: one-body, two-body, or many-body. A few examples of popular interatomic pair potentials include Tersoff (Tersoff, 1989), LJ (Jones, 1924), the embedded atom method (EAM) (Daw & Baskes, 1984), REBO (Jeong et al., 2007), AIREBO (Stuart et al., 2000), CHARMM (MacKerell Jr et al., 2000), and COMPASS (Sun et al., 1998). Potentials are further categorized as bonded (Tersoff, REBO, and EAM) or non-bonded (van der Waals, LJ) electrostatic potentials. The mechanical and thermal behaviour of nanofillers like graphene, carbon nanotubes (CNTs), and hexagonal boron nitride (hBN) is often modelled using Tersoff-type interatomic potentials (Tersoff, 1986, 1988a, 1988b). In Tersoff-type potentials, the potential energy (*E*) is expressed mathematically as:

$$E = \sum_i E_i = \left(\sum_{i \neq j} V_{ij} \right) \Big/ 2, \tag{12.4}$$

$$V_{ij} = f_c(r_{ij})[a_{ij} f_R(r_{ij}) + b_{ij} f_A(r_{ij})]. \tag{12.5}$$

Here, we have the system's total energy (*E*), conveniently broken down into its parts—the site energy (E_i) and the bond energy (V_{ij}). Both indices *i* and *j* extend across all particles in the system while measuring the separation (r_{ij}) between them.

When the wave functions of two atoms overlap, a repulsive potential arises, represented by the function f_R, and an attracting potential arises, represented by the function f_A. These can be expressed as:

$$f_R(r) = A\exp(-\lambda_1 r), \tag{12.6}$$

$$f_A(r) = -B\exp(-\lambda_2 r). \tag{12.7}$$

These atomic interactions are only allowed to take place within a certain cut-off distance, defined by a cut-off function f_c, to keep the computational cost down.

$$f_c(r) = \begin{cases} 1, \quad r < R - D \\ (1/2) - \sin\left(\dfrac{\pi(r - R)}{2D} \right) \Big/ 2, \qquad R - D < r < R + D \\ 0, \quad r > R + D \end{cases} \tag{12.8}$$

The cut-off radius (R) is typically selected such that just the first neighbour shell is considered. The function b_{ij} represents a measure of the bond order. In addition, the terms that act to limit the range of interaction to the first neighbour shell are included in b_{ij}. The function a_{ij} consists solely of such range-limiting terms.

The use of MD-based simulations to forecast the characteristics of these novel materials is becoming more feasible. To further enhance these computations, researchers are also making improvements to these potentials. Brenner (1990) expanded upon the original Tersoff potential stated in the article (Tersoff, 1986). Optimized Tersoff potential parameters have been suggested for increasing the fidelity of these atomistic simulations by many researchers, including Brenner (1990), Matsunaga et al. (2000), Matsunaga and Iwamoto (2001), Albe et al. (1997), Albe and Möller, 1998), Nord et al. (2003), and Sevik et al. (2011).

12.3 POTENTIAL NANOMATERIALS FOR HYDROGEN STORAGE

12.3.1 Graphene

With excellent mechanical, thermal, and electrical properties, Patchkovskii et al. (2005) observed that graphene is a potential nanomaterial for hydrogen storage by physisorption with weak van der Waals (vdW) interaction with hydrogen atoms. Compared to other carbon-based nanomaterials, graphene showed higher binding energy with hydrogen atoms by physisorption. For better adsorption of hydrogen and reduction in the adsorption energy barrier, the functionalization of graphene was most favourable (Patchkovskii et al., 2005). Functionalizing the graphene with aluminium resulted in improved hydrogen storage, i.e., 5.1 wt% (Ao et al., 2009). Further, functionalizing the graphene with oxygen, i.e., forming graphene oxide (GO), also showed improved hydrogen storage at room temperature (Wang et al., 2009). A different phenomenon is utilized to improve the hydrogen storage in graphene nanosheets. In a study by Ataca et al. (2008), the graphene nanosheets are metallized by adsorbing the Li atoms on the graphene. It was observed that each Li atom was able to absorb four hydrogen atoms, resulting in a high storage capacity of graphene of 12.8 wt%. Yadav et al. (2014) incorporated defect engineering into graphene nanosheets for better hydrogen storage. Single vacancy and mixed defects at high concentrations in the graphene nanosheets showed an improvement in hydrogen storage up to 7 wt%.

12.3.2 Hexagonal boron nitride (hBN)

Hexagonal boron nitride has a similar structure and comparable mechanical properties to graphene nanosheets and may be an alternative nanomaterial for hydrogen storage as fuel (Boldrin et al., 2011b). Jhi and Kwon (2004) analysed the binding energy of hBN nanosheets to hydrogen atoms at room temperature and observed a 40% improvement compared to carbon nanotubes (CNTs). Liang et al. (2022) functionalized the hBN by polymer grafting and Ni nanoparticles. They observed an improvement of 7.3 wt% and 6.1 wt% in hydrogen storage compared to pristine hBN. Zhang et al. (2023) also functionalized the hBN with a super alkali cluster (OLi_3) using DFT studies. The Li atom in functionalized hBN behaves as a binding site for hydrogen atoms and observed a 9.73 wt% improvement in hydrogen storage compared to pristine hBN.

The above-discussed nanomaterial will also have its potential application in hydrogen storage if reinforced to the metal-organic frameworks (MOFs).

12.4 HYDROGEN STORAGE PERFORMANCE OF MOF NANOCOMPOSITES

Liu et al. (2022) examined the graphene oxide-reinforced MOFs (MOFGO) utilized for hydrogen storage at room temperature. Hydrogen storage capacity was observed to be greater in MOF-GO nanocomposites with tiny pores than those with larger pores. The enhanced hydrogen storage was determined to be the result of an increase in the number of pores per unit volume or surface area. The analysis revealed that the hydrogen storage efficiency increased from 9.88 mg/go to 11.48 mg/go when the MOFs were reinforced with graphene oxide (GO). For hydrogen storage, Mahdizadeh and Goharshadi (2019) analysed graphitic carbon nitride-reinforced palladium nanocomposites. The researchers documented an increase in the capacity of hydrogen storage to 30 g/L at 177 K, surpassing the threshold established by the Department of Energy (DOE). In their investigation, Deniz et al. (2021) examined the physisorption of hydrogen in lithium-doped fullerene-pillared graphene nanocomposites (Li-FPNG) at ambient temperature. They discovered that the Li-FPNG exhibited three times greater hydrogen storage capacity (3.2 wt%) than the undoped FPNG. An additional investigation conducted by Dimitrakakis et al. (2008) documented a hydrogen storage increase of 7.4 wt% or six times that of undoped FPNG. Numerous studies documented in the scientific literature have demonstrated that MOF nanocomposites have a greater capacity for hydrogen storage than unreinforced MOFs (Ariharan et al., 2018; Baykasoglu et al., 2016; Mert et al., 2020). The impact of ultrafine nanoparticles on the hydrogen storage capacity of magnesium (Mg) was also investigated by Jia et al. (2015). Applying nickel

(Ni) nanoparticles to MgH_2 matrices showed that ultrafine nanoparticles positively influenced Mg's adsorption and desorption of hydrogen atoms.

Additionally, Liu et al. (2007) have attempted to improve the hydrogen storage capacity at room temperature through catalyst modification and carbon doping of MOFs. The hydrogen storage efficiency of MOFs such as Material Institute Lavoisier (MIL) MIL-101 and MIL-53 was enhanced by 1.14 wt% and 0.63 wt%, respectively, compared to pristine MOFs. In their study, Zhou et al. (2015) employed the spillover mechanism to assess the hydrogen storage capacity of zeolitic imidazolate frameworks ZIF-8/GO MOFs by incorporating nanoparticles into pristine ZIF-8 MOFs. The results demonstrated a hydrogen storage capacity that was 2.2 times greater than that of pristine ZIF-8 MOFs. Sumida et al. (2010) evaluated the hydrogen storage capacity of iron-benzene tristetrazolate (Fe-BTT)-based MOFs using a high-throughput method. Excellent outcomes were achieved with respect to hydrogen storage at room temperature, specifically 8.4 gm/L. When evaluating the University of Oslo (UIO-66) MOFs for hydrogen storage at room temperature, Bambalaza et al. (2020) detected a 0.8% hydrogen adsorption. The impact of Li-crown ether on the hydrogen storage capacity of MOFs at room temperature was investigated by Orcajo et al. (2019). The researchers examined metal-organic frameworks (MOFs) with varying pore sizes, specifically Cr-MIL-101, Ni-MOF-74, and Fe-MIL-100. The findings indicated that the incorporation of the Li–ether complex into all three MOFs negatively impacted the hydrogen storage capability of Ni-MOF-74, whereas it enhanced the capacity of the other two MOFs. The variability in adsorption capacities is predominantly determined by the volume of pores, or pore sizes. Additionally, Xia and Liu (2017) investigated the hydrogen adsorption capabilities of MOF-519 and MOF-520, which are aluminium-based MOFs. MOF-519 exhibited favourable hydrogen storage characteristics under low-pressure conditions, whereas MOF-520 demonstrated exceptional hydrogen adsorption capabilities at high pressure due to its large pore volume and large surface area. In contrast to pristine MOF-177, Sreedhar et al. (2018) observed that MOF-177 reinforced with platinum nanoparticles exhibited exceptional hydrogen storage at room temperature, storing 2.5 wt% of hydrogen. Suksaengrat et al. (2016) investigated the hydrogen adsorption capabilities of Mg-MOF-74 using titanium (Ti) as the catalyst. At ambient conditions, they observed an exceptional hydrogen adsorption capacity of 1.2 wt% for Mg-MOF-74 in the presence of Ti catalysis.

12.5 FUTURE PROSPECTS AND CONCLUSION

Hydrogen is a clean fuel that can sustainably fulfil the world's energy needs in the long run. Still, hydrogen storage poses numerous challenges ranging from bulky storage tanks to the transportation and use in moving vehicles. Hence, the use of graphene, hBN, MOFs, and their nanocomposites is gaining traction for hydrogen storage applications, possibly removing the challenges in hydrogen storage. However, lower hydrogen storage per unit weight of these materials is still a hurdle that must be overcome. Therefore, future research efforts and explorations must be focused on improving the per-unit weight storage of hydrogen for these materials.

REFERENCES

Aceves, S. M., Martinez-Frias, J., & Garcia-Villazana, O. (2000). Analytical and experimental evaluation of insulated pressure vessels for cryogenic hydrogen storage. *International Journal of Hydrogen Energy*, *25*(11), 1075–1085.

Ahmed, A., Liu, Y., Purewal, J., Tran, L. D., Wong-Foy, A. G., Veenstra, M., Matzger, A. J., & Siegel, D. J. (2017). Balancing gravimetric and volumetric hydrogen density in MOFs. *Energy & Environmental Science*, *10*(11), 2459–2471.

Ahmed, A., Seth, S., Purewal, J., Wong-Foy, A. G., Veenstra, M., Matzger, A. J., & Siegel, D. J. (2019). Exceptional hydrogen storage achieved by screening nearly half a million metal-organic frameworks. *Nature Communications*, *10*(1), 1568.

Albe, K., & Möller, W. (1998). Modelling of boron nitride: Atomic scale simulations on thin film growth. *Computational Materials Science*, *10*(1), 111–115. https://doi.org/https://doi.org/10.1016/S0927-0256(97)00172-9

Albe, K., Möller, W., & Heinig, K.-H. (1997). Computer simulation and boron nitride. *Radiation Effects and Defects in Solids*, *141*(1–4), 85–97. https://doi.org/10.1080/10420159708211560

Allendorf, M. D., Hulvey, Z., Gennett, T., Ahmed, A., Autrey, T., Camp, J., Cho, E. S., Furukawa, H., Haranczyk, M., & Head-Gordon, M. (2018). An assessment of strategies for the development of solid-state adsorbents for vehicular hydrogen storage. *Energy & Environmental Science*, *11*(10), 2784–2812.

Andrew, R. C., Mapasha, R. E., Ukpong, A. M., & Chetty, N. (2012). Mechanical properties of graphene and boronitrene. *Physical Review B*, *85*(12), 125428. https://doi.org/10.1103/PhysRevB.85.125428

Ao, Z. M., Jiang, Q., Zhang, R. Q., Tan, T. T., & Li, S. (2009). Al doped graphene: a promising material for hydrogen storage at room temperature. *Journal of Applied Physics*, *105*(7), 074307.

Arenal, R., Wang, M.-S., Xu, Z., Loiseau, A., & Golberg, D. (2011). Young modulus, mechanical and electrical properties of isolated individual and bundled single-walled boron nitride nanotubes. *Nanotechnology*, *22*(26), 265704. https://doi.org/10.1088/0957-4484/22/26/265704

Ariharan, A., Viswanathan, B., & Nandhakumar, V. (2018). Nitrogen-incorporated carbon nanotube derived from polystyrene and polypyrrole as hydrogen storage material. *International Journal of Hydrogen Energy*, *43*(10), 5077–5088.

Ataca, C., Aktürk, E., Ciraci, S., & Ustunel, H. (2008). High-capacity hydrogen storage by metallized graphene. *Applied Physics Letters*, *93*(4), 043123.

Bambalaza, S. E., Langmi, H. W., Mokaya, R., Musyoka, N. M., & Khotseng, L. E. (2020). Experimental demonstration of dynamic temperature-dependent behavior of UiO-66 metal–organic framework: compaction of hydroxylated and dehydroxylated forms of UiO-66 for high-pressure hydrogen storage. *ACS Applied Materials & Interfaces*, *12*(22), 24883–24894.

Baykasoglu, C., Ozturk, Z., Kirca, M., Celebi, A. T., Mugan, A., & To, A. C. (2016). Effects of lithium doping on hydrogen storage properties of heat welded random CNT network structures. *International Journal of Hydrogen Energy*, *41*(19), 8246–8255.

Boldrin, L., Scarpa, F., Chowdhury, R., & Adhikari, S. (2011a). Effective mechanical properties of hexagonal boron nitride nanosheets. *Nanotechnology*, *22*(50), 505702. https://doi.org/10.1088/0957-4484/22/50/505702

Boldrin, L., Scarpa, F., Chowdhury, R., & Adhikari, S. (2011b). Effective mechanical properties of hexagonal boron nitride nanosheets. *Nanotechnology*, *22*(50), 505702. https://doi.org/10.1088/0957-4484/22/50/505702

Brenner, D. W. (1990). Empirical potential for hydrocarbons for use in simulating the chemical vapor deposition of diamond films. *Physical Review B*, *42*(15), 9458–9471. https://doi.org/10.1103/PhysRevB.42.9458

Collins, D. J., & Zhou, H.-C. (2007). Hydrogen storage in metal–organic frameworks. *Journal of Materials Chemistry*, *17*(30), 3154–3160.

Daw, M. S., & Baskes, M. I. (1984). Embedded-atom method: Derivation and application to impurities, surfaces, and other defects in metals. *Physical Review B*, *29*(12), 6443–6453. https://doi.org/10.1103/PhysRevB.29.6443

Deniz, C. U., Mert, H., & Baykasoglu, C. (2021). Li-doped fullerene pillared graphene nanocomposites for enhancing hydrogen storage: A computational study. *Computational Materials Science*, *186*, 110023.

Dimitrakakis, G. K., Tylianakis, E., & Froudakis, G. E. (2008). Pillared graphene: A new 3-D network nanostructure for enhanced hydrogen storage. *Nano Letters*, *8*(10), 3166–3170.

Ghoufi, A., Deschamps, J., & Maurin, G. (2012). Theoretical hydrogen cryostorage in doped MIL-101 (Cr) metal–organic frameworks. *The Journal of Physical Chemistry C*, *116*(19), 10504–10509.

Golberg, D., Costa, P. M. F. J., Lourie, O., Mitome, M., Bai, X., Kurashima, K., Zhi, C., Tang, C., & Bando, Y. (2007). Direct force measurements and kinking under elastic deformation of individual multiwalled boron nitride nanotubes. *Nano Letters*, *7*(7), 2146–2151. https://doi.org/10.1021/nl070863r

Jeong, B.-W., Lim, J.-K., & Sinnott, S. B. (2007). Tensile mechanical behavior of hollow and filled carbon nanotubes under tension or combined tension-torsion. *Applied Physics Letters*, *90*(2), 023102.

Jhi, S.-H., & Kwon, Y.-K. (2004). Hydrogen adsorption on boron nitride nanotubes: A path to room-temperature hydrogen storage. *Physical Review B*, *69*(24), 245407.

Jia, Y., Sun, C., Peng, Y., Fang, W., Yan, X., Yang, D., Zou, J., Mao, S. S., & Yao, X. (2015). Metallic Ni nanocatalyst in situ formed from a metal–organic-framework by mechanochemical reaction for hydrogen storage in magnesium. *Journal of Materials Chemistry A*, *3*(16), 8294–8299.

Jones, J. E. (1924). On the determination of molecular fields.—II. From the equation of state of a gas. *Proceedings of the Royal Society of London. Series A, Containing Papers of a Mathematical and Physical Character*, *106*(738), 463–477.

Kudin, K. N., Scuseria, G. E., & Yakobson, B. I. (2001). ${\mathrm{C}}_{2}\ mathrm{F},$ BN, and C nanoshell elasticity from ab initio computations. *Physical Review B*, *64*(23), 235406. https://doi.org/10.1103/PhysRevB.64.235406

Kumar, R., & Parashar, A. (2016). Atomistic modeling of BN nanofillers for mechanical and thermal properties: A review. *Nanoscale*, *8*(1), 22–49. https://doi.org/10.1039/c5nr06917c

Li, C., & Chou, T.-W. (2003). A structural mechanics approach for the analysis of carbon nanotubes. *International Journal of Solids and Structures*, *40*(10), 2487–2499. https://doi.org/https://doi.org/10.1016/S0020-7683(03)00056-8

Liang, H., Cao, B., Zhu, J., Shen, X., Zhu, M., Geng, B., Zhang, P., Zhu, S., Yu, F., & Zhang, R. (2022). Intercalation optimized hexagonal boron nitride nanosheets for high efficiency hydrogen storage. *Applied Surface Science*, *604*, 154118.

Liu, D., Purewal, J. J., Yang, J., Sudik, A., Maurer, S., Mueller, U., Ni, J., & Siegel, D. J. (2012). MOF-5 composites exhibiting improved thermal conductivity. *International Journal of Hydrogen Energy*, *37*(7), 6109–6117.

Liu, Y.-Y., Zeng, J.-L., Zhang, J., Xu, F., & Sun, L.-X. (2007). Improved hydrogen storage in the modified metal-organic frameworks by hydrogen spillover effect. *International Journal of Hydrogen Energy*, *32*(16), 4005–4010.

Liu, Y. Y., Shen, D., Tu, Z., Xing, L., Chung, Y. G., & Li, S. (2022). Room-temperature hydrogen storage performance of metal-organic framework/graphene oxide composites by molecular simulations. *International Journal of Hydrogen Energy*, *47*(97), 41055–41068.

MacKerell Jr, A. D., Banavali, N., & Foloppe, N. (2000). Development and current status of the CHARMM force field for nucleic acids. *Biopolymers: Original Research on Biomolecules*, *56*(4), 257–265.

Mahdizadeh, S. J., & Goharshadi, E. K. (2019). Hydrogen storage on graphitic carbon nitride and its palladium nanocomposites: A multiscale computational approach. *International Journal of Hydrogen Energy*, *44*(16), 8325–8340.

Matsunaga, K., Fisher, C., & Matsubara, H. (2000). Tersoff potential parameters for simulating cubic boron carbonitrides. *Japanese Journal of Applied Physics*, *39*(Part 2, No. 1A/B), L48–L51. https://doi.org/10.1143/jjap.39.l48

Matsunaga, K., & Iwamoto, Y. (2001). Molecular dynamics study of atomic structure and diffusion behavior in amorphous silicon nitride containing boron. *Journal of the American Ceramic Society*, *84*(10), 2213–2219. https://doi.org/https://doi.org/10.1111/j.1151-2916.2001.tb00990.x

Mert, H., Deniz, C. U., & Baykasoglu, C. (2020). Monte Carlo simulations of hydrogen adsorption in fullerene pillared graphene nanocomposites. *Molecular Simulation*, *46*(8), 650–659.

Nord, J., Albe, K., Erhart, P., & Nordlund, K. (2003). Modelling of compound semiconductors: Analytical bond-order potential for gallium, nitrogen and gallium nitride. *Journal of Physics: Condensed Matter*, *15*(32), 5649–5662. https://doi.org/10.1088/0953-8984/15/32/324

Orcajo, G., Montes-Andres, H., Villajos, J. A., Martos, C., Botas, J. A., & Calleja, G. (2019). Li-Crown ether complex inclusion in MOF materials for enhanced H2 volumetric storage capacity at room temperature. *International Journal of Hydrogen Energy*, *44*(35), 19285–19293.

Parashar, A., & Mertiny, P. (2012). Study of mode i fracture of graphene sheets using atomistic based finite element modeling and virtual crack closure technique. *International Journal of Fracture*, *176*(1), 119–126. https://doi.org/10.1007/s10704-012-9718-y

Patchkovskii, S., Tse, J. S., Yurchenko, S. N., Zhechkov, L., Heine, T., & Seifert, G. (2005). Graphene nanostructures as tunable storage media for molecular hydrogen. *Proceedings of the National Academy of Sciences*, *102*(30), 10439–10444.

Peng, Q., Ji, W., & De, S. (2012). Mechanical properties of the hexagonal boron nitride monolayer: Ab initio study. *Computational Materials Science*, *56*, 11–17. https://doi.org/10.1016/j.commatsci.2. https://doi.org/https://doi.org/10.1016/j.commatsci.2011.12.029

Purewal, J., Liu, D., Sudik, A., Veenstra, M., Yang, J., Maurer, S., Müller, U., & Siegel, D. J. (2012). Improved hydrogen storage and thermal conductivity in high-density MOF-5 composites. *The Journal of Physical Chemistry C*, *116*(38), 20199–20212.

Rajasekaran, G., Narayanan, P., & Parashar, A. (2016). Effect of point and line defects on mechanical and thermal properties of graphene: A review. *Critical Reviews in Solid State and Materials Sciences*, *41*(1), 47–71.

Rowsell, J. L. C., & Yaghi, O. M. (2005). Strategies for hydrogen storage in metal–organic frameworks. *Angewandte Chemie International Edition*, *44*(30), 4670–4679.

Schlapbach, L., & Züttel, A. (2001). Hydrogen-storage materials for mobile applications. *Nature*, *414*(6861), 353–358.

Sevik, C., Kinaci, A., Haskins, J. B., & Çağın, T. (2011). Characterization of thermal transport in low-dimensional boron nitride nanostructures. *Physical Review B*, *84*(8), 85409. https://doi.org/10.1103/PhysRevB.84.085409

Shet, S. P., Priya, S. S., Sudhakar, K., & Tahir, M. (2021). An assessment of strategies for the development of solid-state adsorbents for vehicular hydrogen storage. *International Journal of Hydrogen Energy*, *46*(21), 11782–11803. https://doi.org/10.1016/j.ijhydene.2021.01.020

Siegel, D., Hardy, B., & Team, Hsec. (2015). *Engineering an adsorbent-based hydrogen storage system: What have we learned.* DOE Materials-Based Hydrogen Storage Summit, NREL.

Singh, R., Altaee, A., & Gautam, S. (2021). Nanomaterials in the advancement of hydrogen energy storage. *TIDEE: TERI Information Digest on Energy and Environment*, *20*(3), 318.

Sreedhar, I., Kamani, K. M., Kamani, B. M., Reddy, B. M., & Venugopal, A. (2018). A bird's eye view on process and engineering aspects of hydrogen storage. *Renewable and Sustainable Energy Reviews*, *91*, 838–860.

Ströbel, R., Garche, J., Moseley, P. T., Jörissen, L., & Wolf, G. (2006). Hydrogen storage by carbon materials. *Journal of Power Sources*, *159*(2), 781–801.

Stuart, S. J., Tutein, A. B., & Harrison, J. A. (2000). A reactive potential for hydrocarbons with intermolecular interactions. *The Journal of Chemical Physics*, *112*(14), 6472–6486. https://doi.org/10.1063/1.481208

Suh, M. P., Park, H. J., Prasad, T. K., & Lim, D.-W. (2012). Hydrogen storage in metal–organic frameworks. *Chemical Reviews*, *112*(2), 782–835.

Suksaengrat, P., Amornkitbamrung, V., Srepusharawoot, P., & Ahuja, R. (2016). Density functional theory study of Hydrogen adsorption in a Ti-Decorated Mg-Based metal–organic framework-74. *ChemPhysChem*, *17*(6), 879–884.

Sumida, K., Horike, S., Kaye, S. S., Herm, Z. R., Queen, W. L., Brown, C. M., Grandjean, F., Long, G. J., Dailly, A., & Long, J. R. (2010). Hydrogen storage and carbon dioxide capture in an iron-based sodalite-type metal–organic framework (Fe-BTT) discovered via high-throughput methods. *Chemical Science*, *1*(2), 184–191.

Sun, H., Ren, P., & Fried, J. R. (1998). The COMPASS force field: Parameterization and validation for phosphazenes. *Computational and Theoretical Polymer Science*, *8*(1), 229–246. https://doi.org/https://doi.org/10.1016/S1089-3156(98)00042-7

Tersoff, J. (1986). New empirical model for the structural properties of silicon. *Physical Review Letters*, *56*(6), 632–635. https://doi.org/10.1103/PhysRevLett.56.632

Tersoff, J. (1988a). New empirical approach for the structure and energy of covalent systems. *Physical Review B*, *37*(12), 6991–7000. https://doi.org/10.1103/PhysRevB.37.6991

Tersoff, J. (1988b). Empirical interatomic potential for carbon, with applications to amorphous carbon. *Physical Review Letters*, *61*(25), 2879–2882. https://doi.org/10.1103/PhysRevLett.61.2879

Tersoff, J. (1989). Modeling solid-state chemistry: Interatomic potentials for multicomponent systems. *Physical Review B*, *39*(8), 5566–5568. https://doi.org/10.1103/PhysRevB.39.5566

Wang, L., Lee, K., Sun, Y.-Y., Lucking, M., Chen, Z., Zhao, J. J., & Zhang, S. B. (2009). Graphene oxide as an ideal substrate for hydrogen storage. *ACS Nano*, *3*(10), 2995–3000.

Xia, L., & Liu, Q. (2017). Adsorption of H2 on aluminum-based metal-organic frameworks: A computational study. *Computational Materials Science*, *126*, 176–181.

Yadav, S., Zhu, Z., & Singh, C. V. (2014). Defect engineering of graphene for effective hydrogen storage. *International Journal of Hydrogen Energy*, *39*(10), 4981–4995.

Yang, J., Sudik, A., Wolverton, C., & Siegel, D. J. (2010). High capacity hydrogen storage materials: attributes for automotive applications and techniques for materials discovery. *Chemical Society Reviews*, *39*(2), 656–675.

Zhang, N., Shi, Y., Li, J., Zhang, Y., Guo, J., Fu, Z., & Zhang, P. (2023). Potential applications of OLi3-decorated h-BN monosheet for high hydrogen storage. *Applied Physics Letters*, *123*(8), 083902.

Zhou, H., Zhang, J., Zhang, J., Yan, X., Shen, X., & Yuan, A. (2015). High-capacity room-temperature hydrogen storage of zeolitic imidazolate framework/graphene oxide promoted by platinum metal catalyst. *International Journal of Hydrogen Energy*, *40*(36), 12275–12285.

Züttel, A., Borgschulte, A., & Schlapbach, L. (n.d.). *Hydrogen as a future energy carrier.* Wiley-VCH.

Chapter 13

Assessment of materials for energy-efficient low-pressure hydrogen storage

Adegoke C. Ishola, Ibrahim Albayati, Behnaz Sohani, Faiz Iqbal, and Aliyu M. Aliyu

13.1 INTRODUCTION

Rapid rise in the global mean temperature to 3.2°C above pre-industrial levels has been an alarming trend linked to anthropogenic activities (Hou et al., 2021). These activities are believed to have been a major precursor to greenhouse gas emissions in the atmosphere (Ball & Wietschel, 2009; Lai et al., 2015). The accumulation of which is subtly termed "global warming" and is liable for many natural disasters eroding the globe. The impact of these is experienced in diverse forms and has been a source of concern, prompting nations, both developed and emerging, to formulate strategies aimed at curbing global CO_2 and other greenhouse gas emissions (Schlapbach & Züttel, 2001). Thus, governments are concerned about the effects of global warming and continuously strive to establish strategies and policies to decrease carbon dioxide and other greenhouse gases from the atmosphere: a way to not only mitigate future occurrences but also to ensure the global temperature rise is delayed. As the world's largest source of greenhouse emissions, the energy industry is a prime target for green ideas and legislation. It is an integral part of human existence and has been marked as a growth catalyst in different aspects of human history. The energy industry involves the production of useful energy via the utilisation, transportation, and storage of raw materials, transformed resources, and all resource transformation activities related to energy in its entirety. This sector contributes about 73.2% CO_2 (Figure 13.1) to the atmosphere (Winter & Brodd, 2004) through its activities and has thrived for centuries by relying on traditional energy sources such as petroleum and/or coal, known as fossil fuels, which have been extensively relied upon over the years (Schlapbach & Züttel, 2001). The combustion of these fuel sources via the plenteous energy conversion processes produces a significant amount of greenhouse gases (i.e., CO_2, CO, methane, CFCs, etc).

Among a myriad of potential solutions, hydrogen has been identified as a promising energy carrier owing to its clean-burning properties (producing

 DOI: 10.1201/9781003537816-13

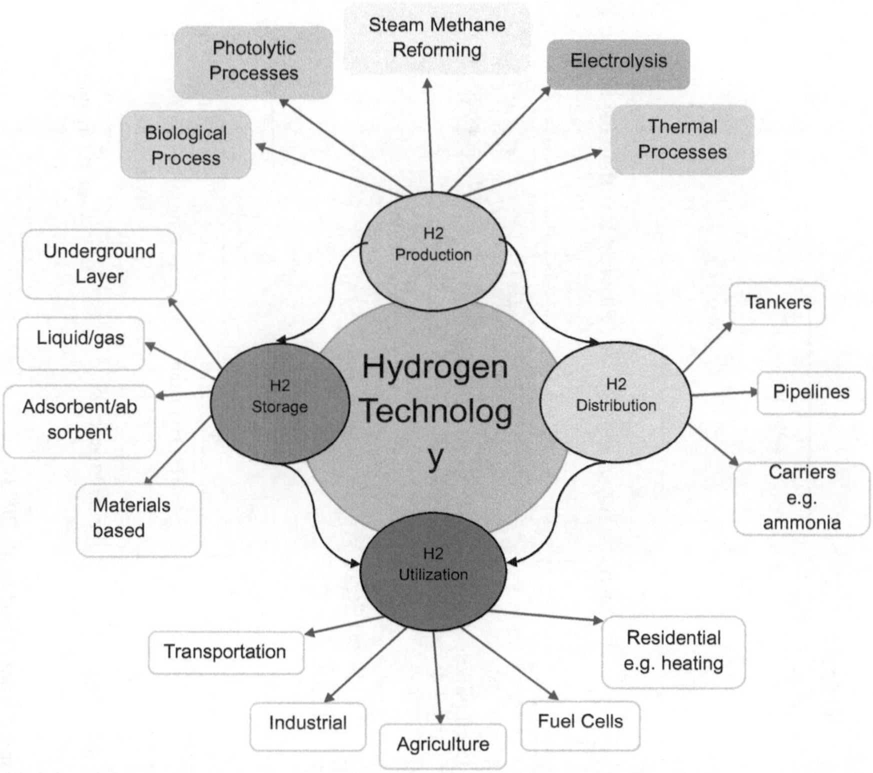

Figure 13.1 Hydrogen energy technologies.

water (H_2O) as its only byproduct), high-energy content (2.82×105 J.mol^{-1}) (Ball & Wietschel, 2009; Hou et al., 2021) as well as its abundancy in nature (Lai et al., 2015; Schlapbach & Züttel, 2001; Winter & Brodd, 2004). Hydrogen holds the potential for decarbonizing a wide array of sectors including industry, transport, and heating (Kojima, 2019). Hydrogen also possesses the potential to tackle the issue of storing energy, a problem that's growing with the rise and improvement of renewable energy sources that works optimally—only in favourable weather conditions (International Energy Agency, 2023). What makes hydrogen even more versatile is that it can be produced from many sources like water, natural gas, and even renewables such as wind or solar energy and can find many applications (both stationary and mobile) in diverse fields (Kojima, 2019; Niaz et al., 2015). Figure 13.2 shows a summary of hydrogen technologies linked to its production, storage, and utilisation. This means that there are many ways to generate and align it into our energy systems (Yue et al., 2021). Yet, shifting to a hydrogen-based economy comes with its own hurdles to overcome. This is

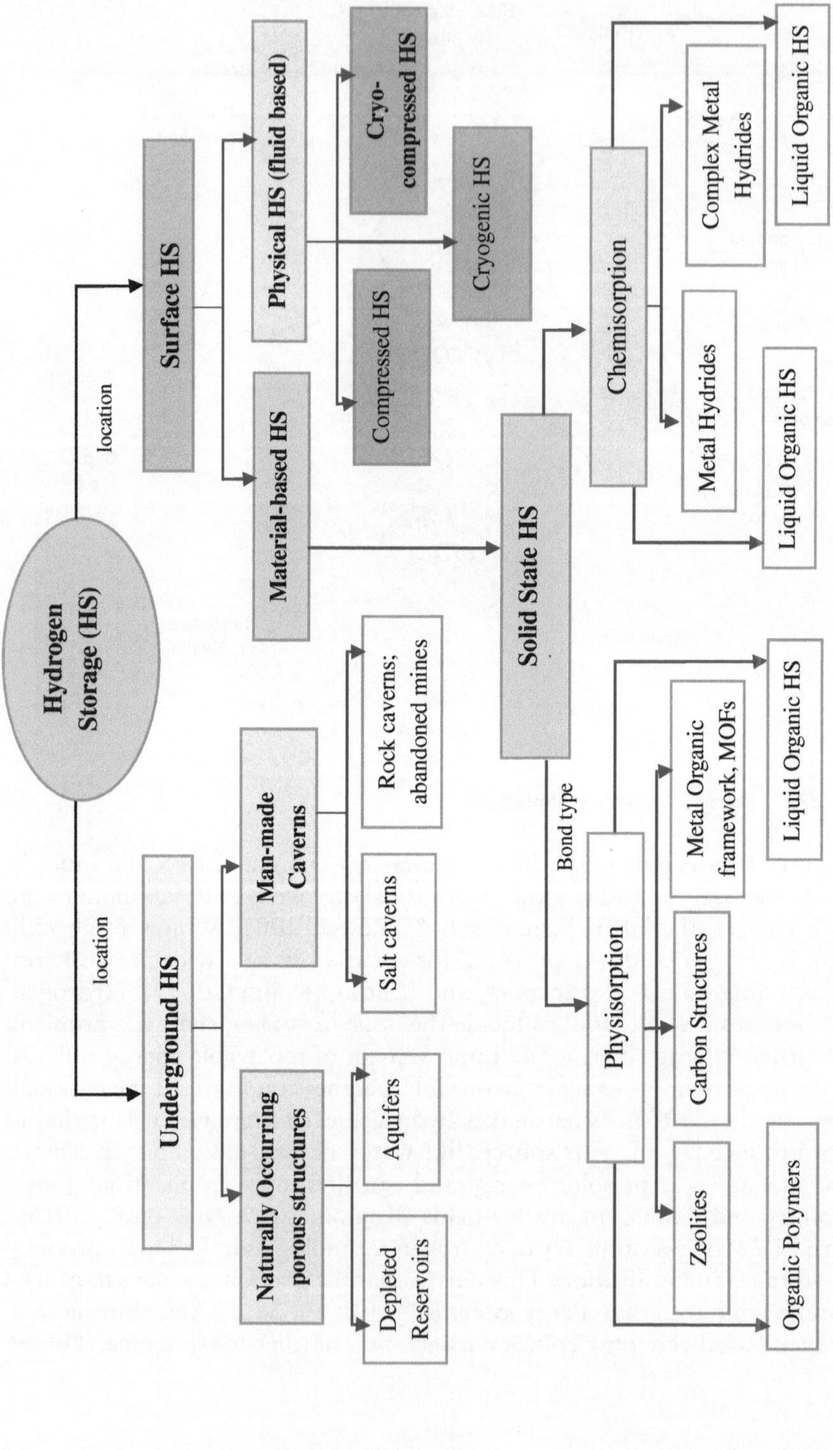

Figure 13.2 Hydrogen storage technology and material types (adapted from Nagar et al., 2023).

because hydrogen possesses very low volumetric and gravimetric densities. At standard ambient temperature and pressure, 1 kg of H_2 (g) has a volume of 11 m^3 (Züttel, 2003), thus massive volume storage will be required for many vital applications if it is to be sustainable. However, hydrogen gas has found applications for some activities by being compressed under high pressure in pressure vessels with pressures usually around 700—800 bar, and these high pressures raise safety concerns (Nagar et al., 2023; Züttel, 2003). Another application of hydrogen at ambient pressure has been in its liquid state under cryogenic temperatures (about –252°C) and in open systems allowing for huge consumption of energy (Barthélémy, 2012). Other storage applications exist such as metal halides, adsorption, absorption, and complex compounds. However, enduring obstacles remain, including issues related to reversibility stemming from hydrogen bonding, elevated pressures, and increased consumption. Consequently, to enhance hydrogen density, various approaches can be employed, such as applying energy for gas compression, reducing temperatures below a critical threshold, or mitigating the repulsive forces between hydrogen molecules through interaction with other substances (Züttel, 2003).

As a result, this chapter will systematically consider literature on efficient low-pressure hydrogen storage materials and technologies to ascertain trends and perhaps project future prospects for the hydrogen economy since the development of the hydrogen economy depends hugely on breakthroughs in hydrogen storage materials.

13.2 HYDROGEN STORAGE METHODS

The storage of hydrogen is an essential part of the hydrogen economy. It is the hinge between hydrogen production and utilisation/transportation. It currently lags in terms of development when compared with hydrogen production and hydrogen utilisation due to inherent complexities within material types, technologies, safety of utilisation, transportation, kinetics, efficiencies, and cost. Despite these challenges, it remains a potential replacement for fossil fuels and the possible key to unlocking a low-carbon economy as the world pursues zero-carbon emissions by 2050 (Nagar et al., 2023; Niaz et al., 2015). It also has the potential to serve as an energy vector for maximising the potential of renewable energy sources, which have been significantly affected by fluctuations in weather conditions. Thus, a series of interests, research, funds, and policies have been put in place to ensure that this potential is unlocked in earnest.

Numerous systems and technologies have thus undergone scrutiny in the quest for ideal materials for hydrogen storage. However, an optimal material for this purpose must exhibit a combination of high volumetric and gravimetric energy densities, compatibility with standard temperature and pressure conditions, rapid energy absorption and release kinetics, inherent safety features, and economic viability. Nevertheless, despite hydrogen gas's superior energy density per unit weight (although it is worth noting that liquid hydrogen exhibits a significantly lower energy density per unit volume when compared to hydrocarbon fuels like gasoline, boasting a density of 8 MJ/l in contrast to gasoline's density of 32 MJ/l, a fourfold difference (Osman et al., 2022)), its volumetric energy content is notably low at 9.9 MJ/m^3 when compared to alternative fuels. Thus, it necessitates a more spacious container for storage. This has made it impossible to have a singular storage solution for integrating hydrogen across a range of sectors. Hydrogen storage solutions must therefore be tailored to align with the unique requirements, capacities, and operational parameters of various segments within the energy supply chain and diverse end-user applications (Amirthan & Perera, 2022). Currently, the major hydrogen storage approaches employed can be widely summarised into two including underground hydrogen storage (UHS) and surface hydrogen storage (SHS) (Amirthan & Perera, 2022; Demirci, 2023). These classifications are summarised as seen in Figure 13.2.

13.2.1 Compressed hydrogen gas storage (CHGS)

Compressed hydrogen gas storage is the most matured, well-established form of hydrogen storage technology, which involves physical storage of compressed hydrogen in high-pressure vessels widely used in diverse industries (Abe et al., 2019; (USDoE, 2010). Its application requires that hydrogen is pressurized between 350 bar and up to 700 bar with about 11–13% of stored hydrogen energy applied towards its storage (Abe et al., 2019). The advanced development of its upstream and mid-level supply chain, encompassing production facilities and refuelling stations, facilitates swift and efficient high-pressure hydrogen refuelling, ensuring its robust compatibility with vehicular systems (USDoE, 2010).

Its application has also spread to the petroleum industry for hydro-processing, crude-oil cracking, and the production of products like ethanol, ammonia, and higher-order alcohols (Okolie et al., 2021). *However, for the most profound application, in order to increase the gravimetric and volumetric capacity of this storage system to acceptable values, it is required to compress hydrogen to high pressures ranging from 700 bar to 1000 bar (Zheng et al., 2012). Nonetheless, pressurising at such elevated pressures increases the probability of leakage (Rivard et al., 2019). Another challenge faced by this technology is the very low hydrogen mass to system mass—which is solely impacted by the light mass of hydrogen molecules as well as the mass of storage substances usually in the form of tanks, etc. The five major pressure vessel types classified based on their material are shown in Figure 13.3 with Table 13.1 showing the properties of each vessel type.

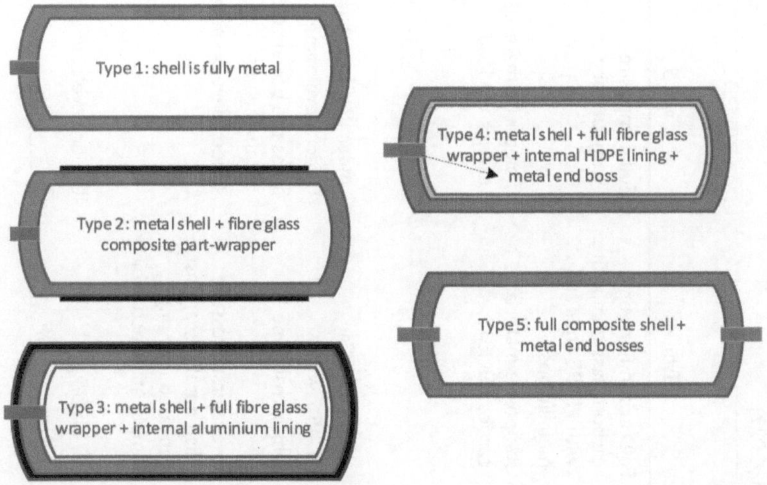

Figure 13.3 Hydrogen pressure vessel types (based on Amirthan and Perera (2022)).

13.2.2 Cryogenic hydrogen storage (CHS)

The storage of hydrogen in its liquid form at extremely low temperatures, usually below the boiling point of dihydrogen ~20 K (Yanxing et al., 2019a), is another technology employed. This storage offers a higher hydrogen content when compared to pressurized compressed gaseous hydrogen. While the density of liquid hydrogen is 71 g/L at the normal boiling point temperature and that of its gaseous counterpart at 700 bar and 288 K is 42 g/L (USDoE, 2010). Therefore, storage in liquid form provides approximately 1.7 times more density than the hydrogen content in its gaseous state (Amirthan & Perera, 2022). Nevertheless, owing to the exceptionally low boiling point of liquid hydrogen, the cooling technology necessary to maintain these temperatures consumes roughly 30% of its overall energy content (USDoE, 2010; Midilli et al., 2005) and could be higher. It is estimated to be about 4–15.0 kWh/kg compared to about 6.0 kWh/kg for compression at 70 MPa (USDoE, 2010; Folkson, 2022). Compact and lighter cryogenic pressure vessels offer superior safety benefits compared to compressed hydrogen ones. Overall, the continuous hydrogen boil-off and the energy-intensive liquefaction process limit the practicality of liquid hydrogen storage to applications where the primary focus is on energy content rather than cost. It is particularly relevant in contexts such as aircraft and space travel, where the emphasis lies on maximizing energy content rather than minimizing storage costs. Liquid hydrogen also holds promise for applications such as the transcontinental shipping of hydrogen via ships (Barthélémy et al., 2017).

Table 13.1 Pressure vessels classification and properties (Amirthan & Perera, 2022), (Xu et al., 2022)

	Pressure vessels				
	Type 1	*Type 2*	*Type 3*	*Type 4*	*Type 5*
Components	All metal construction	Steel vessel with glass fibre composite	Full composite overwrap with metal (aluminium) liner	Full carbon fibre or carbon-glass composite with HDPE liner and metallic boss	Full composite without liner
Structural load	Metal body withstands the whole load	Steel and composite materials share the load equally	The composite structure bears the most load, and the metal liner takes only about 5% mechanical load	Composite material carries the load	Composite material carries the load
Storage pressure	17.5—50 MPa	20—30 MPa	35—70 MPa	Up to 100 MPa	Under development
Gravimetric density	~1.7 wt%	≤2.1 wt%	4.2—5.5 wt%	4.4—5.7 wt%	Under development
Weight	Relatively weight is high	30—40% less than Type I	Around half the weight of Type II	Lighter than Type III	It has the potential to be the lightest
Features	Internal corrosion	Short life due to internal corrosion	High burst pressure, no permeation, galvanic corrosion between liner and fibre	lower burst pressure. High durability against repeated charging. Simple manufacturability	Under development
Cost (differs with application)	83 USD/kg	86 USD/kg	567 USD/kg	633 USD/kg	Under development

Source: Amirthan and Perera (2022) and Xu et al. (2022).

13.2.3 Cryo-compressed hydrogen storage (CcHS)

This storage system is designed to serve as an alternative to liquid hydrogen storage by combining the characteristics of both liquid and solid hydrogen, resulting in a safer storage solution [55]. It refers to the storage of H_2 at cryogenic temperatures in vessels that can be pressurised (between 25 and 30 MPa (Rao et al., 2020; Yanxing et al., 2019b). Cryo-compressed hydrogen is generated through compression at approximately –233°C [53]. It offers a storage density of 80 g/L [55], surpassing the densities of gaseous and liquid hydrogen previously discussed, and has the potential to meet the capacity targets set by the United States Department of Energy (USDOE) [56]. CcHS can achieve 91%, 175%, and 21% improvement in gravimetric capacity, volumetric capacity, and system cost reduction, respectively, when comparing a 50 MPa CcHS to a Type III 35 MPa CHS (Xu et al., 2022).

13.2.4 Material-based storage—solid-state hydrogen storage (SSHS)

Material-based hydrogen storage involves storing H_2 within either the pores of another substance, on its surface, or in combination with the host molecule. It can involve physical or chemical reactions of H_2 with host materials and offers a promising long-term solution for hydrogen storage needs (Amirthan & Perera, 2022). It is anticipated to provide enhanced storage capacities (McWhorter et al., 2011; Schlapbach & Züttel, 2001) and is usually referred to as solid-state hydrogen storage, where hydrogen is stored within a solid matrix typically without the need for high pressure or extremely low temperature like the liquid and gaseous state storage (Lai et al., 2018). Herein, material (M) forms reversible hydride bonds with hydrogen as indicated by the simplified reaction:

$$M + H_2 \leftrightarrow MH_2 \tag{13.1}$$

In the context of hydrogen storage, the solid-state method offers a distinct advantage: it facilitates the containment of substantial hydrogen quantities within a relatively constrained volume, surpassing the capabilities of both compressed gas and liquid-state methodologies (Durbin & Malardier-Jugroot, 2013). Additionally, utilising solid absorbers presents a relatively enhanced safety profile and affords a superior volumetric capacity compared to methodologies employing compressed gas and liquid-state forms (Sazelee & Ismail, 2021). Research has shown that materials considered for this purpose should exhibit a range of essential characteristics, including excellent reversibility, rapid kinetics, operation at or near ambient conditions, high gravimetric and volumetric capacities, and favourable economic considerations (Eftekhari & Fang, 2017; Hwang & Varma, 2014; Mazzolai, 2012; Roszak et al., 2016; Webb, 2015; B. Zhang & Wu, 2017). Material-based hydrogen storage systems are rooted in either physical sorption or chemical sorption mechanisms, providing a diverse range of options for efficient hydrogen storage solutions.

13.3 PHYSISORPTION OR PHYSICAL SORPTION STORAGE

Physisorption or physical sorption storage refers to a storage mechanism in which molecules or atoms (in this case, hydrogen) are attracted to a host surface due to physical interactions such as weak van der Waals forces (e.g., London dispersion forces, dipole–dipole interactions, or hydrogen bonding), the strength of which is very weak (Prachi et al., 2016a), thus, retaining (i.e., H_2) its identity upon adsorption (Aliezer Martinez-Mesa & Gotthard Seifert, 2014). It is a surface phenomenon that makes research into porous materials mandatory (Prachi R. et al., 2016a). Physisorption is a reversible process that occurs at relatively low temperatures and is not accompanied by chemical reactions (Rimza et al., 2022). Materials for hydrogen storage within this classification present various benefits, including swift kinetics, reversibility, affordability, chemical resilience, lightweight nature, substantial surface area, and relatively substantial storage capacity, [65], (Aliezer Martinez-Mesa & Gotthard Seifert, 2014). Nevertheless, their present hydrogen storage capacity falls short of meeting practical application requirements. This limitation is fundamentally linked to the electronic characteristics of the H_2 molecule and the process of hydrogen adsorption. Martinez-Mesa and Seifert (2014) proposed that the binding energy associated with physisorption falls within the range of 4–10 kJ mol^{-1} and concluded that physisorption offers ease of management with expeditious, lossless adsorption–desorption kinetics. Materials in this class include carbon structures, organic polymers, metal-organic frameworks (MOFs), liquid organic hydrogen storage (LOHS), and zeolites. Another material displaying similar traits is phosphorus nanotubes, which showcase a gravimetric storage of 16.19% at 77 K under a pressure of 50 bar (W. Zhang et al., 2023). As operating pressures shift, the internal area of single-walled phosphorus nanotubes (SWPNTs) achieves saturated hydrogen storage. However, as pressure rises, the external area remains highly promising for additional hydrogen storage.

13.3.1 Carbon-based materials

These materials exhibit diverse structures and configurations, which have been extensively investigated for their hydrogen storage capabilities. They possess the flexibility to be readily tailored to suit different applications and are commonly characterised by their cost-effectiveness. The major features enabling hydrogen storage for carbon structures, according to researchers, are the possession of large specific surface areas (Prachi R. et al., 2016a), pore geometry, and capability to be doped (Dewangan et al., 2022; (Prachi R. et al., 2016a). Dewangan et al. (2022) elucidated in their review that the storage potential of carbon materials, encompassing examples such as carbon nanotubes (both single walled, SWNT, and multiwalled, MWNT), activated carbon (AC), carbon aerogels (CA), templated carbon (TC), and graphene, is intrinsically linked to their pore structure and overall surface area. Nevertheless, achieving optimal hydrogen concentration storage in these materials necessitates considerably high pressure and temperature. Studies showed that activated carbon (AC) can store 5 wt% of hydrogen at low temperature, 77 K and 30–60 bar; porous carbon prepared by template and with a surface area of 3200 m^2/g is observed to store 7% at 77 K and 20 bar (Prachi R. et al., 2016a). The variations in storage capacities detailed in various literature often stem from differences in the fabrication techniques used, especially in the production of carbon nanotubes (CNTs). Echoing these findings, studies by Rimza et al. (2022) further underscored that the hydrogen storage efficacies of microporous carbon materials are significantly influenced by their specific surface area or pore volume, evident at both 77 K and ambient temperatures. While these materials showcase commendable cyclic stability, there exists an accompanying premium in terms of cost (Prachi R. et al., 2016a). Additionally, their present hydrogen storage capacity falls short of meeting practical application requirements. This limitation is fundamentally linked to the electronic characteristics of the H_2 molecule and the process of hydrogen adsorption. For example, the hydrogen storage capacity of activated carbons at room temperature reaches approximately 1 wt% under moderate pressures (50–100 bar) and about 3.0 wt% under extremely high pressures, such as 500 bar.

13.3.2 Organic polymers

In the realm of low-pressure hydrogen storage, certain materials have gained prominence due to their unique atomic constitution, predominantly comprising C, H, N, and O. Their appeal is further accentuated by regulated porosity, expansive specific surface areas, robust stability, structural adaptability, and cost-effectiveness. These materials offer the boon of structural modifiability and multifaceted functionalisation. Noteworthy among these organic polymers are COFs (crystalline nanoporous organic polymers rooted in carbon), PAFs (having a diamond-esque tetrahedral configuration), HCPs, CMPs, and PIMs. Both COFs and PAFs are characterised by expansive surface areas, featherweight densities, and commendable thermal and mechanical resilience. Notably, COFs can achieve hydrogen storage capacities spanning from 3.5 to 10.0 wt% under elevated pressures (80–100 bar) but predominantly at sub-zero temperatures (77 K). Analogously,

PAFs have the potential to manifest superior hydrogen capacities, oscillating between 7.0 and 16.03 wt% under ambient temperatures and heightened pressures, a feat achieved post-metal doping with Li, Mg, and Ca. However, their overarching benefits are somewhat offset by prohibitive costs, subpar processability, and compromised mechanical attributes, particularly in their powdery state. In juxtaposition, HCPs, CMPs, and PIMs score favourably in processability and economic feasibility. However, their restricted surface area caps their hydrogen storage prowess. Specifically, HCPs register hydrogen assimilation capacity of under 4 wt% at frigid conditions and a mere 0.5 wt% closer to ambient temperatures. CMPs, under standard atmospheric pressure at 77 K, typically log values below 3 wt%. Intriguingly, Li-doping escalates this to 6.1 wt%. PIMs oscillate between a capacity of 3.7 to 4.1 wt%.

13.3.3 Zeolites

Zeolites are three-dimensional crystalline aluminosilicates with a high internal surface area of up to 1000 m^2g^{-1}. These structures, with expansive internal surface areas, possess organised architectures and consistent pore dimensions (Hassan et al., 2021; Prachi R. et al., 2016a). They are built of TO_4 tetrahedrons sharing all four corners, where T indicates Si^{4+} and Al^{3+} ions. The process of encapsulation compels hydrogen into these zeolite cavities under specific high temperature and pressure conditions, maintaining its confinement even at ambient temperatures. The gas storage capabilities of zeolites, stemming from their adsorption properties, render them appealing for such applications (Prachi R. et al., 2016a). Consequently, zeolites have been researched as prospective materials for hydrogen storage. At standard room conditions, zeolites exhibit hydrogen storage capacities just under 0.1 wt% (Langmi et al., 2022). However, when the temperature is lowered to 77 K, these capacities can achieve levels as high as 2.55 wt% (Langmi et al., 2022; Mohan et al., 2019). NaY zeolite and CaX zeolite attained a maximum gravimetric hydrogen storage capacity of 1.81 wt% at 15 bar and 77 K and of 2.19 wt% at 15 bar pressure and 77 K temperature, respectively (Prachi R. et al., 2016a; Züttel, 2003). It is noteworthy that such capacities can be augmented through the introduction of metals like palladium or via processes such as bridging and spillover.

13.3.4 Metal-organic frameworks (MOFs)

Metal-organic frameworks (MOFs) are organometallic materials comprising a metallic core linked with organic ligands that connect with one another (Nagar et al., 2023). These crystalline structures, characterised by their porous nature, facilitate hydrogen storage by allowing it to permeate their crystalline composition (Niaz et al., 2015). Notably, MOFs can boast specific surface areas reaching an impressive 4800 m^2g^{-1} and are known for their lightweight attributes and heightened enthalpy levels (Langmi et al., 2014) . The absorption process in MOFs is remarkably swift, mere seconds (Yuan et al., 2018), with the rate being a function of both the diffusivity of hydrogen within the MOF and the size of the crystalline structures. MOFs showcase the potential to achieve hydrogen storage capacities of 5–15 wt% at 77 K and high pressure and about 1 wt% at ambient temperatures (Nagar et al., 2023; Prachi R. et al., 2016a). Research has demonstrated that the introduction of nanoparticles, specifically platinum and palladium can bolster these storage capacities in MOFs (Yuan et al., 2018). Moreover, it is posited that the integration of metal ions such as Li^+, Cu^{2+}, and Mg^{2+} could further enhance these capacities under standard conditions (Nagar et al., 2023). Despite being able to store hydrogen up to the specified target by USDOE, MOFs present certain challenges. Their thermal conductivity is notably subpar, making the orchestration of MOF-centric storage systems a demanding task (Rivard et al., 2019). Added to this is the financial burden, stemming from the costs related to primary ingredients like organic ligands and metal salts, processing methodologies that rely on non-recyclable organic solvents and substantial initial capital investments. These factors cumulatively serve as roadblocks to MOFs' widespread industrial adoption (Nagar et al., 2023).

13.3.5 Liquid organic hydrogen carriers (LOHCs)

LOHCs are liquids or low-melting solids, which have the capability to undergo reversible exothermic hydrogenation and endothermic dehydrogenation under certain conditions, facilitated by the presence of catalysts (Chu et al., 2023). In the context of hydrogen storage, LOHCs favour a low temperature and a high-pressure process (Chu et al., 2023). The gravimetric capacities of hydrogen in LOHCs typically vary between 6 wt% and 8 wt%; however, the process consistently absorbs external heat, primarily due to the energy disparity between hydrogen atom dissociation and the C-H bond's activation energy (Forberg et al., 2016). This highlights the need to address the significant energy consumption from temperature variations in the reaction and the diminishing catalytic activity (Chu et al., 2023; Forberg et al., 2016). Additionally, LOHCs stand out due to their cost-effectiveness, enhanced safety features, and notable reversibility. LOHCs are attractive for their high hydrogen storage capacities (Table 13.2), but challenges in their application such as its high temperature dehydrogenation process, including complex catalytic conditions, large energy consumption, low hydrogen releasing speed, and toxic by-products generated from side reactions, have been a major setback (Xu et al., 2022). LOHCs generally require much energy during their dehydrogenation.

Table 13.2 Properties of selected LOHC compounds

Properties	*Toluene–MCH*		*Naphthalene–decalin*		*Benzene–cyclohexane*		*DBT–PDBT*	
	Toluene	*MCH*	*Naphthalene*	*Decalin*	*Benzene*	*Cyclohexane*	*DBT*	*PDBT*
Chemical formula	C_7H_8	C_7H_{14}	$C_{10}H_8$	$C_{10}H_{18}$	C_6H_6	C_6H_{12}	$C_{21}H_{20}$	$C_{21}H_{38}$
Melting point	−95	−127	80.3	Cis −43.0; trans −30.4	5.5	6.5	−30	−34
Boiling point	111	101	218	Cis −94.6; trans −185.5	80	81	278	395
Phase at ambient	Liquid	Liquid	Solid	Liquid	Liquid	Liquid	Liquid	Liquid
Hydrogenation								
Gravimetric density (wt%)	6.16		7.29		7.2		6.2	
Volumetric density (kg/m³)	47.4		65.4		55.9		57	
Heat of reaction (kJ/mol)	204.8		319.5 (cis), 332.5 (trans)		205.9		588.5	
Temperature (K)	200–300°C		150–250°C		150–250°C		180	
Pressure (bar)	10–50		20–50		10–50		10–50	
Dehydrogenation with a selected catalyst								
Temperature (K)	250–350		300–350		330		260–320	
Pressure (bar)	1–5		1–4		1–4		1–5	
			319.5 (cis), 332.5 (trans)					
Advantages	Both liquid in wide range temperature		Relatively high H_2 content		Relatively high H_2 content		Higher intrinsic safety, good thermal stability	
Disadvantages	Irritative, inflammable, volatile		Different phase (difficulties in storage and transportation)		High melting point (possibility of phase change)		High energy for dehydrogenation	

13.4 CHEMISORPTION OR CHEMICAL STORAGE

Chemisorption in hydrogen storage occurs when the hydrogen molecule dissociates into discrete hydrogen atoms on the material's surface, subsequently diffusing and forming a chemical bond with the substrate (Prachi et al., 2016b). These bonds differentiate hydrides into metal hydrides (defined by metallic bonding), ionic hydrides (characterised by ionic bonding), and covalent hydrides (exhibiting covalent bonding). The latter two are frequently linked to complex hydrides (Barthélémy, 2012). The suitability of hydrides for storage applications is contingent upon their absorption thermodynamic characteristics and kinetics. It is noteworthy that liquid hydrides are more amenable for use as fuels at ambient conditions, in contrast to solid hydrides which typically require temperatures nearing 120°C for effective hydrogen discharge (Sazelee & Ismail, 2021). However, both hydrides tend to operate at pressures lower than all other storage types considered. Concurrently, metal hydrides and complex hydrides serve as the most referred examples of materials that undergo chemisorption. Relative to physisorption, chemisorption is perceived as a propitious medium for hydrogen storage in solid state, attributed to its elevated hydrogen capacity. To illustrate, magnesium hydride (MgH_2) can discharge close to 7.6 wt% of hydrogen ideally upon heating to approximately 400°C, while lithium alanate ($LiAlH_4$) can desorb around 7.9 wt% of hydrogen when subjected to temperatures nearing 350°C (Sazelee & Ismail, 2021).

13.4.1 Metal hydrides (material properties)

Metal hydride (MH)–hydrogen storage technology represents a chemical storage approach in which hydrogen is retained in a metal through a chemical absorption mechanism. They are known for their high volumetric efficiency, moderate to high efficiency, and low-pressure desorption/absorption requirements (Niaz et al., 2015). MH possesss the capability to store a higher hydrogen volume safely and efficiently in comparison to gaseous and liquid storage. This has made it possible to be considered for vehicular use and in combination with physical hydrogen storage vessels (Prachi R. et al., 2016b; Rusman & Dahari, 2016). When metals are exposed to hydrogen gas, hydrogen molecules first adhere to the metal's surface. In this process, when a solid solution with a low hydrogen-to-atomic metal ratio (<0.1) is created, it is identified as the ∞-phase. As more energy is introduced to the system and the hydrogen-to-metal atomic ratio surpasses 0.1, the hydride phase emerges and proliferates, known as the β-phase. When both phases exist concurrently, a flat plateau is observable in the pressure–composition isotherm (PCI). The extent of this plateau indicates the quantity of hydrogen that can be stored (Bellosta von Colbe et al., 2019a; W. Liu et al., 2018). The PCI together with the van't Hoff plots is a major tool to ascertain the pressure and temperature working parameters for various metal hydride substances (Klopčič et al., 2023; W. Liu et al., 2018). The curves in the van't Hoff plot can be described with the following equation:

$$\ln\left(\frac{p_{eq}}{p_0}\right) = -\frac{H}{RT} + \frac{S}{R} \tag{13.2}$$

where R is the gas constant, T is the temperature, ΔS and ΔH are reaction entropy and enthalpy of the absorption/desorption reaction, respectively, the reference pressure p_0 (usually atmospheric pressure of approximately 1 bar). p_{eq} is the resulting equilibrium pressure in the middle of the plateau of a PCI curve.

The ΔS ΔS of hydrogen gas dominates the total reaction entropy. Therefore, while most metal hydrides have similar entropy values due to the gas's nature (Lototskyy et al., 2014), their enthalpy, reflecting in the van't Hoff plot's slope, can vary. Klopčič et al. (2023) specified that low-temperature hydrides, with lesser enthalpy values, operate near ambient conditions, making their heat management critical. This is because the equilibrium pressure, where absorption equals desorption, increases with temperature, and as most hydrides release heat during absorption, this can raise the required hydrogen pressure leading to a problem with heat management. The process becomes inefficient, resulting in prolonged filling times (Bellosta von Colbe et al., 2019b). The PCI (pressure–composition isotherm) diagram illustrates these relationships, highlighting the importance of understanding temperature, pressure, and material-specific characteristics for efficient hydrogen storage in metal hydrides. Thus, in the article by Klopčič et al. (2023), it is required to understand these properties in order to be able to control de/hydrogenation of MH at low pressure. Also, one serious issue in the metal hydrides is sensitivity towards oxygen and moisture, which leads to surface contamination when these hydrides are exposed to atmospheric air (Thangarasu et al., 2023).

13.4.2 Common metal hydrides for low-pressure hydrogen storage

Since high-pressure gaseous hydrogen storage and low-temperature liquid hydrogen storage are energy-intensive and have safety challenges (Abdalla et al., 2018; Pukazhselvan et al., 2012), much research has shown that low-pressure storage might be the breakthrough that is required, and the ability of metal hydrides to operate effectively at ambient temperature and pressure makes them more practical candidates from an application perspective. Metal hydrides, as explained earlier, have properties that make them fit for low-pressure hydrogen storage. For brevity, only the commonly discussed MH will be considered for this study.

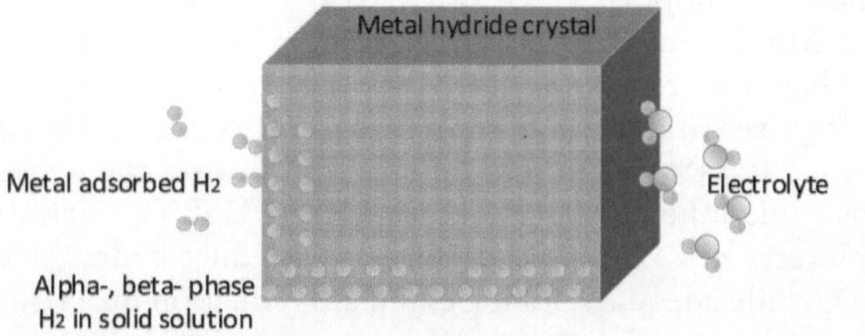

Figure 13.4 Dispersion of hydrogen molecules through a metal substance—forming hydrides. Image based on original by Amirthan and Perera (2022).

13.4.2.1 Magnesium hydride, MgH_2

Considered one of the most promising MHs due to its high natural abundance, operational safety of its hydrides when juxtaposed with other adjacent reactive lightweight metals, non-toxicity, affordability, high-energy density, and high reversible hydrogen storage capacity (7.69 wt%), MgH_2 has been widely researched (Li et al., 2018; Shao et al., 2012; Yang et al., 2023). Though challenges like stable thermodynamics, 76 kJ/mol-H_2 (Yartys et al., 2019), slow kinetics of hydrogen desorption, high delivery temperature (300°C—400°C) (W. Liu et al., 2018), and swift capacity degradation have significantly impeded its practical utilisation (Bolarin et al., 2022). It reacts with hydrogen forming a stable hydride:

$$Mg + H_2 \leftrightarrow MgH_2 + H \tag{13.3}$$

ΔH is the heat of reaction that describes the thermodynamic stability of the hydride. MgH2 has theoretical gravimetric and volumetric values of 7.6 wt% and 110 kgm³, respectively, which is in line with the USDOE target for vehicular applications. Thus, it is considered a potential hydride for needs within the hydrogen economy. However, its thermodynamic stability with enthalpy and entropy values of 74.7 kJ mol^{-1} and 130 J K^{-1} mol^{-1}, respectively, restricts its hydrogen release and absorption rates at ambient temperatures (Vigeholm et al., 1983). Thus, low pressure and a higher temperature appear to be favourable during dehydrogenation (Vigeholm et al., 1983). Different research efforts have therefore been employed to improve the thermodynamic properties of MgH_2—nano-structuring by ball milling, mechanical alloying, catalysis, nanoconfinement, and the use of additives (Wei Liu et al., 2014; J. Zhang et al., 2019; X. L. Zhang et al., 2020; Zhou et al., 2013).

Yang et al. (2023) discussed on approaches to improve the kinetics and thermodynamics of MgH_2 stating that catalytic materials like transition metals, metal oxides, metal halides, metal sulphides, metal hydrides, and various carbon materials could help bolster the performance of MgH_2—a process known as doping. Cui et al. (2014) showed that Ti can act as a catalyst for MgH_2, enabling it to release 6.35 wt% hydrogen in 10 min at 275°C. They also reported that MgH_2 catalysed by Nb_2O_5 showcased the most effective catalytic action among metal oxides. This combination could release 7.0 wt% H_2 in just 60 s at 300°C, and the dehydrogenated compounds could take in 7.0 wt% H_2 in 130 s at 250°C, usually at ambient pressure. Nevertheless, excessive doping could reduce hydrogen storage (Lu et al., 2021), while adequate doping could aid reduction in activation energy for hydrogen release and vice versa (Huang et al., 2021; Peng et al., 2021), as well as reduction in structure stability and reduction in dehydrogenation temperature (Edalati et al., 2018; Ouyang et al., 2020) all of which bolsters the hydrogen capability of MgH_2. Table 13.3 shows some of the research and operational information for the hydrogen capability of MgH_2.

13.4.2.2 Complex hydrides

Complex hydrides are compounds that contain metal cations and intricate anionic hydride species (Orimo et al., 2007; Schueth et al., 2005). These materials possess high hydrogen content, enabling the storage and release of significant amounts of hydrogen (Zhou et al., 2013). The most common examples of complex hydrides include borohydrides and alanates, i.e., $NaBH_4$, $NaAlH_4$, $LiAlH_4$, etc. Alanates and borohydrides undergo reversible hydrogen desorption in the presence of selected catalysts, such as Ti-based catalysts, to obtain an improved desorption capability (Kojima, 2019). Sazelee and Ismail (2021) stated in their work that $LiAlH_4$ decomposes in three steps, with an inception reaction occurring at 150°C, releasing about 5.3 wt% of hydrogen, and an intermediate reaction occurring at 180°C, releasing approximately 2.6 wt% of hydrogen. The final step is said to occur at temperatures >350°C with an equal dehydrogenation capacity of approximately 2.6 wt% of hydrogen. $LiBH_4$ can be reverted, producing end products of LiH and B. To reform $LiBH_4$, hydrogen is absorbed at 963 K and 20 MPa. Additionally, borohydrides offer compelling hydrogen desorption reactions ideal for hydrogen storage, with a notable theoretical hydrogen density exceeding 9.6 wt%. Their stability as complex hydrides is evident from their standard heat of formation values.

13.5 OTHER MATERIAL CATEGORISATIONS

It is worthy to state that throughout the material descriptions, certain special sets of categorisations which were not initially described have been mentioned. This is because despite the motive to have a storage system that can cater for set targets like the USDOE for vehicles, hydrogen storage in the bulk of numerous substances (though feasible) still has not been able to

Table 13.3 Comparative analysis of all storage types

Storage method	*Operational pressure range*	*Storage density range (wt%)*	*Operating temperature range (K)*	*Used up energy for operation (MJ/kg-H_2)*	*Safety*	*Reference*
Compressed hydrogen	300–700 bar (Xu et al., 2022)	5–6	Ambient (298)	~13	Potential risks at high pressures	Yanxing et al. (2019a)
Liquid hydrogen	~1–6 bar (Yanxing et al., 2019a)	36.6 kg/m^3 (Xu et al., 2022)	20.28	~33	Risks due to cryogenic temps	Agarwal and Dondapati (2020); Chu et al. (2023); Yanxing et al. (2019a, 2019b)
Metal hydrides	Ambient	Varies	Ambient to 400 (varies by metal)	15–40 (varies by metal)	Generally safe	Cento et al. (2007); Jia et al. (2022); Langmi et al. (2014); Murray et al. (2009); Silvera and Dias (2021)
Chemical hydrides	Ambient	Varies	Ambient to 400 (depends on hydride)	50–200 (depends on hydride)	Controlled release; safe with precautions	Mohan et al. (2019)
Carbon-based materials	<20 bar	Up to 7.5	77 (for optimal uptake)	Not specified	Safe under normal conditions	[5], Rehman et al. (2023); Rimza et al. (2022)

cater for required gravimetric and volumetric capacities for efficient usage (Lai et al., 2018). This led to the innovation of novel methods that provide precise management of how hydrogen interacts with these materials, enabling hydrogen absorption and release at or near ambient temperatures (below 100°C, akin to the temperature of a proton exchange membrane fuel cell) within a brief timeframe (minutes) and at moderate hydrogen pressures (below 100 bar) (Kondo-François Aguey-Zinsou & Ares-Fernández, 2010). This means that the ionic and hydrogen diffusion within the material must be relatively fast (Kondo-François Aguey-Zinsou & Ares-Fernández, 2010). As suggested by Lai et al. (2018), it is expedient to apply an alternative approach through particle resizing to exploit the nanosizing effect on both the hydrogen binding strength and diffusion paths for improved thermodynamics and fast kinetics (Lai et al., 2018; Kondo-François Aguey-Zinsou & Ares-Fernández, 2008). The article by Lai et al. (2018) further showed using the relation:

$$\Delta G = \Delta G0 + RT \ In \ \frac{p}{p_0} + \frac{3\ V_M \Delta_{M \to MH_2}(\gamma, r)}{r} \tag{13.4}$$

where r = particle size, p = pressure, T = temperature, R = Universal gas constant, VM = molar volume, and γ = nanoparticle surface energy. In a state of equilibrium (where $\Delta G = 0$), using the van't Hoff relation, a decrease in particle size can result in a decreased reaction enthalpy, provided that it $V_M \Delta_{M \to MH_2}$ is positive.

$$\boldsymbol{In}\ p_{eq} = \frac{1}{RT}\left(\Delta H0 + \frac{3\ V_M \Delta_{M \to MH_2}(\gamma, r)}{r}\right) - \frac{\Delta S}{R} \tag{13.5}$$

Consequently, this leads to a higher equilibrium pressure (p_{eq}), structural rearrangements, and change in atomic distances, affecting the stability of the hydride phase. Nanosizing can lead to closer anion–cation proximity, strengthening polar bonds and subsequently lowering hydrogen release temperatures (Grochala & Edwards, 2004). Decreasing particle size addresses hydrogenation and dehydrogenation challenges by diminishing the distances for solid-state diffusion and enhancing both surface area and surface reactivity (Schneemann et al., 2018; Schulz et al., 1999). This idea has led to diverse means of producing nanosized materials, such as mechanical or grinding or ball milling substances into finer grains (Varin et al., 2009). Ball milling is the most prevalent technique for producing nanoscale materials. In this process, high-speed balls persistently strike the materials, reducing particle sizes and amplifying specific surface areas. This method introduces defects and generates metastable phases, which enhance the material's hydrogen uptake and release capabilities.

Research by Ouyang, Liu, et al. (2020;) and Yang et al. (2023) has led to further classification of storage substances in the following. (a) Macroporous substances: materials with pore diameters larger than 50 nm, which can facilitate fast bulk H_2 transport by improving the kinetics of absorption/desorption properties of a storage material. Although they may experience limited hydrogen storage capacities, they play a vital role in real-world applications. (b) Mesoporous substances: these have pore diameters ranging from 2 nm to 50 nm. Due to their intermediate pore size, these materials exhibit high surface areas and pore volumes, making them attractive for hydrogen storage. The large surface area provides numerous adsorption sites, enabling substantial hydrogen uptake, especially under cryogenic temperatures. (c) Microporous substances: materials with pore sizes <2 nm. Zeolites, activated carbons, and some metal-organic frameworks (MOFs) are classic examples of microporous materials (Murray et al., 2009). They facilitate physisorption of hydrogen molecules within their confined spaces. (d) Nanoporous substances: have pore sizes typically less than 100 nm. They encompass both microporous and mesoporous substances. Their structure provides vast surface areas and confinement effects essential for hydrogen storage. Certain MOFs and covalent organic frameworks (COFs) are examples of nanoporous structures optimised for hydrogen uptake (Langmi et al., 2014). Through the integration of catalysis and nanoscale particle manipulation, remarkable enhancements in kinetics, thermodynamics, and reversibility have been realised (Palm et al., 2018; Tuul & Palm, 2021; Cento et al., 2007).

13.6 COMPARATIVE ANALYSIS OF HYDROGEN STORAGE MATERIALS

Table 13.3 highlights, on a high level, the major differences that exist between all the discussed storage approaches. From the table, we can see that metal hydride and complex hydride including alloy, which is not included here, operate at a low varying pressure usually between ambient pressure, 1 bar and about 50 bar depending on the application and the choice of material being used. On the other hand, however, the low pressure is balanced by a relatively high temperature required for desorption to occur. Alternatively, the compressed gas hydrogen storage exhibits a high-pressure and a low-temperature operation for its operation. The gravimetry from this storage is the lowest among all other types. In the same vein, the cryogenic storage medium, which tends to have a much fairer gravimetry and density, is faced with potential leakage as well as hydrogen loss in the long run of its usage. The table, therefore, gives a quick peek at the usage operations of all material types.

13.7 MARKET ANALYSIS AND TRENDS FOR HYDROGEN STORAGE MATERIALS

The current hydrogen market is dominated by the physical storage of hydrogen in liquid and gaseous forms. However, storage in other modules, which have been discussed, is also expected to grow at a significant rate. Knowing the role hydrogen material advancements might play and the urgent need to achieve net-zero by 2050, the market has attracted stakeholders who are key players towards achieving the inherent potential of storage materials. Due to limitations with available information and articles, the hydrogen storage market by type is not available for easy access. Thus, this section of the article would only be limited to data sourced from presumably trusted online sources with some level of assumption being made to allow for calculations and forecasts. According to a report (Market Research Community, 2023), the global hydrogen storage market is valued at US$15.4 billion in 2022 and projected to reach US$28.4 billion by 2030 while growing at a compound annual growth rate (CAGR) of 7.95% from 2023 to 2030. By assuming that the rate would stay the same, the market value can be further forecasted up to the year 2050—the expected year for the full actualisation of the net-zero economy. The value of the hydrogen storage market is, therefore, calculated and estimated to be valued at US$131.15 billion by 2050. This clearly shows that the hydrogen storage market is likely to grow up to 850% of what it was in 2022. While the accuracy of this assumption hinges on the premise that all factors stay consistent as indicated by Market Research Community (2023). However, there is a likelihood to be wrong that if major economies are to meet various net-zero targets set by their government, there would likely be a sharper spike in the market size. Figure 13.5 therefore shows the forecasted values of the hydrogen storage market. The major driver for this growth is the increased demand for hydrogen storage in different sectors as well as the adoption of high usability of hydrogen storage in the power generation and energy storage industry.

Also, the worldwide market for hydrogen storage tanks, which is linked to storage materials for high-pressure vessels, achieved a valuation of approximately USD 164.22 million in 2022 and is anticipated to expand to around USD 1.56 billion by 2030, experiencing a robust compound annual growth rate of approximately 32.87% during the period from 2023 to 2030. Figure 13.6 shows the calculated expected market value for global hydrogen storage tanks up to the year 2035, with an assumption that the set target for achieving a net-zero economy is pursued collectively and that the compound annual growth rate remains the same up to 2035. However, there is a possibility that this value might be lower or higher depending on whether other hydrogen storage forms gain more acceptance or not at the end of the year considered.

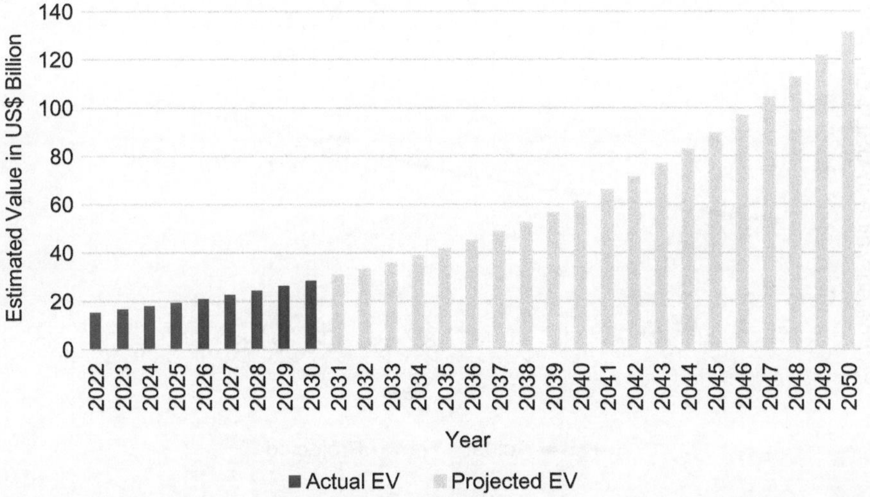

Figure 13.5 Graph of global hydrogen storage market for the year 2022–2023 and extrapolated to cover through to 2050 (data source: Market Research Community 2023).

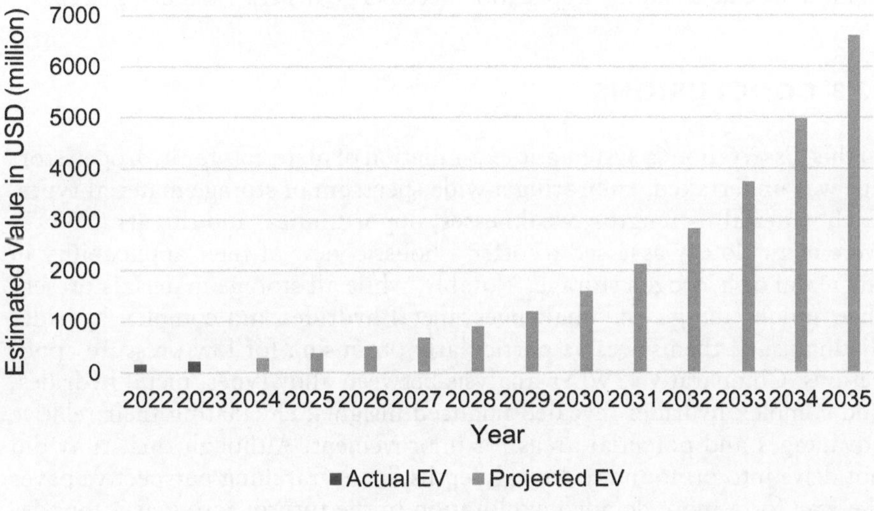

Figure 13.6 Graph showing the estimated value for the global hydrogen tanks storage market for the year 2022–2023 and extrapolated to cover through to 2035 (data source: The Insight Partners, 2021).

Figure 13.7 shows that the market size for hydrogen storage alloys stood at US$2.41 billion in 2019 and was projected to reach US$4.54 billion by 2028, growing at a CAGR of 7.3%. It is worthy of note that North America held the dominant position as the primary market for hydrogen storage alloys

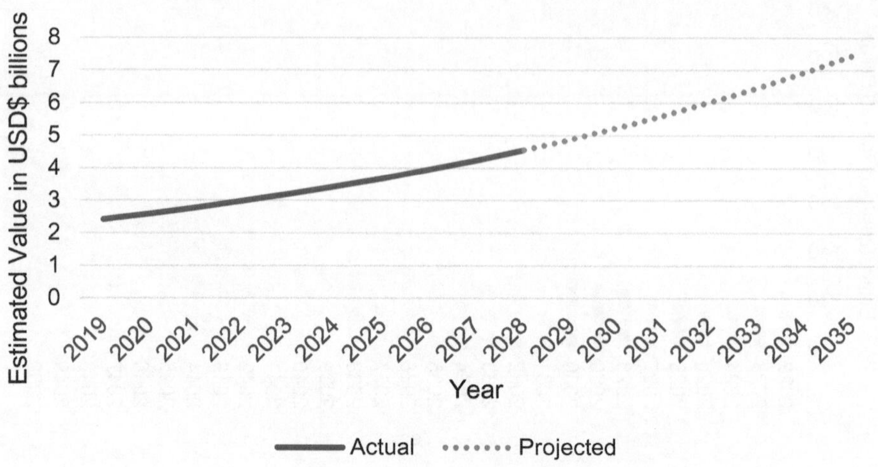

Figure 13.7 Global hydrogen storage alloys market size forecasted up to 2035.

(The Insight Partners, 2021). A significant catalyst for the expansion of the hydrogen storage alloys market in the region lies in their tailor-made designs to meet specific customer application needs (The Insight Partners, 2021).

13.8 CONCLUSIONS

In this dissertation, a systematic examination of materials for hydrogen storage was undertaken, embracing a wide spectrum of storage material types. Each material's strengths, weaknesses, opportunities, and threats (SWOT) were meticulously assessed to offer a holistic view of their applicability in the realm of hydrogen storage. Notably, while all storage materials present their unique merits and challenges, metal hydrides and complex hydrides distinguished themselves as particularly promising for low-pressure applications. Comparative SWOT analysis between alloy types, metal hydrides, and complex hydrides revealed nuanced insights, elucidating their relative advantages and potential areas for improvement. Although this study did not delve into profound technical depths, its overarching perspective paves the way for a more detailed exploration in the future, acting as a foundational guide for researchers and industry experts seeking to harness the full potential of these materials in the evolving energy landscape.

REFERENCES

Abdalla, A. M., Hossain, S., Nisfindy, O. B., Azad, A. T., Dawood, M., & Azad, A. K. (2018). Hydrogen production, storage, transportation and key challenges with applications: A review. *Energy Conversion and Management*, *165*, 602–627. https://doi.org/10.1016/j.enconman.2018.03.088

Abe, J. O., Popoola, A. P. I., Ajenifuja, E., & Popoola, O. M. (2019). Hydrogen energy, economy and storage: Review and recommendation. *International Journal of Hydrogen Energy*, *44*(29), 15072–15086. https://doi.org/10.1016/j.ijhydene.2019.04.068

Agarwal, R., & Dondapati, R. S. (2020). Numerical investigation on hydrodynamic characteristics of two-phase flow with liquid hydrogen through cryogenic feed lines at terrestrial and microgravity. *Applied Thermal Engineering*, *173*, 115240. https://doi.org/10.1016/j.applthermaleng.2020.115240

Aguey-Zinsou, K.-F., & Ares-Fernández, J.-R. (2008). Synthesis of colloidal magnesium: A near room temperature store for Hydrogen. *Chemistry of Materials*, *20*(2), 376–378. https://doi.org/10.1021/cm702897f

Aguey-Zinsou, K.-F., & Ares-Fernández, J.-R. (2010). Hydrogen in magnesium: new perspectives toward functional stores. *Energy & Environmental Science*, *3*(5), 526. https://doi.org/10.1039/b921645f

Aliezer, M.-M., & Gotthard, S. (2014). Adsorption of molecular hydrogen on nanostructured surfaces. *Revista Cubana de Física, 31*(1), 32–34..

Amirthan, T., & Perera, M. S. A. (2022). The role of storage systems in hydrogen economy: A review. *Journal of Natural Gas Science and Engineering*, *108*, 104843. https://doi.org/10.1016/j.jngse.2022.104843

Ball, M., & Wietschel, M. (2009). The future of hydrogen – opportunities and challenges☆. *International Journal of Hydrogen Energy*, *34*(2), 615–627. https://doi.org/10.1016/j.ijhydene.2008.11.014

Barthélémy, H. (2012). Hydrogen storage – Industrial prospectives. *International Journal of Hydrogen Energy*, *37*(22), 17364–17372. https://doi.org/10.1016/j.ijhydene.2012.04.121

Barthelemy, H., Weber, M., & Barbier, F. (2017). Hydrogen storage: Recent improvements and industrial perspectives. *International Journal of Hydrogen Energy*, *42*(11), 7254–7262. https://doi.org/10.1016/j.ijhydene.2016.03.178

Bellosta von Colbe, J., Ares, J.-R., Barale, J., Baricco, M., Buckley, C., Capurso, G., … Dornheim, M. (2019a). Application of hydrides in hydrogen storage and compression: Achievements, outlook and perspectives. *International Journal of Hydrogen Energy*, *44*(15), 7780–7808. https://doi.org/10.1016/j.ijhydene.2019.01.104

Bellosta von Colbe, J., Ares, J.-R., Barale, J., Baricco, M., Buckley, C., Capurso, G., … Dornheim, M. (2019b). Application of hydrides in hydrogen storage and compression: Achievements, outlook and perspectives. *International Journal of Hydrogen Energy*, *44*(15), 7780–7808. https://doi.org/10.1016/j.ijhydene.2019.01.104

Bolarin, J. A., Zou, R., Li, Z., Zhang, Z., & Cao, H. (2022). MXenes for magnesium-based hydrides: A review. *Applied Materials Today*, *29*, 101570. https://doi.org/10.1016/j.apmt.2022.101570

Cento, C., Gislon, P., Bilgili, M., Masci, A., Zheng, Q., & Prosini, P. P. (2007). How carbon affects hydrogen desorption in NaAlH4 and Ti-doped NaAlH4. *Journal of Alloys and Compounds*, *437*(1–2), 360–366. https://doi.org/10.1016/j.jallcom.2006.09.025

Chu, C., Wu, K., Luo, B., Cao, Q., & Zhang, H. (2023). Hydrogen storage by liquid organic hydrogen carriers: Catalyst, renewable carrier, and technology – A review. *Carbon Resources Conversion*, *6*(4), 334–351. https://doi.org/10.1016/j.crcon.2023.03.007

Cui, J., Liu, J., Wang, H., Ouyang, L., Sun, D., Zhu, M., & Yao, X. (2014). Mg–TM (TM: Ti, Nb, V, Co, Mo or Ni) core–shell like nanostructures: synthesis, hydrogen storage performance and catalytic mechanism. *Journal of Materials Chemistry A*, *2*(25), 9645–9655. https://doi.org/10.1039/C4TA00221K

Demirci, U. B. (2023). Exploring the technological maturity of hydrogen production by hydrolysis of sodium borohydride. *International Journal of Hydrogen Energy*, *48*(76), 29682–29698. https://doi.org/10.1016/j.ijhydene.2023.04.176

Dewangan, S. K., Mohan, M., Kumar, V., Sharma, A., & Ahn, B. (2022). A comprehensive review of the prospects for future hydrogen storage in materials-application and outstanding issues. *International Journal of Energy Research*, *46*(12), 16150–16177. https://doi.org/10.1002/er.8322

Durbin, D. J., & Malardier-Jugroot, C. (2013). Review of hydrogen storage techniques for on board vehicle applications. *International Journal of Hydrogen Energy*, *38*(34), 14595–14617. https://doi.org/10.1016/j.ijhydene.2013.07.058

Edalati, K., Uehiro, R., Ikeda, Y., Li, H.-W., Emami, H., Filinchuk, Y., … Horita, Z. (2018). Design and synthesis of a magnesium alloy for room temperature hydrogen storage. *Acta Materialia*, *149*, 88–96. https://doi.org/10.1016/j.actamat.2018.02.033

Eftekhari, A., & Fang, B. (2017). Electrochemical hydrogen storage: Opportunities for fuel storage, batteries, fuel cells, and supercapacitors. *International Journal of Hydrogen Energy*, *42*(40), 25143–25165. https://doi.org/10.1016/j.ijhydene.2017.08.103

Folkson, R. (2022). Hydrogen as an energy vector for transportation vehicles. In *Alternative fuels and advanced vehicle technologies for improved environmental performance* (pp. 151–171). Elsevier. https://doi.org/10.1016/B978-0-323-90979-2.00013-5

Forberg, D., Schwob, T., Zaheer, M., Friedrich, M., Miyajima, N., & Kempe, R. (2016). Single-catalyst high-weight% hydrogen storage in an N-heterocycle synthesized from lignin hydrogenolysis products and ammonia. *Nature Communications*, 7(1), 13201. https://doi.org/10.1038/ncomms13201

Grochala, W., & Edwards, P. P. (2004). Thermal decomposition of the non-interstitial hydrides for the storage and production of Hydrogen. *Chemical Reviews*, *104*(3), 1283–1316. https://doi.org/10.1021/cr030691s

Hassan, I. A., Ramadan, H. S., Saleh, M. A., & Hissel, D. (2021). Hydrogen storage technologies for stationary and mobile applications: Review, analysis and perspectives. *Renewable and Sustainable Energy Reviews*, *149*, 111311. https://doi.org/10.1016/j.rser.2021.111311

Hou, Q., Yang, X., & Zhang, J. (2021). Review on Hydrogen storage performance of MgH_2: Development and trends. *ChemistrySelect*, *6*(7), 1589–1606. https://doi.org/10.1002/slct.202004476

Huang, T., Huang, X., Hu, C., Wang, J., Liu, H., Ma, Z., … Ding, W. (2021). Enhancing hydrogen storage properties of MgH2 through addition of Ni/CoMoO4 nanorods. *Materials Today Energy*, *19*, 100613. https://doi.org/10.1016/j.mtener.2020.100613

Hwang, H. T., & Varma, A. (2014). Hydrogen storage for fuel cell vehicles. *Current Opinion in Chemical Engineering*, *5*, 42–48. https://doi.org/10.1016/j.coche.2014.04.004

International Energy Agency. (2023). *CO2 emissions in 2022 – Analysis*. International Energy Agency.

Jia, T., Gu, Y., & Li, F. (2022). Progress and potential of metal-organic frameworks (MOFs) for gas storage and separation: A review. *Journal of Environmental Chemical Engineering*, *10*(5), 108300. https://doi.org/10.1016/j.jece.2022.108300

Klopčič, N., Grimmer, I., Winkler, F., Sartory, M., & Trattner, A. (2023). A review on metal hydride materials for hydrogen storage. *Journal of Energy Storage*, 72, 108456. https://doi.org/10.1016/j.est.2023.108456

Kojima, Y. (2019). Hydrogen storage materials for hydrogen and energy carriers. *International Journal of Hydrogen Energy*, *44*(33), 18179–18192. https://doi.org/10.1016/j.ijhydene.2019.05.119

Lai, Q., Paskevicius, M., Sheppard, D. A., Buckley, C. E., Thornton, A. W., Hill, M. R., … Aguey-Zinsou, K.-F. (2015). Hydrogen storage materials for mobile and stationary applications: Current state of the art. *ChemSusChem*, *8*(17), 2789–2825. https://doi.org/10.1002/cssc.201500231

Lai, Q., Wang, T., Sun, Y., & Aguey-Zinsou, K. (2018). Rational design of nanosized light elements for Hydrogen storage: Classes, synthesis, characterisation, and properties. *Advanced Materials Technologies*, *3*(9). https://doi.org/10.1002/admt.201700298

Langmi, H. W., Engelbrecht, N., Modisha, P. M., & Bessarabov, D. (2022). Hydrogen storage. In *Electrochemical power sources: Fundamentals, systems, and applications* (pp. 455–486). Elsevier. https://doi.org/10.1016/B978-0-12-819424-9.00006-9

Langmi, H. W., Ren, J., North, B., Mathe, M., & Bessarabov, D. (2014). Hydrogen storage in metal-organic frameworks: A review. *Electrochimica Acta*, *128*, 368–392. https://doi.org/10.1016/j.electacta.2013.10.190

Li, B., Li, J., Shao, H., & He, L. (2018). Mg-based hydrogen absorbing materials for thermal energy storage—A review. *Applied Sciences*, *8*(8), 1375. https://doi.org/10.3390/app8081375

Liu, W., Setijadi, E. J., & Aguey-Zinsou, K.-F. (2014). Tuning the thermodynamic properties of MgH_2 at the nanoscale via a catalyst or destabilizing element coating strategy. *The Journal of Physical Chemistry C*, *118*(48), 27781–27792. https://doi.org/10.1021/jp5097663

Liu, W., Setijadi, E. J., Crema, L., Bartali, R., Laidani, N., Aguey-Zinsou, K. F., & Speranza, G. (2018). Carbon nanostructures/Mg hybrid materials for hydrogen storage. *Diamond and Related Materials*, *82*, 19–24. https://doi.org/10.1016/j.diamond.2017.12.003

Lototskyy, M. V., Yartys, V. A., Pollet, B. G., & Bowman, R. C. (2014). Metal hydride hydrogen compressors: A review. *International Journal of Hydrogen Energy*, *39*(11), 5818–5851. https://doi.org/10.1016/j.ijhydene.2014.01.158

Lu, Z.-Y., Yu, H.-J., Lu, X., Song, M.-C., Wu, F.-Y., Zheng, J.-G., … Zhang, L.-T. (2021). Two-dimensional vanadium nanosheets as a remarkably effective catalyst for hydrogen storage in MgH2. *Rare Metals*, *40*(11), 3195–3204. https://doi.org/10.1007/s12598-021-01764-7

Market Research Community. (2023). *Hydrogen storage market size, share & trends analysis, by technology (compression, liquefaction, material based), by type of storage (cylinder, merchant, on-site, on-board), by region and forecast period 2023 – 2030*. Market Research Community.

Mazzolai, G. (2012). Perspectives and challenges for solid state hydrogen storage in automotive applications. *Recent Patents on Materials Science*, *5*(2), 137–148. https://doi.org/10.2174/1874464811205020137

McWhorter, S., Read, C., Ordaz, G., & Stetson, N. (2011). Materials-based hydrogen storage: Attributes for near-term, early market PEM fuel cells. *Current Opinion in Solid State and Materials Science*, *15*(2), 29–38. https://doi.org/10.1016/j.cossms.2011.02.001

Midilli, A., Ay, M., Dincer, I., & Rosen, M. A. (2005). On hydrogen and hydrogen energy strategies. *Renewable and Sustainable Energy Reviews*, *9*(3), 255–271. https://doi.org/10.1016/j.rser.2004.05.003

Mohan, M., Sharma, V. K., Kumar, E. A., & Gayathri, V. (2019). Hydrogen storage in carbon materials—A review. *Energy Storage*, *1*(2), e35. https://doi.org/10.1002/est2.35

Murray, L. J., Dincă, M., & Long, J. R. (2009). Hydrogen storage in metal–organic frameworks. *Chemical Society Reviews*, *38*(5), 1294. https://doi.org/10.1039/b802256a

Nagar, R., Srivastava, S., Hudson, S. L., Amaya, S. L., Tanna, A., Sharma, M., … Srinivasan, S. S. (2023). Recent developments in state-of-the-art hydrogen energy technologies – Review of hydrogen storage materials. *Solar Compass*, *5*, 100033. https://doi.org/10.1016/j.solcom.2023.100033

Niaz, S., Manzoor, T., & Pandith, A. H. (2015). Hydrogen storage: Materials, methods and perspectives. *Renewable and Sustainable Energy Reviews*, *50*, 457–469. https://doi.org/10.1016/j.rser.2015.05.011

Okolie, J. A., Patra, B. R., Mukherjee, A., Nanda, S., Dalai, A. K., & Kozinski, J. A. (2021). Futuristic applications of hydrogen in energy, biorefining, aerospace, pharmaceuticals and metallurgy. *International Journal of Hydrogen Energy*, *46*(13), 8885–8905. https://doi.org/10.1016/j.ijhydene.2021.01.014

Orimo, S., Nakamori, Y., Eliseo, J. R., Züttel, A., & Jensen, C. M. (2007). Complex hydrides for hydrogen storage. *Chemical Reviews*, *107*(10), 4111–4132. https://doi.org/10.1021/cr0501846

Osman, A. I., Mehta, N., Elgarahy, A. M., Hefny, M., Al-Hinai, A., Al-Muhtaseb, A. H., & Rooney, D. W. (2022). Hydrogen production, storage, utilisation and environmental impacts: A review. *Environmental Chemistry Letters*, *20*(1), 153–188. https://doi.org/10.1007/s10311-021-01322-8

Ouyang, L., Chen, K., Jiang, J., Yang, X.-S., & Zhu, M. (2020). Hydrogen storage in light-metal based systems: A review. *Journal of Alloys and Compounds*, *829*, 154597. https://doi.org/10.1016/j.jallcom.2020.154597

Ouyang, L., Liu, F., Wang, H., Liu, J., Yang, X.-S., Sun, L., & Zhu, M. (2020). Magnesium-based hydrogen storage compounds: A review. *Journal of Alloys and Compounds*, *832*, 154865. https://doi.org/10.1016/j.jallcom.2020.154865

Palm, R., Kurig, H., Aruväli, J., & Lust, E. (2018). NaAlH 4 /microporous carbon composite materials for reversible hydrogen storage. *Microporous and Mesoporous Materials*, *264*, 8–12. https://doi.org/10.1016/j.micromeso.2017.12.027

Peng, D., Zhang, Y., & Han, S. (2021). Fabrication of multiple-phase magnesium-based hydrides with enhanced hydrogen storage properties by activating NiS@C and Mg powder. *ACS Sustainable Chemistry & Engineering*, *9*(2), 998–1007. https://doi.org/10.1021/acssuschemeng.0c08507

Prachi R., P., Mahesh M., W., & Aneesh C., G. (2016a). A review on solid state hydrogen storage material. *Advances in Energy and Power*, *4*(2), 11–22. https://doi.org/10.13189/aep.2016.040202

Prachi R., P., Mahesh M., W., & Aneesh C., G. (2016b). A review on solid state hydrogen storage material. *Advances in Energy and Power*, *4*(2), 11–22. https://doi.org/10.13189/aep.2016.040202

Pukazhselvan, D., Kumar, V., & Singh, S. K. (2012). High capacity hydrogen storage: Basic aspects, new developments and milestones. *Nano Energy*, *1*(4), 566–589. https://doi.org/10.1016/j.nanoen.2012.05.004

Rao, A. G., Yin, F., & Werij, H. (2020). Energy transition in aviation: The role of cryogenic fuels. *Aerospace*, 7(12), 181. https://doi.org/10.3390/aerospace7120181

Rehman, A., Nazir, G., Heo, K., Hussain, S., Ikram, M., Mahmood, Q., … Abd-Rabboh, H. S. M. (2023). Single step strategy to prepare highly microporous carbons derived from melamine and terephthalaldehyde for high-performance material-based hydrogen storage. *Journal of Energy Storage*, *66*, 107468. https://doi.org/10.1016/j.est.2023.107468

Rimza, T., Saha, S., Dhand, C., Dwivedi, N., Patel, S. S., Singh, S., & Kumar, P. (2022). Carbon-based sorbents for hydrogen storage: Challenges and sustainability at operating conditions for renewable energy. *ChemSusChem*, *15*(11). https://doi.org/10.1002/cssc.202200281

Rivard, E., Trudeau, M., & Zaghib, K. (2019). Hydrogen storage for mobility: A review. *Materials*, *12*(12), 1973. https://doi.org/10.3390/ma12121973

Roszak, R., Firlej, L., Roszak, S., Pfeifer, P., & Kuchta, B. (2016). Hydrogen storage by adsorption in porous materials: Is it possible? *Colloids and Surfaces A: Physicochemical and Engineering Aspects*, *496*, 69–76. https://doi.org/10.1016/j.colsurfa.2015.10.046

Rusman, N. A. A., & Dahari, M. (2016). A review on the current progress of metal hydrides material for solid-state hydrogen storage applications. *International Journal of Hydrogen Energy*, *41*(28), 12108–12126. https://doi.org/10.1016/j.ijhydene.2016.05.244

Sazelee, N. A., & Ismail, M. (2021). Recent advances in catalyst-enhanced LiAlH4 for solid-state hydrogen storage: A review. *International Journal of Hydrogen Energy*, *46*(13), 9123–9141. https://doi.org/10.1016/j.ijhydene.2020.12.208

Schlapbach, L., & Züttel, A. (2001). Hydrogen-storage materials for mobile applications. *Nature*, *414*(6861), 353–358. https://doi.org/10.1038/35104634

Schneemann, A., White, J. L., Kang, S., Jeong, S., Wan, L. F., Cho, E. S., … Stavila, V. (2018). Nanostructured metal hydrides for hydrogen storage. *Chemical Reviews*, *118*(22), 10775–10839. https://doi.org/10.1021/acs.chemrev.8b00313

Schueth, F., Bogdanovic, B., & Felderhoff, M. (2005). Light metal hydrides and complex hydrides for hydrogen storage. *ChemInform*, *36*(2). https://doi.org/10.1002/chin.200502229

Schulz, R., Huot, J., Liang, G., Boily, S., Lalande, G., Denis, M. C., & Dodelet, J. P. (1999). Recent developments in the applications of nanocrystalline materials to hydrogen technologies. *Materials Science and Engineering: A*, *267*(2), 240–245. https://doi.org/10.1016/S0921-5093(99)00098-2

Shao, H., Xin, G., Zheng, J., Li, X., & Akiba, E. (2012). Nanotechnology in Mg-based materials for hydrogen storage. *Nano Energy*, *1*(4), 590–601. https://doi.org/10.1016/j.nanoen.2012.05.005

Silvera, I. F., & Dias, R. (2021). Phases of the hydrogen isotopes under pressure: metallic hydrogen. *Advances in Physics: X*, *6*(1). https://doi.org/10.1080/23746149.2021.1961607

Thangarasu, S., Palanisamy, G., Im, Y. M., & Oh, T. H. (2023). An alternative platform of solid-state hydrides with polymers as composite/encapsulation for hydrogen storage applications: Effects in intermetallic and complex hydrides. *International Journal of Hydrogen Energy*, 48(56), 21429–21450. https://doi.org/10.1016/j.ijhydene.2022.02.115

The Insight Partners. (2021). *Hydrogen storage alloys market size and global analysis - 2028*. The Insight Partners.

Tuul, K., & Palm, R. (2021). Influence of Nanoconfinement on the Hydrogen Release Processes from Sodium Alanate. *Reactions*, 2(1), 1–9. https://doi.org/10.3390/reactions2010001

USDoE. (2010). *Hydrogen sorption center of excellence (HSCoE) final report*. USDoE.

Varin, R. A., Czujko, T., Wronski, Z. S., & Zaranski, Z. (2009). Nanomaterials for hydrogen storage produced by ball milling. *Canadian Metallurgical Quarterly*, *48*(1), 11–25. https://doi.org/10.1179/cmq.2009.48.1.11

Vigeholm, B., Kjøller, J., Larsen, B., & Pedersen, A. S. (1983). Formation and decomposition of magnesium hydride. *Journal of the Less Common Metals*, *89*(1), 135–144. https://doi.org/10.1016/0022-5088(83)90259-X

Webb, C. J. (2015). A review of catalyst-enhanced magnesium hydride as a hydrogen storage material. *Journal of Physics and Chemistry of Solids*, *84*, 96–106. https://doi.org/10.1016/j.jpcs.2014.06.014

Winter, M., & Brodd, R. J. (2004). What are batteries, fuel cells, and supercapacitors? *Chemical Reviews*, *104*(10), 4245–4270. https://doi.org/10.1021/cr020730k

Xu, Z., Zhao, N., Hillmansen, S., Roberts, C., & Yan, Y. (2022). Techno-economic analysis of hydrogen storage technologies for railway engineering: A review. *Energies*, *15*(17), 6467. https://doi.org/10.3390/en15176467

Yang, H., Ding, Z., Li, Y.-T., Li, S.-Y., Wu, P.-K., Hou, Q.-H., ... Shaw, L. L. (2023). Recent advances in kinetic and thermodynamic regulation of magnesium hydride for hydrogen storage. *Rare Metals*, *42*(9), 2906–2927. https://doi.org/10.1007/s12598-023-02306-z

Yanxing, Z., Maoqiong, G., Yuan, Z., Xueqiang, D., & Jun, S. (2019a). Thermodynamics analysis of hydrogen storage based on compressed gaseous hydrogen, liquid hydrogen and cryo-compressed hydrogen. *International Journal of Hydrogen Energy*, *44*(31), 16833–16840. https://doi.org/10.1016/j.ijhydene.2019.04.207

Yanxing, Z., Maoqiong, G., Yuan, Z., Xueqiang, D., & Jun, S. (2019b). Thermodynamics analysis of hydrogen storage based on compressed gaseous hydrogen, liquid hydrogen and cryo-compressed hydrogen. *International Journal of Hydrogen Energy*, *44*(31), 16833–16840. https://doi.org/10.1016/j.ijhydene.2019.04.207

Yartys, V. A., Lototskyy, M. V., Akiba, E., Albert, R., Antonov, V. E., Ares, J. R., ... Zhu, M. (2019). Magnesium based materials for hydrogen based energy storage: Past, present and future. *International Journal of Hydrogen Energy*, *44*(15), 7809–7859. https://doi.org/10.1016/j.ijhydene.2018.12.212

Yuan, S., Feng, L., Wang, K., Pang, J., Bosch, M., Lollar, C., ... Zhou, H. (2018). Stable metal–organic frameworks: Design, synthesis, and applications. *Advanced Materials*, *30*(37). https://doi.org/10.1002/adma.201704303

Yue, M., Lambert, H., Pahon, E., Roche, R., Jemei, S., & Hissel, D. (2021). Hydrogen energy systems: A critical review of technologies, applications, trends and challenges. *Renewable and Sustainable Energy Reviews*, *146*, 111180. https://doi.org/10.1016/j.rser.2021.111180

Zhang, B., & Wu, Y. (2017). Recent advances in improving performances of the lightweight complex hydrides Li-Mg-N-H system. *Progress in Natural Science: Materials International*, *27*(1), 21–33. https://doi.org/10.1016/j.pnsc.2017.01.005

Zhang, J., Li, Z., Wu, Y., Guo, X., Ye, J., Yuan, B., ... Jiang, L. (2019). Recent advances on the thermal destabilisation of Mg-based hydrogen storage materials. *RSC Advances*, *9*(1), 408–428. https://doi.org/10.1039/C8RA05596C

Zhang, W., Cui, Y., Zhu, C., Huang, B., Yan, S., Lou, Y., & Zhang, P. (2023). Sequential hydrogen storage in phosphorene nanotubes: A molecular dynamics study. *International Journal of Hydrogen Energy*, *48*(62), 23909–23916. https://doi.org/10.1016/j.ijhydene.2023.03.212

Zhang, X. L., Liu, Y. F., Zhang, X., Hu, J. J., Gao, M. X., & Pan, H. G. (2020). Empowering hydrogen storage performance of MgH2 by nanoengineering and nanocatalysis. *Materials Today Nano*, *9*, 100064. https://doi.org/10.1016/j.mtnano.2019.100064

Zheng, J., Liu, X., Xu, P., Liu, P., Zhao, Y., & Yang, J. (2012). Development of high pressure gaseous hydrogen storage technologies. *International Journal of Hydrogen Energy*, *37*(1), 1048–1057. https://doi.org/10.1016/j.ijhydene.2011.02.125

Zhou, C., Fang, Z. Z., Lu, J., & Zhang, X. (2013). Thermodynamic and kinetic destabilisation of magnesium hydride using Mg–in solid solution alloys. *Journal of the American Chemical Society*, *135*(30), 10982–10985. https://doi.org/10.1021/ja4058794

Züttel, A. (2003). Materials for hydrogen storage. *Materials Today*, *6*(9), 24–33. https://doi.org/10.1016/S1369-7021(03)00922-2

Chapter 14

Sustainable supply chain practices in the emerging hydrogen transportation industry

Durwesh Jhodkar, Somadatta Karanjekar, Bharat Chede, Akhilesh Kumar Choudhary, and Tapus Bajpai

14.1 INTRODUCTION

Hydrogen supply chain is growing as a result of growing demand and technological advancements, with continuous research aimed at improving systems for delivery, storage, and distribution of hydrogen. This rise highlights the importance of hydrogen as a vital component in accomplishing decarbonization targets and tackling climate change issues. It also signals a shift toward cleaner energy sources. Nowadays, CO_2 emissions and fossil fuel consumption are increasing drastically. Green hydrogen plays a vital role in attaining a decarbonized and sustainable future to reduce fossil fuel demands (Dayhim et al., 2014). Green hydrogen is generated through a process called electrolysis. Its primary source of energy is renewable energy, such as solar, wind, or hydroelectric power. Due to minimal carbon emissions and the production of electricity obtained from renewable sources, the produced hydrogen is known as "Green hydrogen" (Gorji, 2023). Building hydrogen infrastructure, such as manufacturing facilities, storage facilities, and distribution methods, is highly risky and expensive. From early techniques like compressed gas and liquid storage to cutting-edge alternatives like pipeline networks and long pipe trailers, the transportation of hydrogen has changed. These developments ease the shift to a sustainable hydrogen economy by addressing issues with efficiency, safety, and infrastructure (Wickham et al., 2022).

Many other researchers have shown their interest in developing sustainability in the hydrogen supply chain. The production of hydrogen needs to follow several steps to reach the refueling station pump. This includes the development of a supply chain with five main components: production, transmission, separation/purification, distribution, and refueling stations (Wickham et al., 2022) as shown in Figure 14.1. Hydrogen is delivered to customers in a sustainable and dependable manner, promoting the development of hydrogen infrastructure and permitting the integration of hydrogen into a variety of uses, such as energy storage, transportation, and industry.

 DOI: 10.1201/9781003537816-14

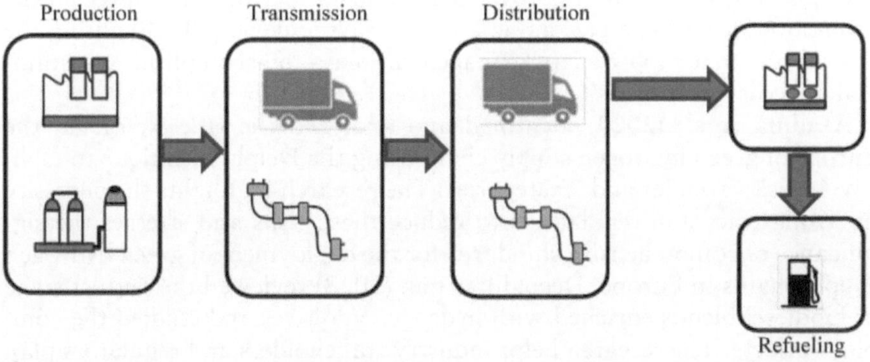

Figure 14.1 Supply chain of hydrogen potential for the transport industry (Wickham et al., 2022).

Riera et al. (2023) reviewed the optimization models of hydrogen supply chains. In this study, various other topics for future research are identified, such as methods to address sophisticated models, effects on climate change, integrated sector modeling, improving spatial and temporal resolutions, and projections for hydrogen demand. Franzmann et al. (2023) have outlined green hydrogen production until 2050 across 28 countries. This study provides spatially and temporally explicit cost potential curves for green hydrogen. Results indicate that countries with significant solar power such as the Middle East and South Africa have high hydrogen potential and cost efficiencies (<2.30 EUR/kg). For minimal cost effect, decentralized PV-based production is preferred as it guarantees sustainable water sources. Additionally, democratic regions might encourage global green energy systems to improve geostrategic security.

Durakovic et al. (2023) have explored the green hydrogen deployment on generation and transmission investments in Europe and the North Sea region until 2060. Hydrogen deployment frequently decreases grid stress during peak demand, potentially stabilizing or lowering electricity costs despite increased demand, even though its effects on pricing vary by area. The results indicate a 50% increase in the capacity of overall power generation. Sgarbossa et al. (2023) discussed the various aspects of hydrogen supply chains (HSC) in a systematic literature review. This study addresses complexities and the planning matrix for the development of renewable HSCs within the evolving low-carbon economy, emphasizing the need for adaptive supply chain strategies. Squadrito et al. (2023) have critically discussed the latest literature and institutional reports and identified the current status of green hydrogen technologies and their associated drawbacks, geopolitical implications, and benefits (Elvira et al., 2022). This study focuses on hydrogen technology in Mexican maritime transportation in order to address sustainability goals. The report proposes strategies to lower transportation costs and improve environmental performance through the promotion of reusable packaging and fuel-saving devices, highlighting the crucial role that supply chain management plays in accomplishing sustainability goals.

Azadnia et al. (2023) identified and analyzed the risk factors in the European green hydrogen supply chain using the Delphi technique to identify 43 risk variables and 7 categories. The research highlights the necessity of strong rules and regulations to reduce these risks and stresses the significance of comprehensive standards for the deployment of green hydrogen supply chains in Europe. Deepalika et al. (2023) reviewed the performance of biodiesel blends enriched with hydrogen additives and studied the combined effects. The research helps industry stakeholders and regulators plan for sustainable energy by offering insights into the hydrogen supply chain landscape of the future through the integration of geographically aggregated demand models.

While there is enthusiasm for creating a hydrogen economy for mobility, barriers remain due to inadequate infrastructure and competition from fossil fuels. Hydrogen's competitiveness must be assessed by analyzing various scenarios that consider safety, environmental, and financial concerns (De-León Almaraz et al., 2014). To address various challenges and uncertainties in hydrogen supply chains and production, various types of optimization models have been created. Almansoori et al. (2016) introduced a model for designing hydrogen supply chains under emission constraints. Seo et al. (2020) proposed a centralized storage model for hydrogen supply chains, which was found to reduce costs and improve efficiency. These models collectively contribute to the development of systematic methodologies for hydrogen expansion, with potential applications in different regions and countries.

In this study, various published articles have been analyzed thoroughly, which are essential for supply chain practices in the hydrogen industry. In this work, hydrogen transportation has been the focus to identify the essential factors in it. Different methods of hydrogen transportation have evaluated in this work such as long pipe trailer transportation, pipeline transportation, and liquid hydrogen transportation. The outcomes of this study reveal several key findings and insights into the logistical challenges and opportunities regarding supply chain practices in the hydrogen storage and transportation industry.

14.2 HYDROGEN STORAGE AND TRANSPORTATION

Hydrogen is a promising energy solution, with a clean supply chain and high energy conversion efficiency. The introduction of a hydrogen supply chain using existing gas pipelines can significantly improve the generation

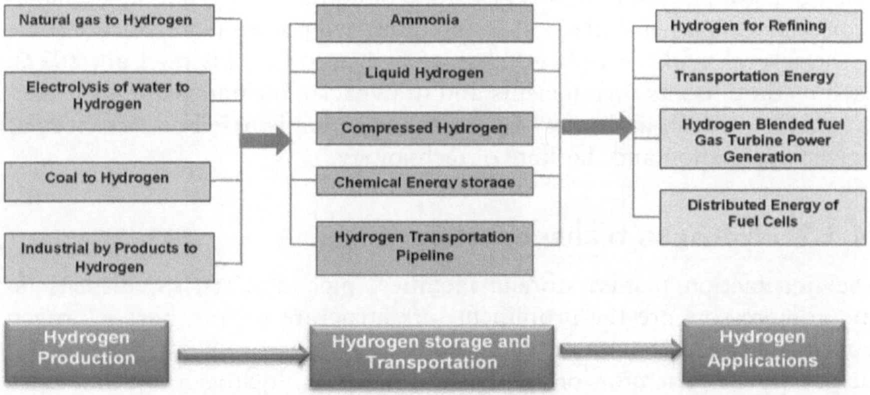

Figure 14.2 Various industrial technologies for hydrogen production, storage, transportation, and applications (source: Zhang & Jiang, 2023).

of clean energy, with potential reductions in CO_2, SOx, NOx, and $PM_{2.5}$ emissions. A clean, sustainable energy future is promised by the growing hydrogen industry, which involves transportation and storage of hydrogen. For it to be widely adopted, safe storage and efficient transportation are required (Maryam, 2017). Compression, liquefaction, or chemical conversion into carriers like ammonia are common techniques utilized to transport hydrogen (Squadrito et al., 2023). Storage techniques vary depending on scale and application (see Figure 14.2). Large amounts of hydrogen can be stored for a long time in depleted gas fields, salt domes, and underground caverns. On a smaller scale, possibilities for secure storage in diverse environments include solid-state materials, cryogenic tanks, and high-pressure tanks (Weichenhain, 2021). In this study, the ongoing research and technological advancements related to hydrogen storage and transportation have been discussed. The latest advancements are cost-effective and sustainable.

14.2.1 Hydrogen storage

Hydrogen storage technology is essential for the safe, efficient, and compact hydrogen gas storage, which is crucial for the shift to sustainable energy. It includes methods such as compressed gas, liquefaction, metal hydrides, and advanced porous materials, unlocking hydrogen's potential as a clean energy carrier in various sectors (Dayhim et al., 2014). Global collaboration, energy security, climate change, economic competitiveness, technological innovation, policy and regulation, and globalization all have an impact on hydrogen storage. Research and development efforts are focused on improving hydrogen storage systems to raise their capacity, effectiveness, durability, and safety (Franzmann et al., 2023). International collaboration and investment in research initiatives drive technological innovation in hydrogen storage. Compressed gas storage, liquid hydrogen storage, metal hydride storage, chemical storage, and improved porous materials are some of the several methods utilized to store hydrogen (Wickham et al., 2022). Each method has its own benefits and drawbacks, including storage capacity, efficiency, cost, and safety issues; the selection is based on the needs of a given application and the state of technology.

14.2.2 Hydrogen transportation

The liquefaction plants, storage facilities, pipelines, trucks, dispensers, and compressors are the prominent infrastructure to transport hydrogen from the site of production to the point of end use such as an industrial facility, power generator, or fueling station. Green hydrogen revolutionizes transportation by offering a renewable energy source for vehicles, trains, and ships, while tackling the pressing issue of reducing carbon emissions and promoting cleaner mobility solutions. Implementing sustainable supply chain practices is crucial for effectively incorporating hydrogen into the transportation industry (Azadnia et al., 2023; Husarek et al., 2021). Commercially available transportation technology exists in the United States, with companies providing bulk hydrogen. The growth in hydrogen demand will necessitate expanding regional infrastructure and developing new technologies like chemical carriers for high-density hydrogen delivery and high-throughput fueling systems for heavy-duty transportation.

In the initial phase of the hydrogen market, demand is low and transportation distance is restricted. The cost-effective method involves using high-pressure gas transport with cylinders onboard. As the market progresses, the demand area expands to include gaseous and low-temperature liquid transportation. In the future, advancements will focus on high-density storage, a pipeline network, and creating new standards for storage and distribution (Gorji, 2023; Zhang & Jiang, 2023).

14.2.1.1 Long pipe trailer transportation

For transporting hydrogen, it must either be liquefied or pressurized and delivered as a compressed gas to be transported in large quantities. However, hydrogen is the smallest and lightest element present on Earth (Aceves et al., 2010). The weight of gasoline of one gallon is about 2.75 kg, whereas one gallon of hydrogen (at 0°C and 1 atm pressure) only weighs 0.00075 kg. Transporting hydrogen has several difficulties, especially when considering mobility (Xinhua et al., 2006). A long pipe trailer is a custom-made transport vehicle for moving hydrogen gas across long distances. It includes linked pipes on a trailer frame for secure transfer from production sites to distribution locations (Husarek et al., 2021). With storage tanks and safety measures, it provides an affordable method for delivering hydrogen to distant regions. Issues such as hydrogen embrittlement and regulations are managed to guarantee safe operation and promote the expansion of hydrogen infrastructure. Reynolds et al. (1975) note that the pipeline transportation of hydrogen is significantly more expensive than that of natural gas, with the cost of cooling hydrogen for transport outweighing the savings in pipeline cost. Using a trailer to transport long pipes at 20 MPa is recommended for distances under 200 km. Expenses increase due to disassembly, labor, and fuel costs. Domestic regulations cap pressure for long pipe trailers at 20 MPa, while globally, trailers with pressures up to 50 MPa are employed. Relaxing pressure standards could reduce costs. At 100 km, 50 MPa trailers cost 65% less than 20 MPa trailers. By 2025, China is expected to widely adopt 50 MPa long pipe trailers (Zhang & Jiang, 2023). This suggests that while long pipe trailer transportation of hydrogen may be feasible, it may not be the most cost-effective option.

14.2.1.2 Pipeline transportation

Generally, large-scale hydrogen transportation is preferred by pipeline transportation; however, high installation costs affect its utilizations (Jinyang et al., 2012). Gaseous hydrogen transported through pipelines is the preferred way of transporting it from point of production to end use. In the United States, 1600 miles of hydrogen pipelines are currently operated (Rusman & Dahari, 2016). These pipelines are generally owned commercially by various producers and are situated in regions containing a significant number of large hydrogen clients, such as chemical and petroleum refineries, like the Gulf Coast (Jianwei et al., 2017).

Transporting a large volume of gaseous hydrogen via pipeline is an economical way. However, constructing new pipeline infrastructure requires high initial capital expenditures associated with hydrogen delivery infrastructure. Therefore, researchers and engineers are focusing on resolving technical challenges regarding pipeline transmissions. Hydrogen embrittlement risks in steel pipelines necessitate controlling leaks to improve cheaper, reliable, durable compression technology for hydrogen. Using pipelines made of steel or composite materials, hydrogen is transported over long distances efficiently. Nowadays, researchers have developed fiber-reinforced polymer (FRP) pipelines for hydrogen supply because it does not require welding, and the installation cost of FRP is 20% less than those steel pipelines.

Effective monitoring systems for the pipeline transportation enhance safety and reliability, minimizing energy consumption per unit of hydrogen transported. Hydrogen pipeline transportation is efficient, reliable, safe, scalable, and integrates with existing infrastructure, enabling widespread adoption as a clean energy source (Husarek et al., 2021).

14.2.3 Liquid hydrogen transportation

Liquid hydrogen storage involves storing hydrogen at an extremely low temperature (–253°C). This method offers high energy density and is commonly used in space exploration and select transportation applications. However, maintaining such low temperatures requires intricate insulation systems and specialized infrastructure (Rusman & Dahari, 2016). Liquid hydrogen storage is crucial for enabling hydrogen-fueled rockets, vehicles, and extending space missions. Research in cryogenic liquid hydrogen storage systems for aircraft applications is essential (Mital et al., 2006). Continuous improvements include vacuum-based insulation in double walls and multilayer insulation systems, but challenges remain in producing robust and lightweight storage solutions. Materials-based hydrogen storage solutions, including those for cryogenic temperatures, are also being explored (Jianwei et al., 2017). These include new porous materials, composite hydride systems, and the use of ionic liquids. Heat transfer issues in cryogenic liquid hydrogen storage are another important consideration, with a focus on efficient heat transfer during compression and intercooling, as well as improved thermal insulation of cryogenic tanks (Jinsong et al., 2005). Finally, recent advances in liquid-phase chemical hydrogen storage, including the use of catalysts and recycling solutions, are contributing to the development of safe and efficient hydrogen storage systems (Lang & Jia, 2020). Terrestrial usage is substantial but limited because of the expensive and complex infrastructure needed. However, the goal of present research is to enhance the efficacy and safety of liquid hydrogen storage devices.

Research on liquid hydrogen transportation has highlighted its potential as a cost-effective and efficient carrier for large-scale hydrogen energy development (Zhang & Jiang, 2023). Various transportation methods have been explored, with a trade study at Kennedy Space Center identifying mobile semi-trailer tankers and railcars as the most effective options. An innovative approach involves transforming hydrogen into liquid ammonia or ammonia water for transportation, with subsequent re-transformation into hydrogen at the destination (Rupsha Bhattacharyya, 2015). To ensure the purity of liquid hydrogen during transportation, the use of improved thermal insulation in railroad tank cars has been proposed (Carabtree, 2008).

Liquid transportation of hydrogen involves storing and moving hydrogen in its liquid form, which offers advantages like high energy density and purity. This method is crucial for large-scale applications, especially in sectors like aerospace, energy, and maritime shipping. Liquid hydrogen is handled at extremely low temperatures (around 20 K), requiring specialized cryogenic storage and transportation containers (Ichikawa, 2008; Teng He & Qijun Pei, 2015) . The use of liquid hydrogen as a carrier for energy is gaining traction due to its efficiency, long storage capabilities, and economic viability (Alhumaidan et al., 2011). In the context of maritime shipping, storing hydrogen in its liquid phase is seen as a viable solution due to its high energy demands and long-range requirements. Overall, liquid hydrogen transportation is a key component in the transition toward a low-carbon hydrogen economy (Zacharia & Rather, 2015).

14.3 CONCLUSIONS

This chapter introduces the growing importance of hydrogen as a clean energy source and its role in decarbonization efforts. It reviews studies on hydrogen supply chain optimization and discusses transportation techniques including long pipe trailer, pipeline, and liquid hydrogen transportation. The key findings are summarized as follows:

(i) Hydrogen shows promise for clean energy, but challenges in infrastructure and cost-effectiveness remain. Further research is needed for its full potential as a sustainable energy source, requiring collaboration, investment, and standardized regulations.
(ii) Highlights the significance of continuing advancements in hydrogen transportation technologies. Cooperation among stakeholders, funding research projects, and implementing standardized regulations are crucial in unlocking hydrogen's complete potential.
(iii) A well-established hydrogen transportation network is crucial for aiding the shift to a more eco-friendly and long-lasting energy future.
(iv) Extensive adoption of hydrogen as an energy carrier can decrease dependence on fossil fuels and alleviate the effects of climate change.

REFERENCES

Aceves, S. M., Espinosa-Loza, F., Ledesma-Orozco, E., Ross, T. O., Weisberg, A. H., Brunner, T. C., & Kircher, O. (2010). High-density automotive hydrogen storage with cryogenic capable pressure vessels. *International Journal of Hydrogen Energy*, *35*(3), 1219–1226. https://linkinghub.elsevier.com/retrieve/pii/S0360319909018394

Alhumaidan, F., Cresswell, D., & Garforth, A. (2011). Hydrogen storage in liquid organic hydride: Producing hydrogen catalytically from methylcyclohexane. *Energy & Fuels*, *25*(10), 4217–4234. https://doi.org/10.1021/ef200829x

Almansoori, A. B.-T. (2016). Design of optimization model for a hydrogen supply chain under emission constraints - A case study of GermanyTitle. *Energy*, *111*, 414–429.

Almansoori, A., & Betancourt-Torcat, A. (2016). Design of optimization model for a hydrogen supply chain under emission constraints-A case study of Germany. *Energy*, *111*, 414–429.

Azadnia, A. H., McDaid, C., Andwari, A. M., & Hosseini, S. E. (2023 [2022 May]). Green hydrogen supply chain risk analysis: A European hard-to-abate sectors perspective. *Renewable and Sustainable Energy Reviews*, *182*, 113371. Elsevier Ltd. https://doi.org/10.1016/j.rser.2023.113371

Carabtree, R. H. (2008). Hydrogen storage in liquid organic heterocycles. *Energy and Environmental Science*, *1*, 134–138.

Chengguang Lang, Yi Jia, X. Y. (2020). Recent advances in liquid-phase chemical hydrogen storage. *Energy Storage Materials*, *26*, 290–312.

Dayhim, M., Jafari, M. A., & Mazurek, M. (2014). Planning sustainable hydrogen supply chain infrastructure with uncertain demand. *International Journal of Hydrogen Energy*, *39*(13), 6789–6801. Elsevier Ltd. http://dx.doi.org/10.1016/j.ijhydene.2014.02.132

De-León Almaraz, S., Azzaro-Pantel, C., Montastruc, L., & Domenech, S. (2014). Hydrogen supply chain optimization for deployment scenarios in the Midi-Pyrénées region, France. *International Journal of Hydrogen Energy*, *39*(23), 11831–11845.

Deepalika, Kumar, V., & Choudhary, A. K. (2023, September). A comparative review on evaluation of performance, combustion, and emission characteristics of biodiesel blends enriched with hydrogen, additives and their combined effect. *Thermal Science and Engineering Progress*, *46*, 102185. Elsevier Ltd. https://doi.org/10.1016/j.tsep.2023.102185

Durakovic, G., del Granado, P. C., & Tomasgard, A. (2023). Powering Europe with North Sea offshore wind: The impact of hydrogen investments on grid infrastructure and power prices. *Energy*, *263*(PA), 125654. Elsevier Ltd. https://doi.org/10.1016/j.energy.2022.125654

Elvira, K., Marisol, R., Susanne, H., Leif-Magnus, J., Sandoval-Pineda, J. M., & de G, G. H. R. (2022). Hydrogen technology for supply chain sustainability: The Mexican transportation impacts on society. *International Journal of Hydrogen Energy*, *47*(70), 29999–30011. The Author(s). https://doi.org/10.1016/j.ijhydene.2022.03.081

Franzmann, D., Heinrichs, H., Lippkau, F., Addanki, T., Winkler, C., Buchenberg, P., Hamacher, T., et al. (2023). Green hydrogen cost-potentials for global trade. *International Journal of Hydrogen Energy*, *48*(85), 33062–33076. The Author(s). https://doi.org/10.1016/j.ijhydene.2023.05.012

Gorji, S. A. (2023). Challenges and opportunities in green hydrogen supply chain through metaheuristic optimization. *Journal of Computational Design and Engineering*, *10*(3), 1143–1157.

Husarek, D., Schmugge, J., & Niessen, S. (2021). Hydrogen supply chain scenarios for the decarbonisation of a German multi-modal energy system. *International Journal of Hydrogen Energy*, *46*(76), 38008–38025. Elsevier Ltd. https://doi.org/10.1016/j.ijhydene.2021.09.041

Ichikawa, & M. (2008). 18 - Organic liquid carriers for hydrogen storage. In G. B. T.-S.-S. H. S. Walker (Ed.), *Woodhead publishing series in electronic and optical materials* (pp. 500–532). Woodhead Publishing. https://www.sciencedirect.com/science/article/pii/B9781845692704500181

Ren, J., Musyoka, N. M., Langmi, H. W., & Mkhulu Mathe, S. L. (2017). Current research trends and perspectives on materials-based hydrogen storage solutions: A critical review. *International Journal of Hydrogen Energy*, *42*(1), 289–311.

Zhang, J., Fisher, T. S., Ramachandran, P. V., Gore, J. P., & Mudawar, I. (2005). A review of heat transfer issues in hydrogen storage technologies. *ASME Journal of Heat and Mass Transfer*, *127*(12), 1391–1399.

Zheng, J., Liu, X., Xu, P., Liu, P., Zhao, Y., & Yang, J. (2012). Development of high pressure gaseous hydrogen storage technologies. *International Journal of Hydrogen Energy*, *37*(1), 1048–1057.

Maryam, S. (2017). Review of modelling approaches used in the HSC context for the UK. *International Journal of Hydrogen Energy*, *42*(39), 24927–24938. Elsevier Ltd. http://dx.doi.org/10.1016/j.ijhydene.2017.04.303

Mital, S. K., Gyekenyesi, J. Z., Arnold, S. M., Sullivan, R. M., Manderscheid, J. M., & Murthy, P. L. N. (2006). *Review of current state of the art and key design issues with potential solutions for liquid hydrogen cryogenic storage tank structures for aircraft applications. Nasa-Tm-2006-214346.* c:%5CICT%5CEILS%5CVerbrennung, Raketenmotoren f?r IL%5CDevelopment of IL Monopropellants for IN-space Propulsion.pdf TS - RIS

Reynolds, R. A., & Slager, W. L. (1975). Pipeline transportation of hydrogen. In T. N. Veziroğlu (Ed.), *Hydrogen energy* (pp. 533–543). Springer, Boston, MA, USA. https://doi.org/10.1007/978-1-4684-2607-6_37

Riera, J. A., Lima, R. M., & Knio, O. M. (2023). A review of hydrogen production and supply chain modeling and optimization. *International Journal of Hydrogen Energy*, *48*(37), 13731–13755. The Author(s). https://doi.org/10.1016/j.ijhydene.2022.12.242

Rupsha Bhattacharyya, S. M. (2015). Solid state storage of hydrogen and its isotopes: An engineering overview. *Renewable and Sustainable Energy Reviews*, *41*, 872–883.

Rusman, N. A. ., & Dahari, M. (2016). A review on the current progress of metal hydrides material for solid-state hydrogen storage applications. *International Journal of Hydrogen Energy*, *41*(28), 12108–12126.

Seo, S. K., Yun, D. Y., & Lee, C. J. (2020). Design and optimization of a hydrogen supply chain using a centralized storage model. *Applied Energy*, *262*, 1144–52..https://doi.org/10.1016/j.apenergy.2019.11445.

Sgarbossa, F., Arena, S., Tang, O., & Peron, M. (2023 [2022, October]). Renewable hydrogen supply chains: A planning matrix and an agenda for future research. *International Journal of Production Economics*, *255*, 108674. Elsevier B.V. https://doi.org/10.1016/j.ijpe.2022.108674

Squadrito, G., Maggio, G., & Nicita, A. (2023, July). The green hydrogen revolution. *Renewable Energy*, *216*, 119041. Elsevier Ltd. https://doi.org/10.1016/j.renene.2023.119041

Teng He, Qijun Pei, P. C. (2015). Liquid organic hydrogen carriers. *Journal of Energy Chemistry*, *24*(5), 587–594.

Weichenhain, U. (2021). Hydrogen transportation. The key to unlocking the clean hydrogen economy. *Roland Berger*, 1–28. MBH Sederanger 1 80538 Munich Germany.

Wickham, D., Hawkes, A., & Jalil-Vega, F. (2022 [2021, July]). Hydrogen supply chain optimisation for the transport sector – Focus on hydrogen purity and purification requirements. *Applied Energy*, *305*, 117740. Elsevier Ltd. https://doi.org/10.1016/j.apenergy.2021.117740

Xinhua, W., Rugan, C., Yan, Z., Changpin, C., & Qidong, W. (2006). Hydrogen storage alloys for high-pressure suprapure hydrogen compressor. *Journal of Alloys and Compounds*, *420*(1–2), 322–325.

Zacharia, R., & Rather, S. U. (2015). Review of solid state hydrogen storage methods adopting different kinds of novel materials. (Y. Chen, Ed.). *Journal of Nanomaterials*, *2015*, 914845. Hindawi Publishing Corporation. https://doi.org/10.1155/2015/914845

Zhang, G., & Jiang, Z. (2023, June). Overview of hydrogen storage and transportation technology in China. *Unconventional Resources*, *3*, 291–296. Elsevier B.V. https://doi.org/10.1016/j.uncres.2023.07.001

Chapter 15

Hydrogen fuel in sustainable transformation

A comprehensive review

Ratindra Gautam, Asish Kumar, Shivani Chaudhary, C. K. Kaithwas, Ajay Kumar Shukla, and U. B. Singh

15.1 INTRODUCTION

15.1.1 Hydrogen production and its applications to mobility

Many nations are realizing that hydrogen is essential to attaining both long-term climate neutrality and strategic autonomy. Many strategies stress how critical it is to accelerate its adoption in consumption and throughout the supply chain. Whether hydrogen is used directly in hydrogen fuel cells or converted into green fuels, it presents a strong alternative for sustainable development, since transportation accounts for a large portion of CO_2 emissions. This overview examines recent advancements in the use of hydrogen in the primary energy-intensive transportation sectors. Using a systems engineering approach, we analyze every aspect of the hydrogen supply chain, considering its impact on society, the environment, and the economy. We discuss the difficulties that hydrogen will face when it tries to enter various transportation markets in the future (Gonzalez-Garay et al., 2022).

15.1.2 Technologies for sustainable heat generation in food processing

The food business of today processes most of its products using heat. In this industry, the generation of hot air, steam, and water is currently dominated by fossil fuel boilers. Nonetheless, there is a drive to lower the greenhouse gas emissions connected to thermal processing in order to meet sustainability targets. In place of traditional boilers, this research examines alternative technologies such as hydrogen conversion, geothermal energy, and electric alternatives. When compared to fossil fuels, these technologies—such as fuel cells, microturbines, and heat pumps—offer reduced emissions. The adoption of sustainable heating technology, such as electrical and geothermal systems, shows potential despite obstacles like low fossil fuel prices. Furthermore, research is being done on the use of hydrogen produced by

DOI: 10.1201/9781003537816-15

electrolysis as a transportable energy source for the production of heat and electricity, particularly in areas with restricted access to electricity (Grossmann et al., 2022).

15.1.3 Organic hydrogen carriers in liquid form (OHCL)

Sustainable, renewable energy systems will replace carbon-based ones due to the need to cut CO_2 emissions. As a secondary energy carrier, hydrogen will be essential, especially for large-scale energy storage and for enabling the spread of renewable energy sources around the world. But the constraints of standard hydrogen infrastructure have prevented its widespread deployment. Organic hydrogen carriers in liquid form (OHCL), a recent development in hydrogen handling, present a viable remedy. Through catalytic cycles, hydrogen may be stored and released efficiently in OHCL systems, which are made up of organic molecules that are both hydrogen-rich and hydrogen-lean. When released, OHCL systems produce pure hydrogen, in contrast to gas-based storage techniques. Suitable chemicals, catalysis, and future research possibilities are described, together with the present and potential status of OHCL technology, in this account. Mobile hydrogen production for on-board use, stationary energy storage, and hydrogen logistics are applicative. Additional study is required for applications such as hydrogen transportation to filling stations and direct OHCL-fuel cell use, even though commercial demonstrations have demonstrated the viability of OHCL technology for stationary storage. In order to advance OHCL technology for a hydrogen-driven economy, this account seeks to motivate and direct future efforts (Preuster et al., 2017).

15.2 PRODUCTION OF HYDROGEN TECHNOLOGIES

15.2.1 Microalgal hydrogen production

In order to meet the demands for energy around the world, the energy sector must make a significant transition toward sustainability. This review evaluates the technology and energy sources available today. Hydrogen is an essential energy carrier in solar energy, making it a strong contender as a fossil fuel substitute. This review addresses the possibility of photosynthetic hydrogen production for a solar-powered hydrogen economy, highlighting its importance. There are initiatives afoot to scale up industrial systems, and lab-scale research is making encouraging progress. It is thought that incorporating these developments into effective sustainable energy conversion is crucial, as it may result in mature industrial deployment (Bayro-Kaiser & Nelson, 2017).

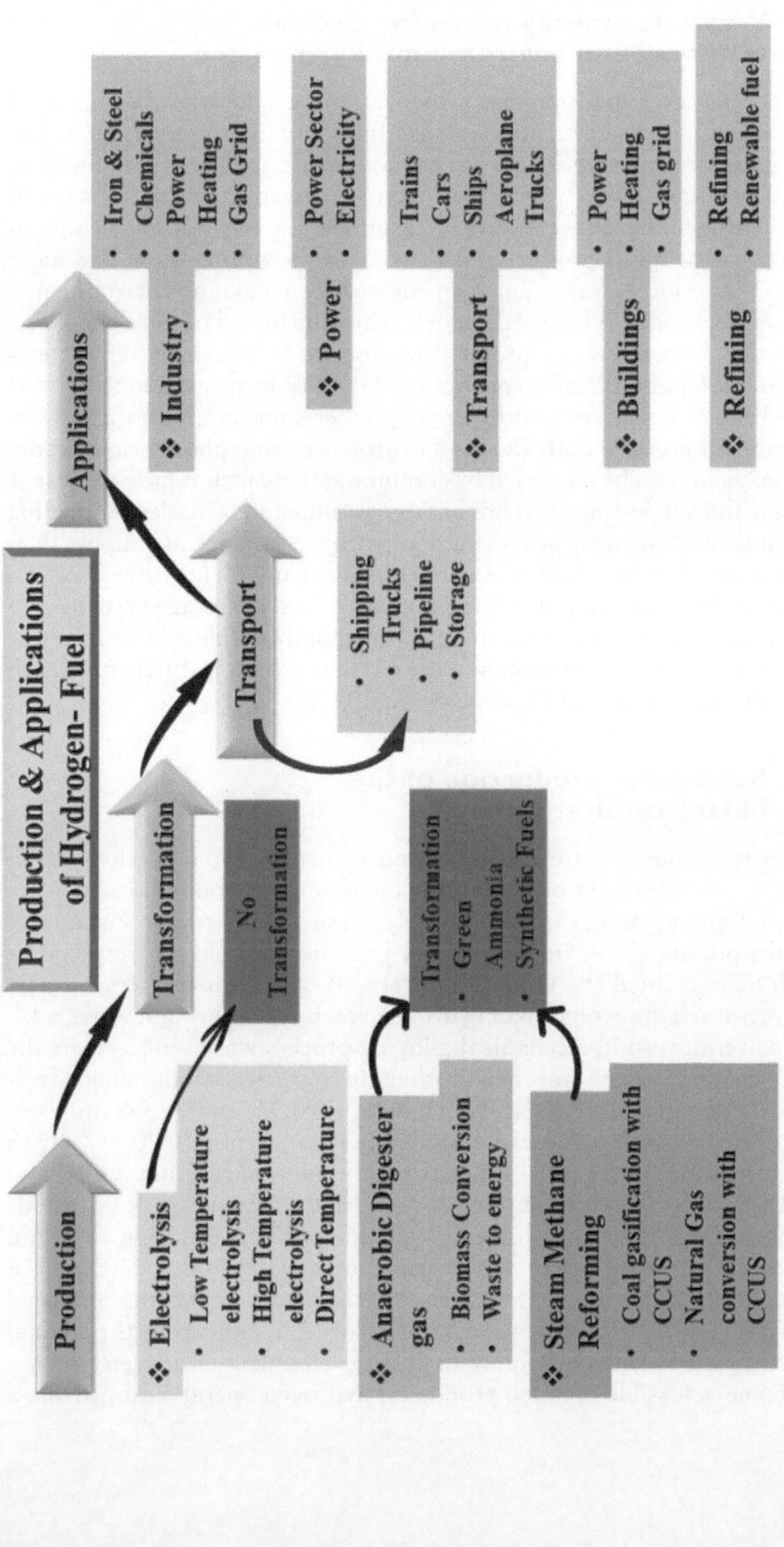

Figure 15.1 Schematic diagram of production and applications of hydrogen fuel.

15.2.2 Waste-to-energy nexus for circular economy and environmental protection

Global population expansion and urbanization have increased demand for energy, which has put conventional fossil fuels under pressure and resulted in price increases and greenhouse gas emissions that are harmful to the environment. In response, focus has switched to renewable, alternative sources such as hydrogen. Because they are environmentally friendly and economical, biological ways of producing hydrogen are becoming more and more popular. They address waste management, energy needs, and environmental challenges all at once. These techniques, which include dark and photo fermentation, use organic waste products to produce hydrogen. While bacteria like *Clostridium* and *Bacillus* spp. are used in dark fermentation to convert carbohydrate-rich sources into hydrogen, under some circumstances, non-sulfur bacteria produce both CO_2 and hydrogen during photofermentation. Hydrogen yields can be increased by combining these techniques: two-stage sequential and single-stage systems are two prominent examples. According to economic assessments, producing biohydrogen is more affordable than producing conventional fuel; possible enhancements include lowering the cost of photobioreactors and putting in place effective storage systems. For biological hydrogen production to be commercially viable and to meet the aims of sustainable development and the circular economy, further research and development is needed (Sharma et al., 2020).

15.2.3 Solid-state production of the binary catalyst $CdFe_2O_4$

Clean energy solutions are urgently needed in the current period of rising energy needs brought on by industrialization and population increase. Traditional energy sources, most notably fossil fuels, are not sustainable since they pollute the environment and pose health risks. In response, a research team created the unique catalyst $CdFe_2O_4$, which offers a viable energy vector for the evolution of hydrogen and oxygen through water splitting. A solventless solid mechanical alloying process was used to create the $CdFe_2O_4$ catalyst, which was then thoroughly characterized utilizing a variety of methods, including FTIR, XRD, XPS, FESEM, and XEDS. In water electrolysis, the catalyst showed a moderate overpotential (470 mV), with an onset potential of 1.6 V/RHE for oxygen evolution reaction (OER) and 0.2 V/RHE for hydrogen evolution process (HER). The electrode potentials required for OER (in an alkaline medium) and HER (in an acidic medium) to achieve 10 mA/cm^{-2} were 1.70 V/RHE (η = 0.47) and −0.30 V/RHE (η = −0.30 V), in that order. Furthermore, the catalyst that was produced exhibited remarkable stability for a duration of 12 h, indicating its potential for real-world implementation. With this development, water electrolysis will become a feasible way for producing hydrogen energy and provide a low-energy alternative to the electrolytic process. A promising catalyst for the synthesis of clean hydrogen fuel, $CdFe_2O_4$ has the potential to reduce global dependence on fossil fuels for energy production (Asiri et al., 2022).

15.2.4 Hydrogen evolution and oxidation

Clean energy systems require efficient hydrogen electrocatalysts. The development of catalysts requires an understanding of the processes and dynamics of hydrogen reactions. Interest in pH-dependent hydrogen evolution and oxidation reactions is growing. This section summarizes the HER/HOR reaction mechanism and offers a thorough kinetic analysis. It also discusses recent developments in affordable electrocatalyst technology and how our understanding of pH-dependent hydrogen electrocatalysis has advanced recently (Tian et al., 2019).

15.2.5 Creation of platinum nanoparticles for use in a reaction to generate hydrogen

The search for alternative energy sources has been fueled by the depletion of fossil resources. Although geothermal, wind, solar, and tidal energy all have potential, they are not without drawbacks. On the other hand, hydrogen gas is a very attractive alternative because it can be used to create energy, with the only result being water. This work presents a unique composite material (PtFCS) that is decorated with platinum nanoparticles and made from fused carbon spheres generated from sugar. TEM, EDS, FTIR, and XRD were some of the methods used to describe these nanoparticles, which were created by reducing chloroplatinic acid in a sodium borohydride solution with sodium citrate acting as a capping agent. At a rate of 0.0438 mL/mgcat•min, the PtFCS composite efficiently catalyzed the hydrolysis of $NaBH_4$, performing best in an acidic environment (pH 6). The catalyst demonstrated steady effectiveness over several applications, with activation energy, indicating its promise as a sustainable catalyst for the production of hydrogen fuel (E. Biehler, Q. Quach, & T. M. J. N. Abdel-Fattah, 2023).

15.2.6 Electrochemical hydrogenation with gaseous ammonia

Ammonia is a fuel that is both a powerful hydrogen-storage medium and a carbon-free, sustainable fuel. It is imperative to devise inventive methods to capitalize on its potential immediately. In this work, we present an electrochemical approach to hydrogenate utilizing carbon electrodes and ammonia as the hydrogen source to produce alkenes, alkynes, and ketones. With amazing tolerance, this technique can handle a broad range of heterocycles and functional groups, such as allyl carbamate, benzyl, and sulfides.

The mechanism enables the intended changes by means of quick stepwise proton and electron transfer processes (Li et al., 2019).

15.2.7 An artificial photosynthetic leaf-to-farm challenge for solar hydrogen generation

One potential method for turning sunlight into sustainable and renewable hydrogen energy is solar water splitting. Three primary approaches to this transition are examined in this review: photovoltaic-electrolysis (PV-EC), photoelectrochemical (PEC), and photocatalytic (PC). The article assesses solar-to-hydrogen production devices and systems with an eye on promoting the green hydrogen economy. It highlights important criteria, including material efficiency, cost, stability, and techno-economic aspects. Often referred to as "an artificial photosynthetic leaf-to-farm challenge," it offers a thorough overview, including principles, materials, technologies, system designs, multinational projects, and scaling-up challenges (Kim et al., 2019).

15.2.8 Cooperative hydrogen and oxidative organic synthesis coupling

Selective organic synthesis and H_2 production may be cooperatively photoredox linked by the use of photoexcited electrons and holes over semiconductor-based catalysts. This is an improvement over conventional photocatalytic H_2 synthesis. We discuss some of the most recent advancements in this field in this part, including the reforming of lignocellulosic biomass, the oxidative coupling of amines, cross-coupling mechanisms, the dehydrogenation of cyclic alkanes, and the selective oxidation of alcohols. Finally, we address the challenges that now exist as well as the prospects for this emerging profession in the future. By providing insights for the thoughtful design of dual-functional photoredox systems, this review hopes to further the development of environmentally responsible and economically viable solar fuel generation as well as value-added fine chemical synthesis (Yao et al., 2021).

15.3 HYDROGEN FUEL IN SUSTAINABLE TRANSFORMATION

15.3.1 Conversion of H_2S into hydrogen fuel and elemental sulfur

A mechanistic insight into the photocatalytic transformation of hydrogen sulfide pollution into hydrogen fuel and elemental sulfur using an exceptional heterojunction nanocomposite solar-energy material hydrogen sulfide (H_2S) poses threats to both living organisms and metallic structures, arising from natural geothermal or bacterial processes and industrial activities, notably in petroleum and gas sectors. A promising solution lies in the photocatalytic conversion of H_2S into hydrogen fuel and elemental sulfur, offering both environmental and energy benefits. The demand for cost-effective, eco-friendly photocatalysts or solar-energy materials is significant. In this study, we synthesized a series of xBi_2S_3•yMnS nanocomposite photocatalysts through a simple hydrothermal method. The catalysts showed effectiveness in transforming H_2S into hydrogen fuel and elemental sulfur in alkaline (pH = 11) conditions at ambient temperature and atmospheric pressure. A molar ratio of $xy = 2$ was shown to yield optimal performance, exhibiting less charge recombination, improved photon absorption, and increased surface area. Significantly, the only oxidation product detected was disulfide (S_{22}^{-}), which readily precipitated as elemental sulfur when the pH was lowered to about five through medium acidification (Lashgari et al., 2019).

15.3.2 Lignocellulose catalytic conversion into chemicals and fuel products in ionic liquids

A sustainable future and a biobased economy depend heavily on the creative use of abundant and renewable lignocellulosic biomass. Ionic liquids (ILs), acknowledged as environmentally benign solvents, are attracting interest for their ability to catalyze the transformation of lignocellulosic feedstocks into diverse products. Leveraging unique IL properties enhances catalytic processes, potentially reducing environmental impact. Additionally, IL-based systems offer a platform for diverse product development. This review delves into the catalytic conversion of lignocellulose components (cellulose, hemicellulose, lignin) into valuable chemicals and fuels using ILs. It also outlines challenges and opportunities in this significant field (Zhang et al., 2017).

15.3.3 Conversion of struvite from wastewater into a molecule that stores hydrogen for fuel ammonia borane

Ammonia borane (NH_3BH_3) serves as a hydrogen carrier, offering a carbon-free renewable energy solution due to its high hydrogen content and air stability. We explore a novel method for producing ammonia borane using ammonia derived from wastewater, aligning with global sustainability goals. In addition to saving water, recycling wastewater allows for the recovery of important nutrients like phosphorus, magnesium, and nitrogen in the form of struvite (NH_4MgPO_4•$6H_2O$). Our method produces high-purity ammonia borane by first turning urine into struvite and then reacting it with alkali borohydrides. This technique offers an effective way to recycle nitrogen waste because of its gentle reaction conditions and low purification processes. By integrating nitrogen removal from wastewater with the production of a renewable, carbon-free energy source, both the energy and wastewater sectors stand to benefit, contributing to a more sustainable environment (Dhingra et al., 2023).

15.3.4 Technological approaches to landfill management: Aiming for carbon neutrality and biomethane recovery

Ineffective landfill management must be addressed right away because it puts human health and the environment at serious risk. Negligently maintained landfills release toxic gases, contaminate water supplies, and deplete essential resources. Linking landfill operations to the sustainable development goals (SDGs) emphasizes how important it is to hit several goals at once. It is critical that landfill practices adapt quickly in order to address urgent issues including climate change, reaching carbon neutrality, guaranteeing food security, and optimizing resource recovery. The significant influence of landfill management on climate change has been demonstrated by a wealth of research, as evidenced by over 191 published articles (1998–2023). Emerging sustainable landfill management approaches, including oxidation, biological, and physico-chemical treatments, are discussed in this chapter and their respective practical applicability is evaluated. Prioritizing resource recovery over traditional treatment methods, it highlights the worldwide significance of landfill management in the pursuit of carbon neutrality. Achieving carbon neutrality by 2050 requires minimizing inputs such as energy, chemicals, and water in nutrient recovery. Techniques like reusing water, recovering energy, and carefully choosing materials for manufacturing are crucial. Considering practicality, the possibility of water technologies for removing macronutrients from landfill leachate is investigated. Recycling and composting are two examples of integrated waste management techniques that are essential to trash reduction and reducing environmental damage. Water technologies provide economic benefits by purifying leachate and making it easier to recover important materials including ammonium, heavy metals, nutrients, and salts. The potential for landfill resource recovery is further increased by using microbial fuel cells to convert methane and hydrogen into bioenergy and electricity generation. Future studies should focus on environmentally friendly and economically viable landfill leachate treatment methods. It is possible to lessen the negative effects improper trash disposal has on the environment and human health by implementing improved landfill management techniques (Al-Hazmi et al., 2024).

15.3.5 Controlling Lewis acidic sites in 1T-2H MoS_2 catalysts for photothermal heating-driven H_2 production from lignocellulosic biomass

It has been effectively established that 1T-2H MoS_2 catalysts with tunable Lewis acidic sites in an alkaline aqueous environment may produce hydrogen (H_2) through photothermal catalysis from lignocellulosic biomass under solar driving. Through the development of an edge-terminated 1T-2H phase structure and an adjustable layer count, these catalysts control the number of Lewis acidic sites. The 1T-2H MoS_2 catalyst proves to be highly effective in reducing polar wood chips, bamboo, rice straw, corncobs, and rice hulls into H_2 by utilizing the plentiful Lewis acidic sites for lignocellulosic biomass oxygenolysis. The catalyst is noteworthy for achieving a peak H_2 generation rate of 3661 $\mu mol \bullet h^{-1} \bullet g^{-1}$ and a solar-to-H_2 (STH) efficiency of 0.18% in the polar wood chip system when illuminated by a 300 W Xe lamp. The method this research proposes to use solar energy to directly convert renewable lignocellulosic biomass into H_2 fuel is both economically and sustainably feasible (Ma et al., 2023).

15.3.6 Tyrosine is the source of all five CO and CN^- ligands in cell-free H-cluster formation and [FeFe] hydrogenase activation

Models for imitating natural hydrogen-evolving processes and [FeFe] hydrogenases are useful catalysts for the sustainable synthesis of hydrogen. Three CO and two CN^- ligands, together with a dithiolate bridge, make up the H-cluster, a special iron-based cofactor required for hydrogen production. Although accessory proteins (HydE, HydF, and HydG) are crucial for H-cluster assembly, it is yet unknown what exactly their tasks include and the biosynthesis mechanism. Here, we report efficient cell-free methods to analyze [FeFe] hydrogenase activation and H-cluster production. We show that the CO and CN^- ligands originate from the carboxylate and amino substituents of tyrosine, respectively, by using isotopic labeling in conjunction with FTIR spectroscopy. These rebuilt in vitro systems provide a flexible tool to explore biosynthetic mechanisms. Understanding protein–protein interactions and the intricate processes involved in H-cluster building and hydrogenase maturation has advanced significantly as a result of this research (Kuchenreuther et al., 2011).

15.3.7 Graphene doped with gold: An extremely stable and effective electrocatalyst for the reduction of oxygen

Hydrogen fuel cells present a viable answer to the pressing need for sustainable energy. However, increasing the efficiency and affordability of platinum-based cathodes is necessary for their broad implementation. Our study shows that adding gold atoms to graphene vacancies results in a stable and cost-effective electrocatalyst. This catalyst outperforms platinum-based ones in catalyzing the oxygen reduction reaction (ORR). By passivating defected graphene with gold atoms, we optimize the catalytic activity while ensuring high stability. Our findings challenge assumptions about material screening, showing that the linear relation among binding energies of reaction intermediates doesn't hold for this material. We anticipate that gold-doped graphene will overcome barriers in hydrogen fuel cell technology, paving the way for sustainable transportation and portable power sources (Stolbov & Alcântara Ortigoza, 2015).

15.3.8 Lewis acid enhancement of base metal catalysts in the reversible hydrogenation of carbon dioxide to formic acid and methanol

Addressing environmental concerns stemming from fossil fuel usage necessitates the exploration of new sustainable energy sources. While hydrogen (H_2) shows promise as a clean energy option, challenges in its storage and transportation as a gas have led to interest in chemical H_2 storage (CHS). Formic acid (FA) and methanol (MeOH) are renewable liquid CHS candidates due to their hydrogen content. However, current commercial synthesis methods for FA and MeOH rely on nonrenewable resources. Direct metal-catalyzed CO_2 hydrogenation using affordable and easily accessible carbon sources is a viable sustainable method. Because they are abundant and reasonably priced, homogeneous transition metal catalysts are preferred, especially those that are based on first-row transition metals. Recent developments for CO_2 hydrogenation to FA and MeOH and FA and MeOH dehydrogenation in iron and cobalt systems are reviewed. Additives and ligands are essential for maintaining the stability of these catalysts and enhancing catalytic activity. Alkali metal Lewis acid cocatalysts are highlighted for their ability to enhance activity and stability. Mechanistic pathways for these transformations are examined, revealing similarities among catalysts. Limitations of current catalytic systems are identified, with suggestions for future improvements in stability, activity, and practicality (Bernskoetter & Hazari, 2017).

15.3.9 Hydrogen evolution and polyethylene terephthalate plastic reformation simultaneously by electrocatalysis using a self-supported amorphous cobalt iron phosphide electrode

A viable way to address energy constraints and plastic pollution is to combine the electrocatalytic hydrogen evolution process (HER) with the transformation of polymers into valuable molecules. Effective bifunctional catalytic electrodes and carefully thought-out electrolysis systems are needed for this strategy. By electrodepositing on nickel foam, we created a self-supported CoFe-P/NF electrode that effectively catalyzes HER and the selective electrooxidation of PET plastic hydrolysate into formate. The electrode demonstrated tolerance to PET monomers, good stability for HER, and a low overpotential (η 100 of 168 mV at J = 100 mA cm^{-2}). Additionally, it demonstrated a high faradaic efficiency (FE, 90%) and productivity (0.1 mmol cm^{-2} h^{-1}) for the generation of formate from the ethylene glycol (EG) component in PET hydrolysate at 1.5 V. Compared to water electrolysis, the CoFe-P/NF electrode allowed for the coproduction of H_2 and formate at a lower voltage and energy consumption. This work highlights the possibility of electrolyzing PET hydrolysate with properly engineered catalytic electrodes to produce formate and H_2 fuel. These electrodes have potential uses in the electrosynthesis of several compounds and energy devices (Chang et al., 2024).

15.4 APPLICATIONS OF HYDROGEN PRODUCTIONS

15.4.1 Hydrogen fuel and biodegradable polymer monomers are produced at high current densities via electrocatalytic upcycling of biomass and plastic wastes

An efficient upcycling method for resource optimization is the conversion of biomass and plastic waste using electrocatalysis into useful chemicals and fuels. It is still crucial to overcome obstacles like raising current density and guaranteeing excellent selectivity for compounds that are in great demand in the market. In this work, we provide an electrocatalytic technique for converting waste polyethylene terephthalate (ethylene glycol) into glycolic acid and glycerol, a biodiesel byproduct, into lactic acid. These goods are important monomers in the synthesis of biodegradable polymers. We achieved significant current densities at moderate potentials (317.7 mA/cm^2 at 0.95 V vs RHE and 326.2 mA/cm^2 at 1.15 V vs RHE, respectively) and outstanding selectivity of 77% for lactic acid and 91% for glycolic acid using a nickel hydroxide-supported gold electrocatalyst (Au/$Ni(OH)_2$). Our results show that ethylene glycol and glycerol can be concentrated at the Au/$Ni(OH)_2$ interface, which can lead to high current densities and improved local concentrations. We used a membrane-free flow electrolyzer to upcycle PET and triglyceride bottles to show that it was feasible. The results showed that it is possible to produce valuable chemicals and H_2 fuel from waste materials in a sustainable manner. We were able to produce 13.7 g of glycolic acid and 11.2 g of lactic acid combined with 9.3 L of H_2 (Yan et al., 2023).

15.4.2 Thermal and structural analysis of the conversion of lignocellulosic biomass to improve sustainability in the progressive organic refinery paradigm for waste-to-energy applications

The depletion of fossil fuel reserves and the resulting damage to the environment caused by human activity demand that sustainable waste-to-energy technologies advance quickly. The production of energy from biowaste in organic bio-refinery systems is consistent with the sustainability paradigm that considers both the environment and energy. According to the World Bioenergy Association, biomass could provide half of the world's primary energy demands. This study examines the burning properties of cotton stalks acidified in diluted hydrochloric acid (HCl) for biomass conversion in organic bio-refineries with the goal of changing the energy landscape in favor of cleaner energy generation. Advanced thermochemical techniques, including pyrolysis and combustion, were employed at various leaching times (0 to 180 min) under controlled conditions. The demineralization process and its effects were evaluated using analytical techniques such as proximal, ultimate, gross calorific value (GCV), thermogravimetric, kinetic, XRD, FTIR, and SEM-EDS. The findings show that while lowering the carbon footprint by 54.80%, longer leaching (180 min) considerably raised the fixed carbon content and higher heating value by 20.23% and 12.48%, respectively. These findings were corroborated by GCV and CHNS analysis. Thermogravimetric analysis showed a trend of increasing conversion efficiency, with the sample that underwent prolonged leaching showing the largest rise of 98% (3H.TT.DEM). The production of functional groups and crystalline modifications were verified by FTIR and XRD investigations, respectively. The most demineralized sample (3H.TT.DEM) had a balanced element distribution, according to EDS analysis; however, SEM analysis demonstrated an increase in pore size and quantity inside the cotton stalk structure post-acidulation. In order to improve waste biorefinery systems, this research proposes sustainable biomass transformation, which will greatly contribute to Sustainable Development Goals 7, 12, and 13 (Qureshi et al., 2024).

15.4.3 Creation of an effective and long-lasting Si/GaN photocathode through photoelectrochemical self-improvement

The advancement of solar fuel production depends on the creation of an effective and durable photoelectrode. In this work, we demonstrate the self-improving behavior of a silicon/gallium nitride photocathode that produces hydrogen with approximately 100% Faradaic efficiency. Through a comprehensive analysis involving various spectroscopic and microscopic methods, along with density functional theory calculations, we elucidate the chemical transformation underlying this self-improvement. More specifically, by partially replacing nitrogen with oxygen, a gallium oxynitride layer formed on the gallium nitride grains, boosting catalytic sites for hydrogen evolution. The conversion of gallium nitride into gallium oxynitride, which improves catalytic activity and guarantees long-term operation, is highlighted in this study. Applications for oxynitride layers as protective catalytic coatings for hydrogen evolution appear promising (Zeng et al., 2021).

15.4.4 Decorated gold nanoparticles AuNP on fused graphene-like materials for use in hydrogen production

Since hydrogen burns cleanly, it has become a popular fuel option for those looking for a sustainable alternative to fossil fuels. A catalyst is required for maximum efficiency in producing hydrogen from sodium borohydride, as the process is quite sluggish. For the hydrolysis of sodium borohydride, this work explores a new catalyst made of fused graphene-like materials (AuFGLM) and gold nanoparticles. Fused graphene-like material (FGLM) was produced in an environmentally friendly manner by pressure-processing a dextrose solution. TEM, SEM, EDS, FTIR, XRD, and Raman spectroscopy were among the characterization methods that provided information about the composition and characteristics of the nanoparticles and composites. A circular carbon backbone and an average diameter of roughly 23 nm for the gold nanoparticles were seen in TEM imaging. Functional groups such as O—H, C—C, and C = O were confirmed by FTIR. According to EDS analysis, the composite had a gold content of about 6.3%. The crystal structures of AuFGLM and FGLM were determined using P-XRD. Catalytic experiments conducted in a range of settings showed that elevated temperatures, acidic pH values, and higher $NaBH_4$ dosages promoted the best results. In later experiments, the catalyst demonstrated reusability with increased hydrogen generation. The activation energy was found to be 45.5 kJ mol^{-1}. With its competitive activation energy when compared to its counterparts, this innovative catalyst has great promise for producing hydrogen sustainably from $NaBH_4$ hydrolysis as a fuel source (E. Biehler, Q. Quach, & T. M. J. M. Abdel-Fattah, 2023).

15.4.5 *Escherichia coli* strains that produce more hydrogen and ethanol in a glycerol-based medium were identified by screening in single-knockout mutant collections

Human activities, particularly the use of fossil fuels, are causing the Earth's temperature to warm. A sustainable substitute is provided by bioenergy, which includes biodiesel, biohydrogen, and bioethanol. However, there is an environmental problem due to the excess glycerol produced as a result of the widespread manufacture of biodiesel. Researchers suggest using waste glycerol to address this by using microbial transformation to make biofuels like ethanol and hydrogen, with *Escherichia coli* being a promising choice. An experimental protocol was created for this work in order to test *E. coli* single mutants for higher production of hydrogen and/or ethanol in comparison to the wild type. Twelve new strains showed increased yields for at least one target product out of the 150 mutants that were examined. Three modified *E. coli* mutant strains were produced by combining mutations in genes like *gnd* and *tdcE*, which demonstrated notable improvements and redirected carbon flux toward desired products. This approach sheds light on metabolic pathways that can be altered to increase the yields of hydrogen and ethanol in several *E. coli* mutants that use glycerol as their main carbon source (Valle Gallardo et al., 2015).

15.4.6 Manganese oxide is more stable when synthesized electrochemically using light during the water oxidation process

To produce hydrogen fuel through electrochemical water splitting, new methods are needed to improve the longevity and efficiency of earth-abundant catalysts. Manganese oxide films with significant disorder and a range of oxidation states have been produced by previous procedures using electrodeposition, which produced active but unstable catalysts for the oxygen evolution reaction (OER) at high current densities. In this study, we show that manganese oxide can be electrodeposited multipotentially under light to form nanostructured films that are far more stable for the OER than films formed in the dark under comparable conditions. For 18 h at a pH of 13, these light deposition films maintain a current density of 10 mA/cm^2 at 2.2 V vs the reversible hydrogen electrode. Still, light does not improve the stability or activity of films of manganese oxide formed at a constant potential. After 1 h of electrolysis, films formed in the dark using multipotential deposition completely lose their activity.

According to electrochemical and structural analysis, photoexcitation during film growth stabilizes the nanostructured film during electrocatalysis by reducing Mn ions and changing the composition and structure of intercalated potassium ions and water molecules within the disordered layers of MnOx sheets that resemble birnessite. These data highlight the possibility of combining several external stimuli, such as light and an external current, to cause structural alterations that improve the stability and activity of earth-available catalysts for important chemical reactions like the oxidation of water (Qin et al., 2023).

15.4.7 Water-splitting electrocatalysts based on hierarchically nanostructured cobalt phosphide/chalcogenide: Dynamics and control of active sites

Effective electrocatalysts are essential for producing hydrogen fuel sustainably through water splitting. Comprehending intricate catalytic systems is still a difficult task. As an alternative electrocatalyst, we utilize a self-templated method to fabricate hierarchically nanostructured Fe-doped cobalt phosphide nanoboxes (Co@CoFe-P NBs). We monitor catalytically active intermediates and structural dynamics during water splitting using operando Raman and X-ray absorption spectroscopy. Fe substitution facilitates the adsorption/desorption of HER intermediates and lowers energy barriers, which improves the efficacy of the hydrogen evolution reaction (HER) by promoting structural reconstruction into P-Co-O-Fe-P topologies. Converted configurations into active CoIV-O-FeIV moieties give rise to competitive oxygen evolution reaction (OER) activity. These results are supported by ex situ investigations, which emphasize the benefits of phosphide-based electrocatalysts. These realizations are essential for making the most of non-oxide, inexpensive electrocatalysts for upcoming energy storage and conversion projects (Zhao et al., 2022).

15.4.8 A few crucial elements for effective electrocatalytic water splitting: Identification of intermediates, morphology alterations, and active site coordination environment

The creation of clean hydrogen fuel through water splitting presents a viable route, but the slow kinetics of existing electrocatalytic techniques provides difficulties. As a result, effective electrocatalyst development is essential. In order to optimize water-splitting electrocatalysts, active sites and intermediate species must be identified. This chapter summarizes recent advancements in this regard. Changes in valence state and morphology in two-dimensional materials are examples of how structural modifications impact catalytic activity. The approaches for spectroscopy and in situ/ex situ electron microscopy are particularly highlighted as useful for identifying molecular active intermediates and describing catalytic sites. Additionally, the review emphasizes the necessity of developing novel in situ methodologies and electrokinetic analytic strategies to direct future research initiatives (Hu et al., 2022).

REFERENCES

Al-Hazmi, H. E., Hassan, G. K., Kurniawan, T. A., Śniatała, B., Joseph, T. M., Majtacz, J., . . . Saeb, M. R. J. J. O. E. M. (2024). Technological solutions to landfill management: Towards recovery of biomethane and carbon neutrality. *Journal of Environmental Management*, *354*, 120414.

Asiri, A. M., Adeosun, W. A., Khan, S. B., Alamry, K. A., Marwani, H. M., Zakeeruddin, S. M., & Grätzel, M. J. S. R. (2022). Solid-state synthesis of CdFe2O4 binary catalyst for potential application in renewable hydrogen fuel generation. *Scientific Reports*, *12*(1), 1632.

Bayro-Kaiser, V., & Nelson, N. J. P. R. (2017). Microalgal hydrogen production: prospects of an essential technology for a clean and sustainable energy economy. *Photosynthesis Research*, *133*, 49–62.

Bernskoetter, W. H., & Hazari, N. J. A. O. C. R. (2017). Reversible hydrogenation of carbon dioxide to formic acid and methanol: Lewis acid enhancement of base metal catalysts. *Accounts of Chemical Research*, *50*(4), 1049–1058.

Biehler, E., Quach, Q., & Abdel-Fattah, T. M. J. M. (2023). Gold nanoparticles AuNP decorated on fused graphene-like materials for application in a hydrogen generation. *Materials*, *16*(13), 4779.

Biehler, E., Quach, Q., & Abdel-Fattah, T. M. J. N. (2023). Synthesis of platinum nanoparticles supported on fused nanosized carbon spheres derived from sustainable source for application in a hydrogen generation reaction. *Nanomaterials*, *13*(13), 1994.

Chang, J., Wang, L., Wu, D., Xu, F., Jiang, K., Guo, Y., . . . Science, I. (2024). Concurrent electrocatalytic hydrogen evolution and polyethylene terephthalate plastics reforming by self-supported amorphous cobalt iron phosphide electrode. *Journal of Colloid and Interface Science*, *655*, 555–564.

Dingra, N., Witty, M., Celis, M., Boppana, N., & Ayudhya, T. J. F. I. C. (2023). Transformation of struvite from wastewater to a hydrogen fuel storage compound ammonia borane. *Frontiers in Chemistry*, *11,* 1269845.

Gonzalez-Garay, A., Bui, M., Freire Ordóñez, D., High, M., Oxley, A., Moustafa, N., . . . Engineering, B. (2022). Hydrogen production and its applications to mobility. *Annual Review of Chemical and Biomolecular Engineering*, *13*, 501–528.

Grossmann, L., Hinrichs, J., Weiss, J. J. C. R. I. F. S., & Safety, F. (2022). Technologies for sustainable heat generation in food processing. *Comprehensive Reviews in Food Science and Food Safety*, *21*(6), 4971–5003.

Hu, C., Hu, Y., Zhu, A., Li, M., Wei, J., Zhang, Y., & Xie, W. J. C. A. E. J. (2022). Several key factors for efficient electrocatalytic water splitting: Active site coordination environment, morphology changes and intermediates identification. *Chemistry–A European Journal*, *28*(36), e202200138.

Kim, J. H., Hansora, D., Sharma, P., Jang, J.-W., & Lee, J. S. J. C. S. R. (2019). Toward practical solar hydrogen production–an artificial photosynthetic leaf-to-farm challenge. *Chemical Society Reviews*, *48*(7), 1908–1971.

Kuchenreuther, J. M., George, S. J., Grady-Smith, C. S., Cramer, S. P., & Swartz, J. R. J. P. O. (2011). Cell-free H-cluster synthesis and [FeFe] hydrogenase activation: all five CO and CN– ligands derive from tyrosine. *PLoS One*, *6*(5), e20346.

Lashgari, M., Ghanimati, M. J. J. O. C., & science, i. (2019). An excellent heterojunction nanocomposite solar-energy material for photocatalytic transformation of hydrogen sulfide pollutant to hydrogen fuel and elemental sulfur: A mechanistic insight. *Journal of Colloid and Interface Science*, *555*, 187–194.

Li, J., He, L., Liu, X., Cheng, X., & Li, G. J. A. C. (2019). Electrochemical hydrogenation with gaseous ammonia. *Angewandte Chemie*, *131*(6), 1773–1777.

Ma, C., Cheng, M., Liu, Q.-Y., Yuan, Y.-J., Zhang, F.-G., Li, N., . . . Zou, Z. J. N. L. (2023). Regulating lewis acidic sites of 1T-2H MoS2 catalysts for solar-driven photothermal catalytic H2 production from lignocellulosic biomass. *Nano Letters*, *24*(1), 331–338.

Preuster, P., Papp, C., & Wasserscheid, P. J. A. O. C. R. (2017). Liquid organic hydrogen carriers (LOHCs): Toward a hydrogen-free hydrogen economy. *Accounts of Chemical Research*, *50*(1), 74–85.

Qin, C., Luo, J., Zhang, D., Brennan, L., Tian, S., Berry, A., . . . Sadtler, B. J. A. N. A. (2023). Light-mediated electrochemical synthesis of manganese oxide enhances its stability for water oxidation. *ACS Nanoscience Au*, *3*(4), 310–322.

Qureshi, T., Farooq, M., Imran, S., Munir, M. A., Javed, M. A., Sohoo, I., . . . Asim, M. J. E. R. (2024). Structural and thermal investigation of lignocellulosic biomass conversion for enhancing sustainable imperative in progressive organic refinery paradigm for waste-to-energy applications. *Environmental Research*, *246*, 118129.

Sharma, S., Basu, S., Shetti, N. P., & Aminabhavi, T. M. J. S. O. T. T. E. (2020). Waste-to-energy nexus for circular economy and environmental protection: Recent trends in hydrogen energy. *Science of the Total Environment*, *713*, 136633.

Stolbov, S., & Alcántara Ortigoza, M. J. T. J. O. C. P. (2015). Gold-doped graphene: A highly stable and active electrocatalysts for the oxygen reduction reaction. *The Journal of Chemical Physics*, *142*(15), 154703.

Tian, X., Zhao, P., & Sheng, W. J. A. M. (2019). Hydrogen evolution and oxidation: Mechanistic studies and material advances. *Advanced Materials*, *31*(31), 1808066.

Valle Gallardo, A., Cabrera Revuelta, G., Cantero Moreno, D., & Bolívar Pérez, J. (2015). Identification of enhanced hydrogen and ethanol Escherichia coli producer strains in a glycerol-based medium by screening in single-knock out mutant collections. *Microbial Cell Factories*, *14*, 1–14.

Yan, Y., Zhou, H., Xu, S.-M., Yang, J., Hao, P., Cai, X., . . . Shao, M. J. J. O. T. A. C. S. (2023). Electrocatalytic upcycling of biomass and plastic wastes to biodegradable polymer monomers and hydrogen fuel at high current densities. *Journal of the American Chemical Society*, *145*(11), 6144–6155.

Yao, M., Wang, Y., Li, X., Sheng, Y., Huo, H., Xi, L., . . . Zhang, W. J. S. D. (2021). Materials informatics platform with three dimensional structures, workflow and thermoelectric applications. *Scientific Data*, *8*(1), 236.

Zeng, G., Pham, T. A., Vanka, S., Liu, G., Song, C., Cooper, J. K., . . . Toma, F. M. J. N. M. (2021). Development of a photoelectrochemically self-improving Si/GaN photocathode for efficient and durable H2 production. *Nature Materials*, *20*(8), 1130–1135.

Zhang, Z., Song, J., & Han, B. J. C. r. (2017). Catalytic transformation of lignocellulose into chemicals and fuel products in ionic liquids. *Chemical Reviews*, *117*(10), 6834–6880.

Zhao, Y., Dongfang, N., Triana, C. A., Huang, C., Erni, R., Wan, W., . . . Science, E. (2022). Dynamics and control of active sites in hierarchically nanostructured cobalt phosphide/chalcogenide-based electrocatalysts for water splitting. *Energy & Environmental Science*, *15*(2), 727–739.

Chapter 16

A review on hybridization and energy management strategies for hydrogen-based fuel cell electric vehicle applications

Shiv Prakash Singh, Suneel Raju Pendem, Brij Kishor, Dogga Raveendhra, and M Venkatesh Naik

16.1 INTRODUCTION

Modern society depends heavily on burning traditional energy sources such as petroleum, coal, and natural gas to meet daily energy demands, resulting in environmental problems like air pollution and greenhouse gas emissions. The growing concern over depleting these resources motivates scientists and engineers to explore renewable energy options, including solar, wind, tidal, hydroelectric, hydrogen, geothermal, and biomass power. However, sources like wind and solar are subject to unpredictable fluctuations. In contrast, hydrogen energy is notable for being clean and easily transportable and storable, making it a crucial component of future energy solutions, especially in fuel cell (FC) electric vehicles. The transportation sector is the largest contributor to greenhouse gas emissions, accounting for up to 29%, as shown in Figure 16.1. The US Department of Energy (DoE) indicates that FCEVs have the potential to significantly reduce greenhouse gas emissions compared to conventional combustion engine vehicles. Additionally, a discussion by the Argonne National Laboratory suggests that fuel cell electric vehicles could cut greenhouse gas emissions by 50% compared to other fuel-based cars and by 75% compared to diesel vehicles.

United Technologies Corporation (UTC) has been developing fuel cell systems for space and defense since 1958. They created the first functional fuel cell application to generate electricity and clean water for the Apollo space missions. Since 1966, all manned US space missions, including space shuttle flights, have utilized UTC's fuel cells. In 1991, UTC Fuel Cells introduced the PureCell model 200 power plant, which produces 200 kW of electricity and 900,000 Btu/h of heat for combined heat and power (CHP) applications. General Motors is also engaged in fuel cell development for both stationary power and automotive uses. In 2004, they started operations in Texas with Dow Chemical Company, using fuel cell systems to produce 35 MW of electricity. In France, Axane is at the forefront of fuel cell technology development, targeting three main commercial markets: (a) portable multi-application generators (500 W–10 kW), (b) stationary applications

DOI: 10.1201/9781003537816-16

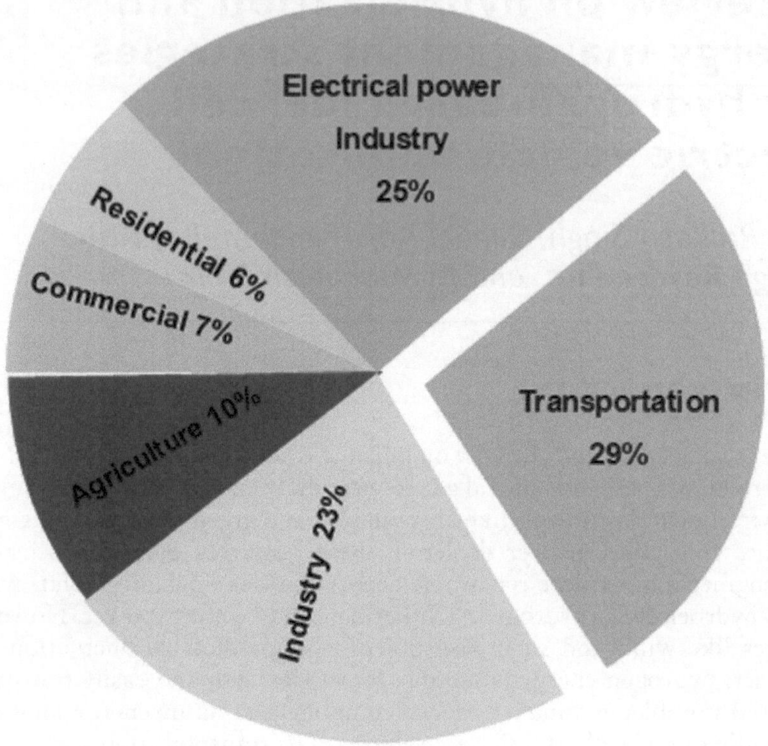

Figure 16.1 Greenhouse gas emission percentage in year 2021 in the United States (Hu et al., 2019).

(more than 10 kW), and (c) mobile applications for small hybrid vehicles (5–20 kW).

FCEVs have the potential to be a crucial technology in the transition to a low-carbon economy, enhancing accessibility, affordability, and energy security. Table 16.1 provides a comprehensive overview of the FCEVs available on the market, including their features, pricing, and manufacturers. In recent years, many passenger car manufacturers have been actively involved in developing FCEVs. While companies like Toyota, Honda, and General Motors produce their own fuel cell stacks, others such as Ford, Mazda, Daimler Chrysler, Hyundai, Fiat, and VW source fuel cell stacks from specialized industry vendors like Ballard Power Systems, Plug Power, and Cummins. Fuel cells generate DC output and are always connected to circuits through power conditioning units such as DC to DC and DC to AC converters. This connection is based on power conversion and control, which falls under the field of power electronics. Advances in power electronics technology are currently driven by the need for optimal control in industrial applications and the creation of more reliable and compact switching power supplies for intelligent systems.

Table 16.1 Manufacturers and features of FCEVs that are currently in market

Model	*Range (km)*	*Motor (type)*	*Top speed*	*FC type*	*FC power*
Honda FCX V4	260–310	Induction motor	150 km/h	PEMFC	78 kW
Ford Focus FCV	320	Induction motor	129 km/h	PEMFC	85 kW
Nissan X-Trail FCV	350	Synchronous motor	Not given	PEFC	90 kW
Mercedes-Benz A F Cell	160–180	—	132 km/h	PEMFC	—
Cheverlet Equinox FC	310	Permanent magnet motor	141 km/h	PEMFC	93 kW
Honda FCX Clarity	390	Brushless DC motor	130 km/h	PEMFC	100 kW
Hyundai Tucson FCEV	594	Induction motor	160 km/h	PEMFC	100 kW
Toyota Mirai FCEV	480	Induction motor	160 km/h	PEMFC	114 kW
Nissan e-Bio Fuel cell	600	—	—	SOFC	5 kW

16.2 FUEL CELL TECHNOLOGY

A fuel cell (FC) is a device that directly converts the chemical energy of a fuel into electrical energy. The concept of the fuel cell was first proposed by William Robert Grove in 1839, long before the modern development of FC technology. Grove's early fuel cell generated water and electricity by introducing hydrogen and oxygen to sulfuric acid, using porous platinum electrodes. In this setup, one electrode oxidizes the fuel, such as hydrogen, while the other electrode reduces an oxidant, like oxygen. This process is similar to the electrochemical reactions in batteries that generate electricity. As illustrated in Figure 16.2, these reactions involve the movement of electrons through an external circuit and the exchange of ions through a solid or liquid electrolyte. Although a single FC cell can theoretically achieve a voltage of 1.23 V, the actual voltage under load typically ranges from 0.6 to 0.7 V at rated current. Therefore, fuel cells are composed of multiple individual cells stacked together to achieve the desired voltage and power output.

16.2.1 Operation of fuel cell

A fuel cell is an electrochemical device that employs a fuel and an oxidizing agent, usually oxygen, to produce electrical energy through redox reactions. Unlike batteries, which operate using stored chemicals, fuel cells

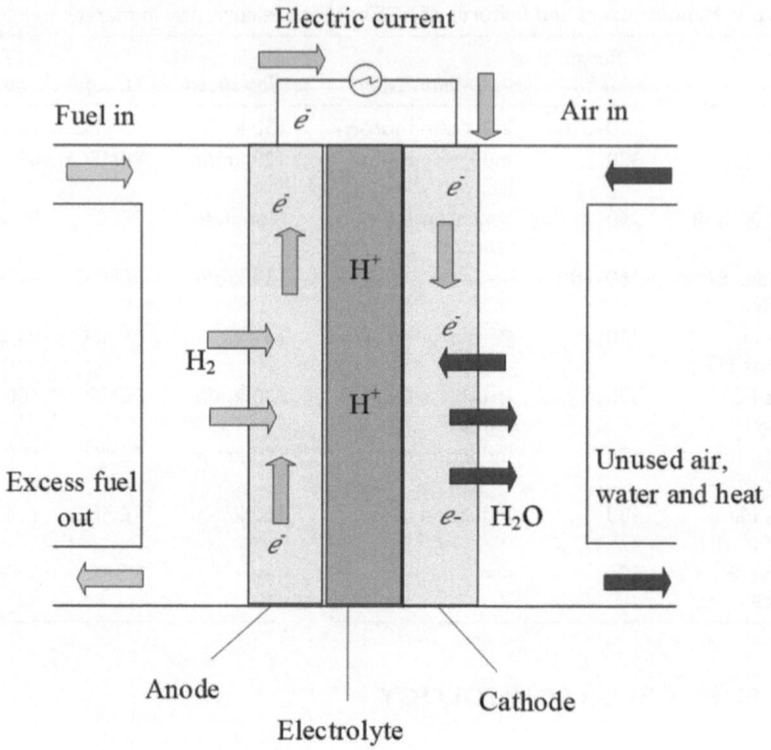

Figure 16.2 Operation of fuel cell.

need a continuous supply of both fuel and oxygen to maintain their chemical reactions and thus generate electricity continuously. While operating a fuel cell is straightforward, its construction is intricate. Fuel is consistently delivered to the anode, while oxidants are supplied to the cathode. Typical fuel sources for fuel cells include hydrogen gases such as methanol, ethanol, and natural gas. These fuel cells undergo simultaneous reactions at the anode and cathode, referred to as a fuel cell reaction. The oxidants utilized can range from chlorine to pure oxygen. The main byproduct of the hydrogen and oxygen reaction in fuel cells is water. The fundamental chemical reactions occurring in a fuel cell are detailed in Equations (16.1)–(16.3). Following the reaction, H^+ ions and electrons (e^-) are generated. H^+ ions move directly toward the cathode, while electrons travel to the cathode through the load. Ultimately, when two oxygen atoms combine with these H^+ ions and electrons, water is produced instead of CO_2. The chemical reactions that occur at the anode and cathode in a PEM electrolyzer are given in Equations (16.4) and (16.5). Comparisons of battery and green hydrogen-operated EVs are given in Table 16.2.

$$\text{At Anode, } H_2 + \frac{1}{2}O_2 \rightarrow H_2O \tag{1}$$

$$\text{At cathode, } \frac{1}{2}O_2 + 2H^+ + 2e^- \rightarrow H_2O \tag{2}$$

$$\text{Complete reaction, } H_2 + \frac{1}{2}O_2 \rightarrow H_2O \tag{3}$$

$$H_2O + \text{Energy} + \text{Heat} \rightarrow 2H^+ + 2e^- + \frac{1}{2}O_2 \tag{4}$$

$$2H^+ + 2e^- \rightarrow H_2 \tag{5}$$

16.2.2 Types of fuel cells

FCs are typically grouped based on their chemical characteristics and operational temperatures, as detailed in Table 16.3. Within the domain of fuel cell electric vehicles (FCEVs), specific types of FCs include proton exchange membrane FC (PEMFC) (Vural et al., 2010), solid-oxide FC (SOFC) (Dimitrova & Maréchal, 2017), direct methanol FC (DMFC) (González-Espasandín et al., 2019), alkaline FC (AFC) (Kordesch et al., 1999), molten carbonate FC (MCFC) (Jafri & Gupta, 2016), and phosphoric acid FC (PAFC) (Kamankesh et al., 2016). Particularly in vehicle applications, PEMFCs that operate at low temperatures and under pressure are the most prevalent types of FCs. This popularity stems from their high power density, lower operational temperatures (60–80°C), and reduced vulnerability to corrosion compared to other FC variants. Fuel cells are primarily classified based on their electrolyte and the type of fuel they use. The following list details the most commonly used fuel types in the market.

Table 16.2 Comparisons of EVs and green hydrogen vehicles

Sr. No.	*Specification*	*Battery EVs*	*Green hydrogen EVs*
1	Definition	Electric cars run on power generated by batteries, which are charged with electricity and used to propel the vehicle.	The operation of hydrogen cars involves utilizing hydrogen fuel to generate power through the conversion of chemical energy to electrical energy.
2	Recharge time	4–5 h	4–7 min
3	Cons	Cheaper than FCEVs	High cost
4	Range	100–200 miles	Up to 300 miles

Table 16.3 Comparisons of fuel cells used in FCEVs

Specifications	*PEMFC*	*PAFC*	*AFC*	*SOFC*	*MCFC*	*DMFC*
Stack power (kW)	<1–250	50–100	1–100	1–3000	300–3000	0.001–100
Cell voltage (V)	1.1	1.1	1	0.8–1	0.7–1	0.2–0.4
Operating temperature (°C)	<100	150–200	90–100	500–1000	600–700	60–200
Efficiency (%)	40–60	40	60	60	50	40
Advantages	Low operating temperature, fast starting, less corrosion, and electrolyte management problem	Higher efficiency with CHP, increased tolerance in fuel impurity	Higher performance, faster cathode reaction and start, lower material cost, low operating temperature	Suitable for CHP, high efficiency, fuel flexibility, hybrid/gas turbine cycle	Suitable for CHP, high efficiency, fuel flexibility	Low cost due to lack of fuel reformer
Disadvantages	Expensive catalyst cost, sensitive to hydrogen impurity	Expensive catalyst cost, long start time	Sensitive to CO_2 levels in oxygen and hydrogen	High operation temperature, corrosion and breakdown, long start time	High operation temperature, long start time	Intermediate adhesion to catalyst surface

16.3 TOPOLOGICAL CLASSIFICATION OF FUEL CELL ELECTRIC VEHICLES

Fuel cell electric vehicles (FCEVs) are equipped with various components, including an electric motor, a motor drive converter, a hydrogen tank, an optional auxiliary DC–DC converter, an FC stack, and an electric motor for power transmission (Carle et al., 2005; Gadalla et al., 2018). The powertrain layout of an FCEV is illustrated in Figure 16.3. During FCEV operation, the FC stack plays a crucial role by supplying energy to the DC bus, ensuring the DC bus voltage remains at the required level (Mallikarjuna & Paulson, 2020). The DC–DC converter connected to the FC is essential as it regulates the DC bus voltage and directs energy produced for vehicle propulsion to the motor drive converter (Ahmadi et al., 2020; (Kim, 2011). Additionally, a DC–AC converter, or inverter, monitors motor torque and speed to maintain safe operation. Ultimately, electric motors convert electrical energy into kinetic energy under the supervision of the drive controller.

16.3.1 Fuel cell hybridization and power supplies

Fuel cell electric vehicles (FCEVs) are categorized based on their energy systems, as illustrated in Figure 16.4. The FC acts as the primary energy generator in FCEVs, supported by various supplementary components for power provision (Tie & Tan, 2013). Hybrid FC applications incorporate a range of energy generation and storage units alongside FC stacks. These include batteries, flywheels, solar panels, ultracapacitors (UCs), and superconducting magnetic energy storage (SMES) devices. Within the context of FCEVs, power supplies are divided into two primary groups: generation

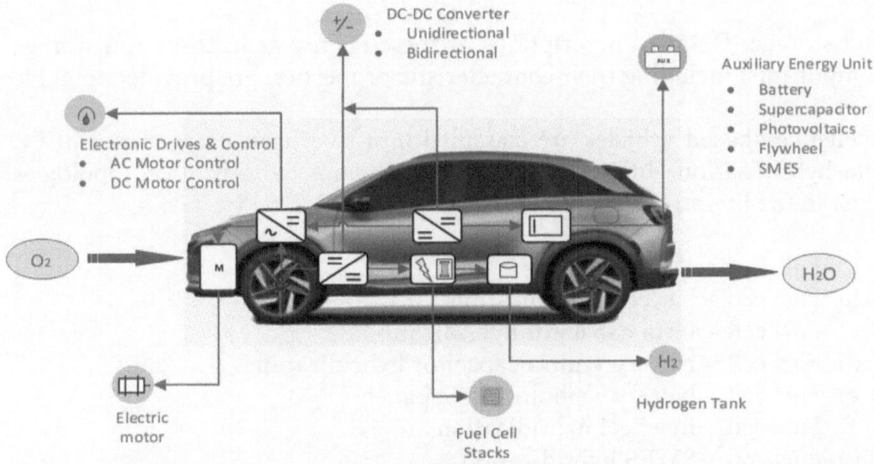

Figure 16.3 Basic block diagram representation of FCEV with auxiliary power distributions.

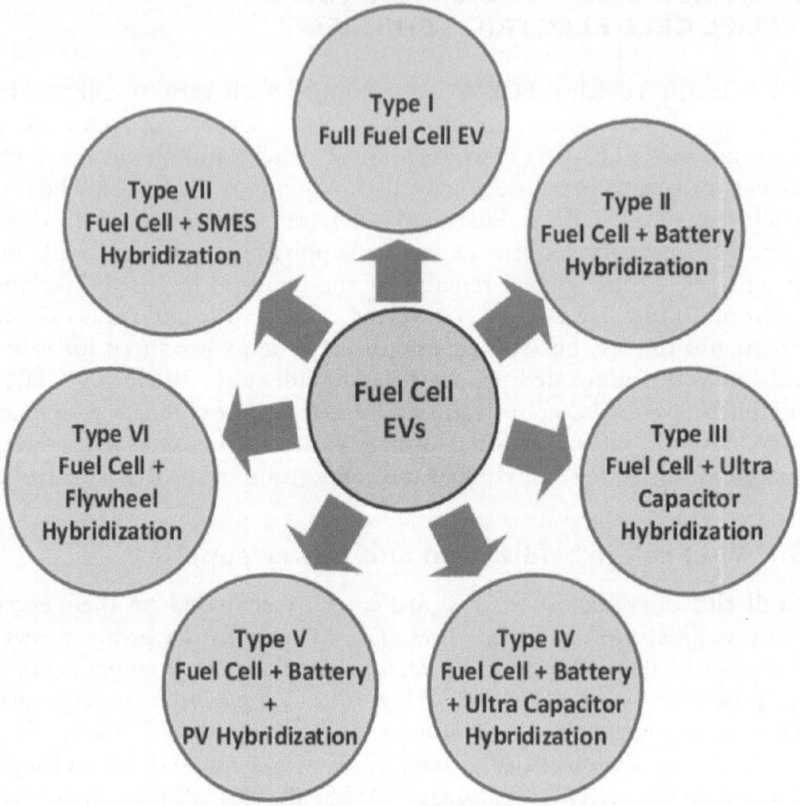

Figure 16.4 Topological classification of FCEVs on the basis of power supplies.

and storage. Detailed descriptions of these energy generation and storage components, including their characteristic properties, are provided in Table 16.4.

Fuel cell-based vehicles are classified into two basic categories: full FC and hybridization with FC. Accordingly, various hybridization topologies exist in the literature, including:

a) Full fuel cell
b) Fuel cell + battery hybridization
c) Fuel cell + ultra capacitor hybridization
d) Fuel cell + battery + ultra capacitor hybridization
e) Fuel cell + battery + photovoltaic panel
f) Fuel cell + flywheel hybridization
g) Fuel cell + SMES hybridization

Table 16.4 Power sources employed in fuel cell electric vehicle

Type	*Unit*	*Energy density (Wh/kg)*	*Lifetime (year)*	*Merits and demerits*
Generation	FC	Very high	20–25 Years	• Highly efficient energy generation • Modular and compact • Smooth power output • Expensive
	PV	Medium	15–20 years	• Clean and silent • Intermittency of power output • Bulky for output
Storage	Battery	High	4–6 years	• Portable and rechargeable • Useful for a limited time • Recharging time
	UC	Very low	10–20 years	• Rapid response • Short-term energy storage
	Flywheel	High	5–10 years	• High-speed charging capability • High power rating • Long charge time
	SMES	Low	25–30 years	• High power output • Short-duration energy storage • High cost

16.3.1.1 Full fuel cell electric vehicles

The full FCEV category denotes vehicles that exclusively utilize a stack of fuel cells as their primary energy source (Saadi et al., 2020). This category's basic configuration includes a fuel tank, FC stack, DC–DC power converter, inverter, and electric motor, as illustrated in Figure 16.5. These vehicles are characterized by their extended driving range, rapid charging times, impressive efficiency, capability to operate in cold conditions, quiet operation due to the absence of moving parts, continuous energy supply, and minimal emissions. These attributes, combined with their straightforward design, make full FCEVs particularly suitable for low-speed vehicles such as buses, trams, airplanes, forklifts, and marine vessels.

16.3.1.2 Fuel cell + battery hybridization

The prevalent hybridization approach for FCEVs integrates FC with Type II battery modules, illustrated in Figure 16.6. Here, the FC is connected to the DC bus via a unidirectional DC–DC converter (UDC), while the battery is linked using a bidirectional DC–DC converter (Ogawa et al., 2009). In

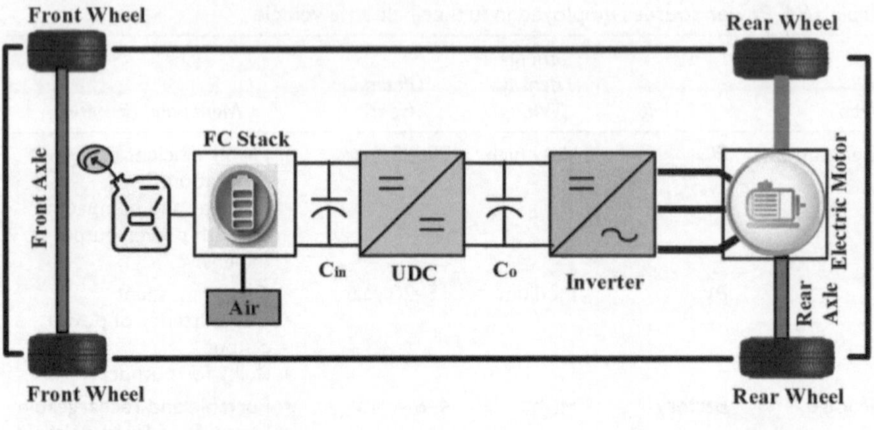

Figure 16.5 Full FCEV architecture

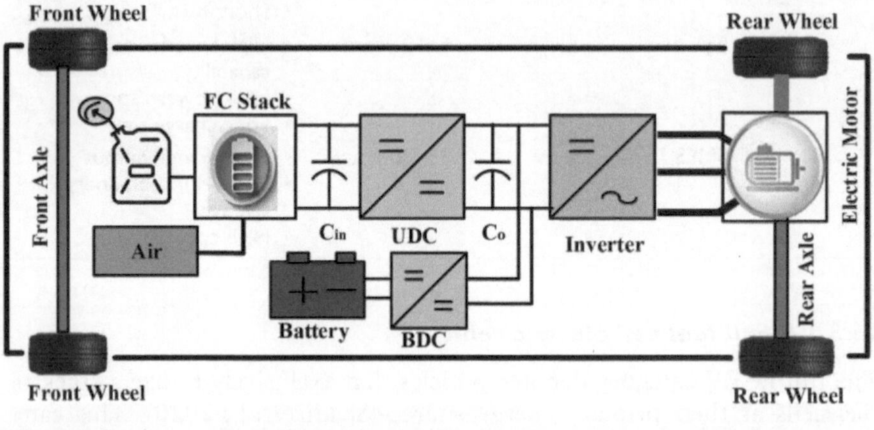

Figure 16.6 Fuel cell with battery arrangement of FCEV.

this configuration, the battery provides power to the electric motor during startup to avoid the FC operating inefficiently at low loads. Subsequently, the FC can sustain the electric motor's operation once the vehicle is running, while also charging the battery as needed.

16.3.1.3 Fuel cell + ultra capacitor hybridization

Type III hybridization, FC + UC architecture shown in Figure 16.7, employs an ultracapacitor in place of a battery (Uzunoglu & Alam, 2007). The UC supports the FC by providing emergency power when needed, but its low energy density restricts sustained energy supply.

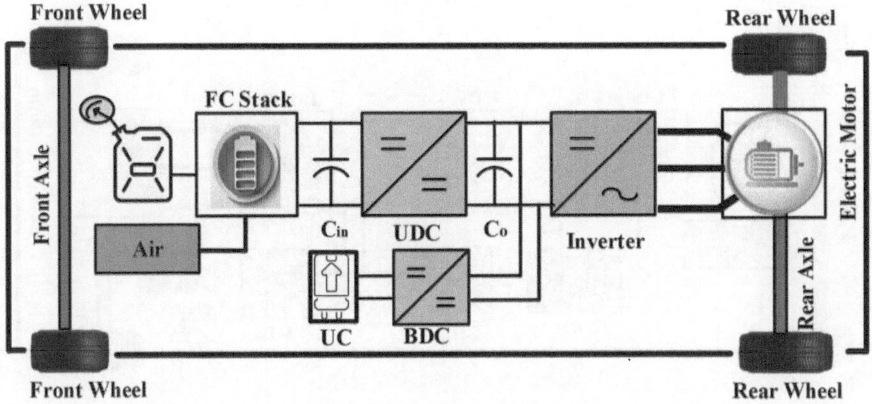

Figure 16.7 Fuel cell with ultracapacitor (UC) arrangement.

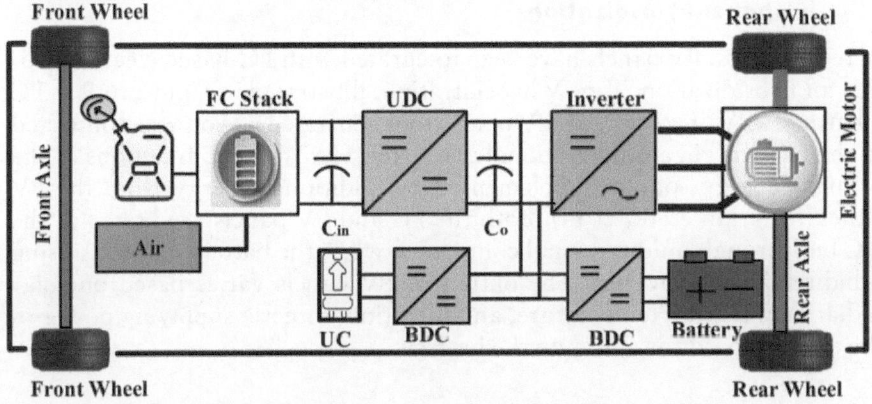

Figure 16.8 Fuel cell with battery and ultracapacitor topology of FCEV.

16.3.1.4 Fuel cell + battery + ultracapacitor hybridization

Unlike earlier designs, Type IV hybridization incorporates two additional components—the battery and ultracapacitor—alongside the primary energy source (FC), illustrated in Figure 16.8. This system employs unidirectional DC–DC converters to connect the FC to the DC bus and bidirectional DC–DC converters (BDCs) to link the battery and UC (Jiang et al., 2005). By integrating the benefits of FC + battery and FC + UC systems, this architecture improves the FC's dynamic response during transient events and ensures continuous energy supply.

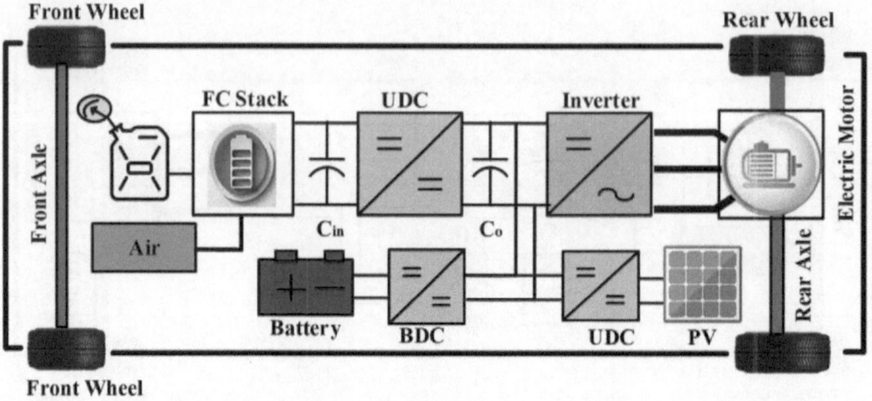

Figure 16.9 Fuel cell, battery, and PV module topology of FCEV.

16.3.1.5 Fuel cell + battery + photovoltaic panel hybridization

In recent years, PV panels have been integrated with FC-based electric vehicles for hybridization. Type V hybridization, illustrated in Figure 16.9 as FC + battery + PV, incorporates PV panels that generate DC voltages connected to the DC bus via a unidirectional converter. Here, the FC functions as the primary energy source, supplemented by additional energy from the PV panel (Mokrani et al., 2014). Both the FC and PV panels are linked to the DC bus through unidirectional converters, while the battery connects using a bidirectional converter. The output of PV panels varies based on solar radiation intensity, temperature, and direction, directly supplying power to the electric motor or charging the battery.

16.3.1.6 Fuel cell + flywheel hybridization

Illustrating FC + flywheel hybridization as shown in Figure 16.10 (Type VI), this configuration expands on previously discussed topologies. Like Type II, the flywheel functions as the energy storage medium instead of batteries, while the FC remains the primary energy source (Huang & Chen, 2017). Flywheels store mechanical energy at high rotational speeds to meet power demands during periods of high energy consumption, converting it into electricity using a generator. Flywheels are noted for their high power ratings, exceptional efficiency, and rapid charging capabilities compared to batteries. They also offer longer lifespans and greater energy storage capacities, are environmentally friendly, and perform effectively across a range of temperatures.

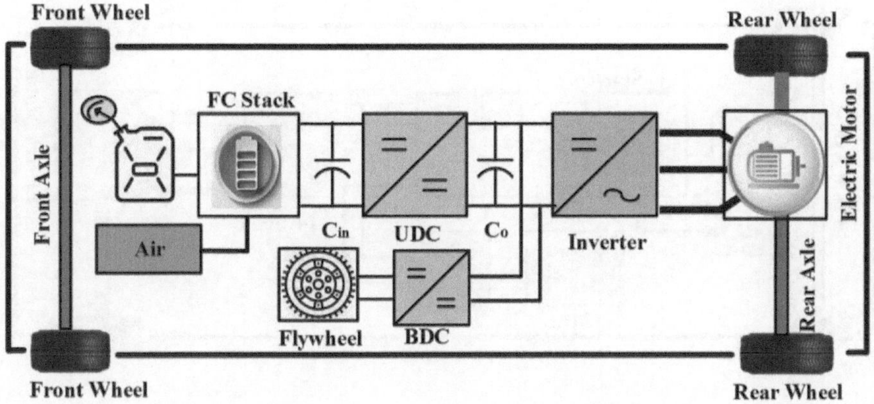

Figure 16.10 Fuel cell, battery, and PV module topology of FCEV.

16.3.1.7 FC + SMES hybridization

The powertrain configuration of FC + SMES hybridization (Type VII), illustrated in Figure 16.10, marks a relatively unexplored territory in hybrid FCEVs (Yang et al., 2020). Nonetheless, there are expectations that SMES units will be integrated with FC stacks in the coming years. SMES stores energy by creating a magnetic field through direct current passing through a superconducting coil. It offers a power conversion ratio of around 95%, high charge/discharge cycles, and faster charging and discharging times compared to conventional storage technologies. However, the current high costs associated with SMES hinder its widespread adoption in FCEVs.

16.4 ENERGY MANAGEMENT STRATEGIES

Energy management strategies in fuel cell electric vehicles involve controlling the DC–DC converter and interfacing fuel cells, batteries, ultracapacitors, and photovoltaic systems. Figure 16.12 categorizes the energy management strategies used for the DC--DC converter controller segment. Additionally, Table 16.5 outlines the advantages and disadvantages of these methods. In fuel cell electric vehicles (FCEVs), energy management from fuel cells to electric motors is governed by two principal rule-based methods: fuzzy rule-based and deterministic rule-based strategies.

16.4.1 Fuzzy rule-based strategies

These strategies uniquely utilize numeric and linguistic data. Fuzzy control methods are notably robust and straightforward to implement. Commonly, the fuzzy rule-based approach is employed for DC-link voltage control

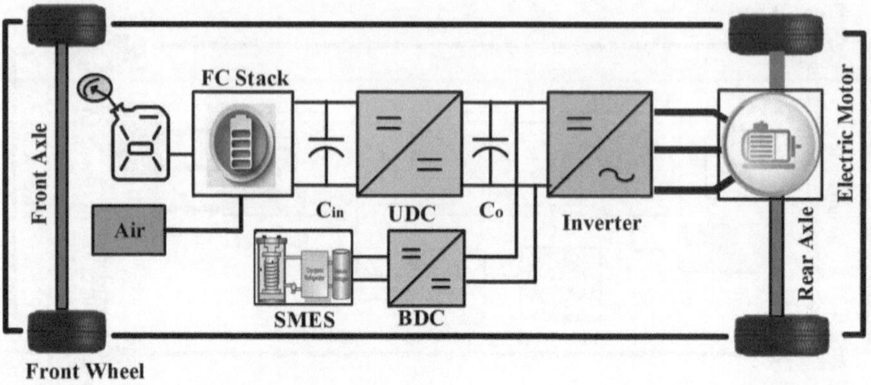

Figure 16.11 Fuel cell with SMES arrangement.

(Li et al., 2012). This strategy functions through three phases: fuzzification, decision-making, and defuzzification, as shown in Figure 16.13. Performance relies on the membership function and fuzzy rules during the logic stage. There are three fuzzy rule-based strategies: optimized, adaptive, and predictive. The optimized method adjusts control mechanisms for fuel usage, emission reduction, state-of-charge maintenance, and improved driving efficiency. These methods optimize the membership function and decision rules using particle swarm optimization, genetic algorithms, bee algorithms, and proportion factor algorithms. The adaptive method incorporates adaptive techniques into the fuzzy rule-based strategy for improved self-regulation. Fuzzy logic is also used in neural networks, machine learning, and decentralized control. The predictive method operates based on forecasting the upcoming state and adapting in real time.

16.4.2 Deterministic rule-based control strategies

These strategies formulate rules based on emission or fuel economy goals for the traction engine (Farhadi et al., 2021). Implementation of these rules typically involves using pre-calculated search tables. Methods encompass state machine control, thermostat control (also known as the on/off method), power follower adjustments, and the gliding average approach. The state machine method operates based on predefined states depicted in a flowchart. Thermostat control maintains fixed torque and speed within specified limits. Power follower adjusts power supply to match the demands of the FCEV. Lastly, the gliding average method provides another deterministic rule-based control strategy utilized in FCEVs.

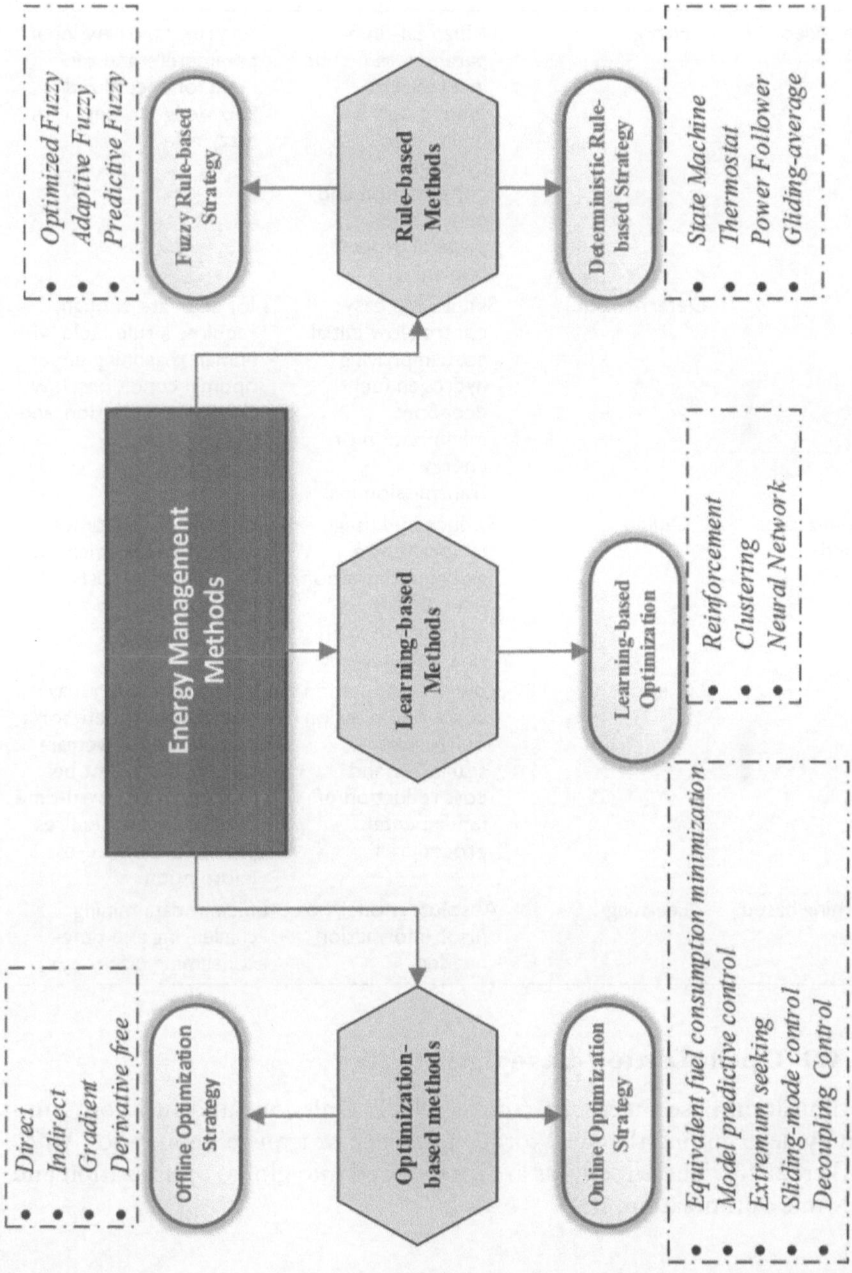

Figure 16.12 Classification of energy management system in FCEVs.

Table 16.5 Advantage and disadvantages of energy management system

Control strategy	*Type*	*Benefits*	*Drawbacks*
Rule based	Fuzzy	Using real-time parameters results in a resilient system that is minimal in hydrogen consumption and effective for physical process reasoning.	Long run-time, few input parameters, and the need for specialized knowledge to derive the fuzzy rule
	Deterministic	Simple and easy control, low initial cost, improving hydrogen fuel economy, minimization of energy transmission loss	Not accurate enough, requires a rule table with human reasoning under optimal conditions, low energy optimization, and gas emission
Optimization based	Online	Reduce fuel usage by localizing a global optimization issue that is appropriate for real-time analysis	Long calculation times and the requirement for an immediate cost function
	Offline	Lower computation costs, optimization of irregular situations, and cost reduction of fundamental procedures	An approximation may not be appropriate for a complicated drivetrain system, it may not be appropriate for real-time analysis, and it requires previous driving cycle information.
Learning based	Learning	Absolute model; no prior information needed	Efficient data mining is a challenging and time-consuming task.

16.4.3 Optimization-based strategies

Optimization-based methods reduce vehicle emissions and fuel consumption by determining the ideal torque reference or transmission ratio (Ali & Söffker, 2018). These methods are categorized into global optimization and real-time optimization.

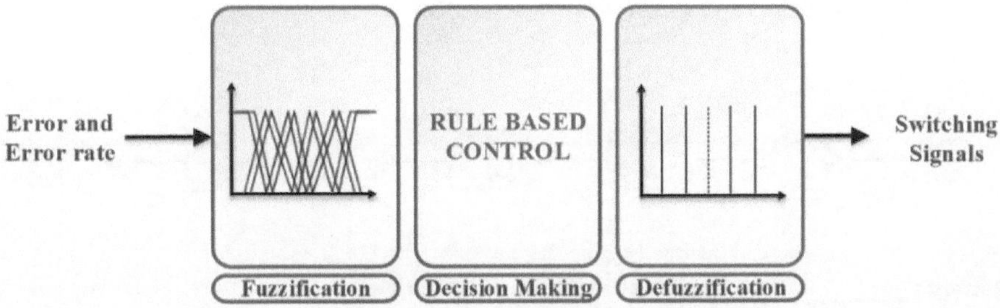

Figure 16.13 The common rule-based method: fuzzy control.

16.4.4 Offline optimization methods

Optimization-based methods are pivotal in reducing vehicle emissions and energy consumption (cost function) through the determination of optimal torque references or transmission ratios. These approaches encompass solutions such as global optimization and real-time optimization. Offline optimization strategies are designed to minimize vehicle losses during prioritized information cycles and are categorized into four types: direct, indirect, gradient, and derivative-free. In FCEV systems, the dynamic programming strategy acts as a typical direct optimization method, employing a multistep decision-making process for both linear and nonlinear systems. Pontryagin's minimum principle, an indirect optimization method, tackles challenges related to global optimization. Gradient optimization techniques, including linear programming and convex programming, are utilized to approximate the nonlinear model of an electric vehicle's fuel consumption, achieving globally optimal solutions (Hou et al., 2023), which proves particularly effective in EMS problems. Convex programming simplifies FCEV system models to ensure they adhere to convexity requirements. Derivative-free methods in FCEV energy management employ meta-heuristic algorithms such as simulated annealing, stochastic control, genetic algorithms, particle swarm optimization, gray wolf optimization, and divided rectangular methods.

16.4.5 Online optimization methods

An online optimization strategy. used for real-time analysis introduces a cost-based function that relies on current state parameters. These techniques encompass decoupling, sliding mode control schemes, extremum searching, model predictive control, and equivalent consumption reduction control. The equivalent consumption minimization method (Mishra et al., 2021) illustrated in Figure 16.14 aims to continually reduce equivalent fuel consumption, transitioning from global to local optimization. Model predictive control utilizes a model to predict future output values based on past

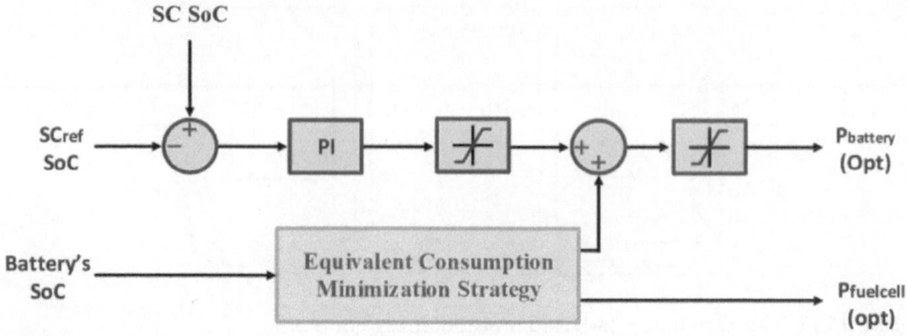

Figure 16.14 The common optimization-based method called the equivalent consumption minimization strategy.

and present system states. Modes of model predictive control include deterministic, stochastic, and telematics approaches. The extremum-seeking approach is applied to identify maximum and minimum values in nonlinear power construction. Sliding mode control is preferred for its simplicity and reliability in monitoring engine/generator torque within optimal ranges. Researchers apply this method in decoupling control to manage shared DC–bus voltage regulation in FC–battery hybridization.

16.4.6 Learning-based method

Learning-based approaches leverage data mining structures to enhance controller behavior in FCEVs. Energy management systems based on learning operate without the need for complete model knowledge to perform system checks. However, creating an accurate database, which significantly impacts control behavior and dimensions, presents challenges that are both time-consuming and demanding. FCEV research incorporates various learning-based methods such as reinforcement learning, clustering learning, and neural network learning to manage energy distribution within the system. In vehicles employing reinforcement learning (Lee & Cha, 2021), as illustrated in Figure 16.15, supervision involves two main components: a learning agent and an environment. These components interact continuously, with the learning agent monitoring the state of the environment and selecting actions that influence it. The reinforcement learning method autonomously determines the optimal action based on data inputs, operating without the need for prior forecasts or predefined instructions.

The clustering learning algorithm builds a system by identifying patterns within the input data. This process involves creating clusters, employing investigative methods to minimize redundancy, and adjusting data to enhance similarity. Another neuron-based learning technique, neural networks, effectively tackles complex challenges in nonlinear FC systems.

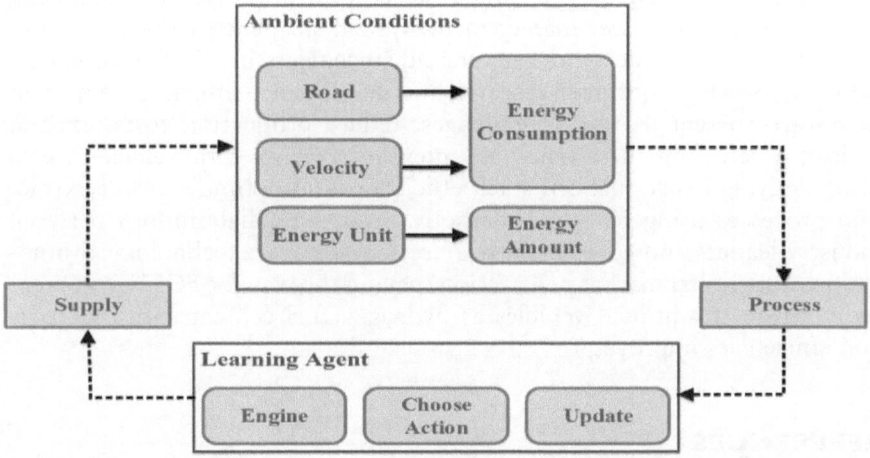

Figure 16.15 The flow chart of the reinforcement learning algorithm.

Nodes within a neural network resemble real multi-link neurons with multiple inputs and outputs. However, a significant drawback of neural networks is their need for a substantial volume of training data to achieve effective training.

16.5 CONCLUSION

The global shift toward alternative energy sources is accelerating, driven by factors such as dwindling fossil fuel reserves, fluctuating oil prices, and the imperative to enhance transportation efficiency while mitigating global warming effects. This transition underscores the pivotal role of electric vehicles (EVs), poised for widespread adoption due to their reduced energy consumption, lower maintenance costs, minimal noise pollution, and superior efficiency driven by electric motors. However, current limitations in battery technology constrain EV suitability primarily to short-distance urban use. This study delves into power transmission configurations, equipment, and control strategies crucial for optimizing efficiency across various components in fuel cell electric vehicles (FCEVs). It explores evolving trends and technical challenges within FCEV technology, distinguishing between full FCEVs and hybrid FCEVs utilizing multiple energy sources. Detailed analyses of energy generation and storage topologies like fuel cells, batteries, ultra-capacitors, photovoltaic panels, flywheels, and superconducting magnetic energy storage (SMES) systems highlight their respective benefits and drawbacks. Techniques for energy management, power transmission, and speed control are scrutinized for their roles in enhancing fuel efficiency, optimizing vehicle architectures, ensuring robust performance, and meeting stringent safety standards. Optimization studies addressing fuel consumption, power management, system size, energy efficiency, and cost-effectiveness underscore ongoing advancements in FCEV technology. Looking forward, continued research and development efforts are expected to resolve current technical challenges, reduce production costs, expand hydrogen refueling infrastructure, integrate FCEVs with vehicle-to-grid technology, enhance fuel cell stack efficiency, and optimize power extraction processes using real-time methods. Strategic collaborations between industry leaders and governments are set to accelerate technological innovations and global market penetration, paving the way for FCEVs to potentially replace traditional vehicles as hydrogen fuel cell capacities increase and efficiencies improve.

REFERENCES

Ahmadi, P., Torabi, S. H., Afsaneh, H., Sadegheih, Y., Ganjehsarabi, H., & Ashjaee, M. (2020). The effects of driving patterns and PEM fuel cell degradation on the lifecycle assessment of hydrogen fuel cell vehicles. *International Journal of Hydrogen Energy*, *45*(5), 3595–3608.

Ali, A. M., & Söffker, D. (2018). Towards optimal power management of hybrid electric vehicles in real-time: A review on methods, challenges, and state-of-the-art solutions. *Energies*, *11*(3), 476.

Brey, S., J. J., Muñoz, D., Mesa, V., & Guerrero, T. (2017). Use of fuel cells and electrolysers in space applications: From energy storage to propulsion/deorbitation. In *Proceedings of the E3S Web of conferences* (Vol. 16, pp. 17004–17012). https://hdl.handle.net/20.500.12412/4718 ISSN: 2267-1242 DOI: 10.1051/e3sconf/20171617004 Date: 201

Carle, G., Axhausen, K. W., Wokaun, A., & Keller, P. (2005). Opportunities and risks during the introduction of fuel cell cars. *Transport Reviews*, *25*(6), 739–760.

Dimitrova, Z., & Maréchal, F. (2017). Environomic design for electric vehicles with an integrated solid oxide fuel cell (SOFC) unit as a range extender. *Renewable Energy*, *112*, 124–142.

Farhadi Gharibeh, H., & Farrokhifar, M. (2021). Online multi-level energy management strategy based on rule-based and optimization-based approaches for fuel cell hybrid electric vehicles. *Applied Sciences*, *11*(9), 3849.

Gadalla, B., Schaltz, E., Zhou, D., & Blaabjerg, F. (2018). Lifetime prediction of boost, z-source and y-source converters in a fuel cell hybrid electric vehicle application. *Electric Power Components and Systems*, *46*(18), 1979–1991.

González-Espasandín, Ó, Leo, T. J., Raso, M. A., & Navarro, E. (2019). Direct methanol fuel cell (DMFC) and H2 proton exchange membrane fuel (PEMFC/H2) cell performance under atmospheric flight conditions of unmanned aerial vehicles. *Renewable Energy*, *130*, 762–773.

Hou, S., Yin, H., Pla, B., Gao, J., & Chen, H. (2023). Real-time energy management strategy of a fuel cell electric vehicle with global optimal learning. *IEEE Transactions on Transportation Electrification*, *9*(4), 5085–5097.

Huang, C. N., & Chen, Y. S. (2017). Design of magnetic flywheel control for performance improvement of fuel cells used in vehicles. *Energy*, *118*, 840–852.

Hu, X., Zou, C., Tang, X., Liu, T., & Hu, L. (2019). Cost-optimal energy management of hybrid electric vehicles using fuel cell/battery health-aware predictive control. *IEEE Transactions on Power Electronics*, *35*(1), 382–392.

Jafri, N. H., & Gupta, S. (2016). An overview of Fuel Cells application in transportation. In *Proceedings of the IEEE transportation electrification conference and expo*, Asia-pacific *(ITEC Asia-Pacific)* (pp. 129–133). doi: 10.1109/ITEC-AP.2016.7512935.

Jiang, X., Wen, X., & Xu, H. (2005). Study on isolated boost full bridge converter in FCEV. In *Proceeding of the international power engineering conference* (pp. 827–830). Vol. 2, doi: 10.1109/IPEC.2005.207022.

Kamankesh, H., Agelidis, V. G., & Kavousi-Fard, A. (2016). Optimal scheduling of renewable micro-grids considering plug-in hybrid electric vehicle charging demand. *Energy*, *100*, 285–297.

Kim, T. (2011). Regenerative braking control of a light fuel cell hybrid electric vehicle. *Electric Power Components and Systems*, *39*(5), 446–60.

Kordesch, K., Gsellmann, J., Cifrain, M., Voss, S., Hacker, V., Aronson, R. R. & Daniel-Ivad, J. (1999). Intermittent use of a low-cost alkaline fuel cell-hybrid system for electric vehicles. *Journal of Power Sources*, *80*(1–2), 190–197.

Lee, H., & Cha, S. W. (2021). Energy management strategy of fuel cell electric vehicles using model-based reinforcement learning with data-driven model update. *IEEE Access*, *9*, 59244–59254.

Li, Q., Chen, W., Li, Y., Liu, S., & Huang, J. (2012). Energy management strategy for fuel cell/battery/ultracapacitor hybrid vehicle based on fuzzy logic. *International Journal of Electrical Power & Energy Systems*, *43*(1), 514–525.

Mallikarjuna Reddy M., B., & Samuel, P. (2020). Analysis, modelling and implementation of multi-phase single-leg DC/DC converter for fuel cell hybrid electric vehicles. *International Journal of Modelling and Simulation*, *40*(4), 279–290.

Mishra, S., Nayak, P. C., Prusty, U. C., & Prusty, R. C. (2021). Model predictive controller-based load frequency control of isolated microgrid system integrated to plugged-in electric vehicle. In *Proceedings of the 1st Odisha international conference on electrical power engineering, communication and computing technology (ODICON)* (pp. 1–5). IEEE. doi: 10.1109/ODICON50556.2021.9428956.

Mokrani, Z., Rekioua, D., & Rekioua, T. (2014). Modeling, control and power management of hybrid photovoltaic fuel cells with battery bank supplying electric vehicle. *International Journal of Hydrogen Energy*, *39*(27), 15178–15187.

Ogawa, K., Yamamoto, T., Hasegawa, H., & Furuya, T. (2009). Development of the fuel-cell/battery hybrid railway vehicle. In *Proceedings of the IEEE vehicle power and propulsion conference* (pp. 1730–1735). IEEE.doi: 10.1109/VPPC.2009.5289693.

Saadi, R., Hammoudi, M. Y., Kraa, O., Ayad, M. Y., & Bahri, M. (2020). A robust control of a 4-leg floating interleaved boost converter for fuel cell electric vehicle application. *Mathematics and Computers in Simulation*, *167*, 32–47.

Tie, S. F., & Tan, C. W. (2013). A review of energy sources and energy management system in electric vehicles. *Renewable and Sustainable Energy Reviews*, *20*, 82–102.

Uzunoglu, M., & Alam, M. S. (2007). Dynamic modeling, design and simulation of a PEM fuel cell/ultra-capacitor hybrid system for vehicular applications. *Energy Conversion and Management*, *48*(5), 1544–1553.

Vural, B., Boynuegri, A. R., Nakir, I., Erdinc, O., Balikci, A., Uzunoglu, M., & Dusmez, S. (2010). Fuel cell and ultra-capacitor hybridization: A prototype test bench-based analysis of different energy management strategies for vehicular applications. *International Journal of Hydrogen Energy*, *35*(20), 11161–11171.

Yang, B., Zhu, T., Zhang, X., Wang, J., Shu, H., Li, S., & Yu, T. (2020). Design and implementation of Battery/SMES hybrid energy storage systems used in electric vehicles: A nonlinear robust fractional-order control approach. *Energy*, *191*, 116510.

Chapter 17

Safety challenges and future scope of green hydrogen for the energy transition

Amit Kumar and Ramna Tripathi

17.1 INTRODUCTION

The demand for energy has expanded considerably due to the world economy and the constantly growing population (Dawood et al., 2020). However, fossil fuels like coal, natural gas, and oil continue to be the main source of energy needed for the economy. The International Energy Agency reports that in 2018 fossil fuels accounted for over 80% of the world's total energy supply: 19.31% and 3.03% of energy were consumed as heat and electricity, respectively; fossil fuels constitute the main energy source for both (IEA, 2022). To overcome this situation, the world needs an alternate solution in the form of clean renewable energy. This process involves the evolution of numerous technologies connected to the production, storage, and transportation of hydrogen in terms of cost and lifecycle environmental effect assessments. Preparing future technology in securing a clean environment hydrogen will definitely make a symbol of international business. Hydrogen is a promising element of the future of the power grid, serving as an adaptable energy source for industry and transportation sectors as shown in Figure 17.1.

Robert Boyle wrote about the reaction that produced hydrogen gas in 1671 when he mixed iron filings with diluted acids. Henry Cavendish named the gas resulting from a metal–acid interaction "inflammable air" in 1766, marking the first identification of hydrogen gas as a distinct substance. With hydrogen accounting for only 4% of global energy use in 2019 and expected to rise to 18% by 2050, it is unlikely that a society built upon a hydrogen economy would materialize anytime soon. Three times as much hydrogen is generated globally each year as was in 1980, with an average of 120 million tons produced annually. Nonetheless, fuel for transportation and fertilizers (among other compounds) account for two-thirds of all hydrogen produced. The comparative table between hydrogen energy and other forms of energy is shown in Table 17.1.

Hydrogen emits no greenhouse gases and can be used for transportation, energy production, and heating (Pielecha et al., 2018). Energy can be exported since it is able to transport and store energy. Its use in the energy

DOI: 10.1201/9781003537816-17

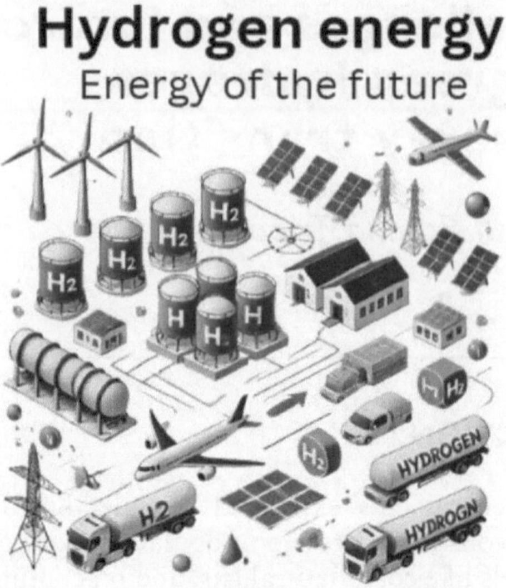

Figure 17.1 Hydrogen energy: A vision of the future.

sector has, however, been rather limited thus far. Nowadays, it comes from fossil fuels and finds application in the glass, electronics, and metallurgy industries. Getting hydrogen has proven to be the largest barrier to its wider application. There are no geological hydrogen resources, in contrast to fossil fuels. Alternatively, splitting water, a process that requires a lot of energy, could produce sustainable hydrogen. Hydrogen thus functions more as an energy carrier than an energy source. When produced using renewable energy sources like sun and wind, hydrogen is an emission-free, renewable fuel. Its relatively low efficiency is the main drawback, as the necessary conversions lose a large portion of the initial energy. Due to its generation from a variety of renewable sources with zero-carbon emissions, hydrogen plays a crucial role in the global energy shift. It is a viable substitute for petroleum-based products because of this green energy fuel, particularly in the car sector. Furthermore, hydrogen may be used to power the turbines in thermoelectric facilities, which would eliminate the need for coal. Hydroelectric, solar, and wind power plants could also use hydrogen to produce secondary energy. In the energy and transportation industries, using hydrogen fuel—which may also be hydrogen—offers quantifiable benefits over traditional modes of operation. The use of conventional batteries, which run on metals like lithium and cobalt that pose serious risks to the environment and human health, is replaced with hydrogen fuel cells. Hydrogen fuel-powered electric cars have a dual benefit in terms of reducing carbon emissions and providing sufficient replenishment for dwindling fossil resources. In the field of mobility, hydrogen fluid dynamics are increasingly intriguing due to its rapid recharging and sustained discharge. Since there are several methods for storing hydrogen, it is logical to categorize the procedure. Because hydrogen has a very high volumetric energy storage density, there will be far less frequent cell discharge.

Table 17.1 Contrasting hydrogen energy with other energy sources

Criteria	*Hydrogen energy*	*Other forms of energy*
Renewable source	Water electrolysis, which uses electricity from solar, wind, or other renewable sources, is one way to manufacture hydrogen, which is abundant.	Coal, oil, natural gas, like other forms of energy, are finite and contribute to environmental pollution and climate change. Nuclear energy has limited fuel resources and poses challenges related to waste disposal and safety.
Environmental impact	Hydrogen burning results in water vapor, making it a clean energy source.	Fossil fuels release greenhouse gases (CO_2, methane) and pollutants, contributing to climate change and air pollution. Nuclear energy generates radioactive waste, posing challenges for disposal and safety.
Energy density	Hydrogen has a high energy density (energy per unit of mass or volume).	Fossil fuels have high energy density, but they come with environmental drawbacks. Renewables like solar and wind have lower energy density and may require large land areas.
Versatility	Hydrogen can be stored and transported easily and used in transportation, industry, electricity generation, and other sectors.	Some energy sources, like fossil fuels, are primarily used for specific applications, and their use may be limited in certain sectors. Renewables may have intermittency issues and storage challenges.
Decentralized production	Hydrogen can be produced on a decentralized basis, allowing for local production and reducing dependency on centralized energy sources.	Fossil fuels and nuclear energy often involve centralized production and distribution, leading to vulnerabilities and inefficiencies.
Storage and transportation	Due to efficient storage and transportation capabilities, hydrogen is a viable option for long-distance transportation and energy storage.	Storage and transportation of fossil fuels can be associated with safety concerns, and renewable energy storage technologies may have limitations and environmental impacts.
Global availability	Hydrogen is abundant globally, and its production can be decentralized, reducing geopolitical dependencies.	Some energy sources may be concentrated in specific regions, leading to geopolitical tensions and supply chain issues.

(Continued)

Table 17.1 (Continued)

Criteria	*Hydrogen energy*	*Other forms of energy*
Technological development	Hydrogen technology is a viable and developing energy source; more research and development could lead to advances.	Other energy sources may have matured technologically, with fewer opportunities for significant breakthroughs.
Economic opportunities	Hydrogen technologies can stimulate economic growth, creating jobs and fostering innovation in various industries.	Dependence on traditional energy sources may limit economic diversification, and the transition to renewables may also create economic opportunities but may face challenges in some regions.
Long-term sustainability	Hydrogen is a long-term sustainable energy solution that can contribute to a low-carbon future.	Some other energy sources may face challenges in terms of long-term sustainability due to either finite resources or environmental concerns.

However, combining the various storage techniques is a challenging undertaking. Classification may be done based on how various hydrogen storage methods interact with one another. Thus, Figure 17.2 depicts the three groups that storage technologies are categorized into:

1. Without any physical or chemical connections to other materials, pure molecular hydrogen can be held as a gas or a liquid.

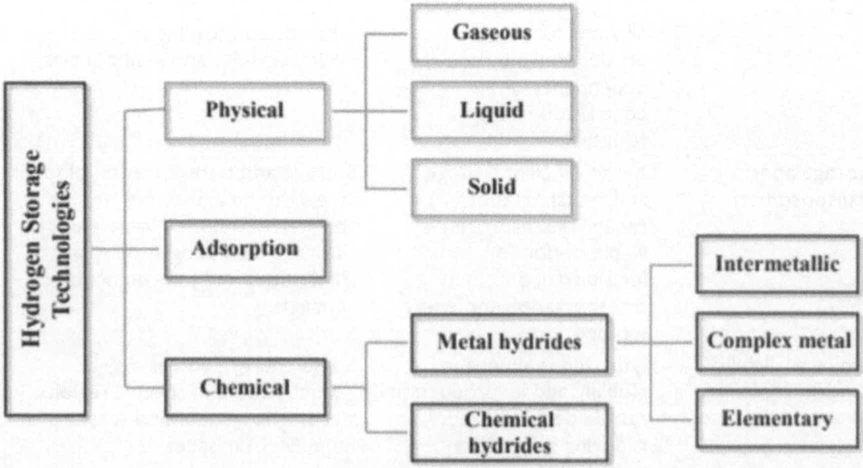

Figure 17.2 Storage technologies (Bairrão et al., 2023).

2. A material held together by a relatively weak force has the ability to adsorb molecular hydrogen.
3. Chemical bonding (absorption by another element) is a possibility for atomic hydrogen.

Distinguishing between chemical and metal hydrides is crucial when it comes to chemical storage methods. The terms metal hydrides and hydroxides indicate that they include metal atoms to which hydrogen may attach directly or through an intricate ion chain, which makes sense given the differences in their chemical storage characteristics. Oxygen, carbon, nitrogen, and hydrogen are examples of nonmetallic atoms that make up chemical hydrides (Abe et al., 2019).

Despite being one of the cleanest fuels, hydrogen is not always produced using the most environmentally friendly methods (Dawood et al., 2020). Table 17.2 displays the hydrogen production process.

Hydrogen is classed into several types according to its production process, including brown, blue, green, yellow, turquoise, and even pink. The several forms of hydrogen creation (see Table 17.3) have drawn a lot of attention over time (Dawood et al., 2020). The most prevalent byproduct of contemporary hydrogen manufacturing techniques is gray hydrogen at the moment. It is obtainable through coal gasification or conventional methane reforming. Because of this unsustainable production process, 830 million tons of CO_2 are released annually, or 2.3% of the world's total emissions of CO_2 (International Energy Agency, 2019). Natural gas reforming releases 8.9 kg of CO_2 for every kilogram of H_2, but coal gasification releases 29.33

Table 17.2 Hydrogen production processes

Method	*Process*	*Raw materials*	*Energy*	*Emission*
Thermal	Reformation	Natural gas	High temperature and steam	CO_2
	Thermochemical hydrolysis	Water	Heat from nuclear energy	No emission
	Gasification	Coal and biomass	Steam, heat, oxygen, and pressure	Some
	Pyrolysis	Biomass	Steam at medium temperature	Some
Electrical	Electrolysis	Water	Electricity	Varies
	Photo-electrochemical	Water	Sunlight	No emission
Biological	Photo-biological	Water and algae	Sunlight	No emission
	Anaerobic digestion	Biomass	Heat	Some
	Fermentation	Biomass	Heat	Some

Table 17.3 Various types of hydrogen generation

Color	*Gray hydrogen*	*Blue hydrogen*	*Turquoise hydrogen*	*Green hydrogen*
Process	Gasification or SMR	SMR or carbon-captured gasification (85–95%)	Pyrolysis	Electrolysis
Source	Coal or methane	Coal or methane	Methane	Renewable energy
CO_2 emissions	9–20 kg of CO_2 is produced for every kilogram of product.	Per kilogram of product, 1.5–4.5 kg of CO_2 is produced.	Per kilogram of product, there is no CO_2 created.	Per kilogram of product, there is no CO_2 created.
Benefits	Minimal cost of production	Using current resources with carbon capture and storage to reduce greenhouse gas emissions.	No carbon dioxide was released.	In line with net-zero carbon dioxide emissions.
Disadvantages	These technologies are not appropriate for a sustainable route (net-zero emission) due to their high CO_2 emissions	Since carbon capture does not satisfy the requirements of a net-zero future and is dependent on changes in the price and availability of fossil fuels, it will never be 100% efficient	Methane's carbon converts to solid carbon black material, although this process is still in the pilot stage and has no industrial applicability.	Presently lacking are policies, infrastructure, and value recognition; the cost of manufacturing is 2–3 times greater than that of gray hydrogen.

kg of CO_2. Currently, fossil fuels are used to make 8.2 million tons of hydrogen, with natural gas reforming providing the majority of these fuels.

Based on the same fundamental techniques as gray hydrogen production, blue hydrogen production is the second most popular approach. However, this substance is made by fusing underground carbon storage and capture technology with fossil fuels. The market for this gas can also be helped by blue hydrogen as it develops and grows. The most promising method of producing hydrogen is "green hydrogen," which involves using renewable energy sources to fuel the process of electrolysis, which separates water molecules into hydrogen and oxygen. Electrolysis needs electrical energy, which is affordable to create from renewable sources.

Still, it is counterproductive to our goal to rely on non-sustainable hydrogen production technologies such as the ones mentioned above. Thus, the only source of clean energy is "Green Hydrogen." India ranks among the world's top emitters of CO_2 due to its huge transportation and industrial sectors. The burning of crude oil for energy purposes is the main cause of greenhouse gas emissions. The consequences of climate change are putting humanity in danger due to the rise in greenhouse gases. The growing inclination of consumers to choose environmentally friendly items in an effort to lessen the effects of climate change may be addressed using green hydrogen. Furthermore, India has imposed a carbon tax, now valued at Rs. 400 per ton, which connects gasoline pricing to emissions.

Fuel costs will rise as a result, forcing consumers to transition to greener energy sources. The world's lightest and most plentiful element is hydrogen, a non-toxic gas. It is capable of existing in both liquid and solid states. Hydrogen is an ideal fuel for the industrial and transportation industries due to its high energy content. Since hydrogen and oxygen gases are the process' final products, electrolysis is a clean process because it produces no greenhouse gases into the atmosphere. Green hydrogen is therefore the best fuel to replace crude oil. However, the cost of producing 1 kg of hydrogen via green hydrogen production is a little higher, ranging from Rs. 350 to Rs. 400/kg (Dawood et al., 2020); it also takes 39 kWh of power. Green hydrogen is the cleanest form of hydrogen, but it is much harder to produce. The difficulties of producing hydrogen are compounded by the intricacy and difficulty of transporting and storing it. Green hydrogen's cost and efficiency are now its main drawbacks (Appleby, 1994). Furthermore, conversion efficiency is very poor. Improvements in production technology are on the horizon and will address the problem. Green hydrogen can currently only be produced via electrolysis, which divides water into hydrogen and oxygen molecules.

Various types of electrolysis technology are explained by Yu et al. (2021). The PEM method and the alkaline electrolysis method are shown in Figures 17.3 and 17.4, respectively.

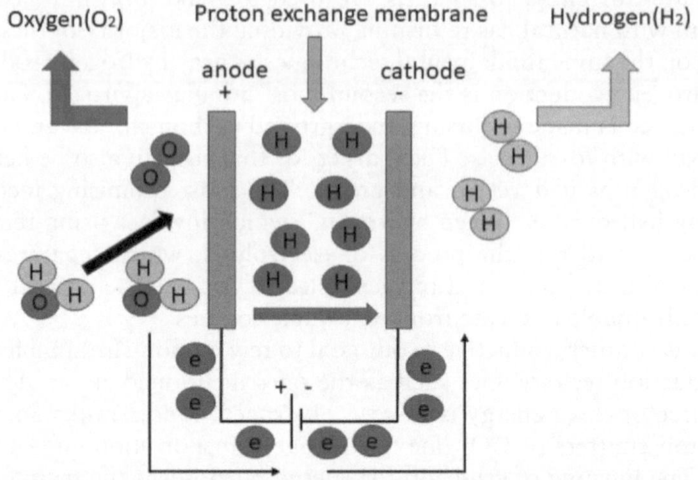

Figure 17.3 The alkaline electrolysis method (Chowdhury et al., 2018).

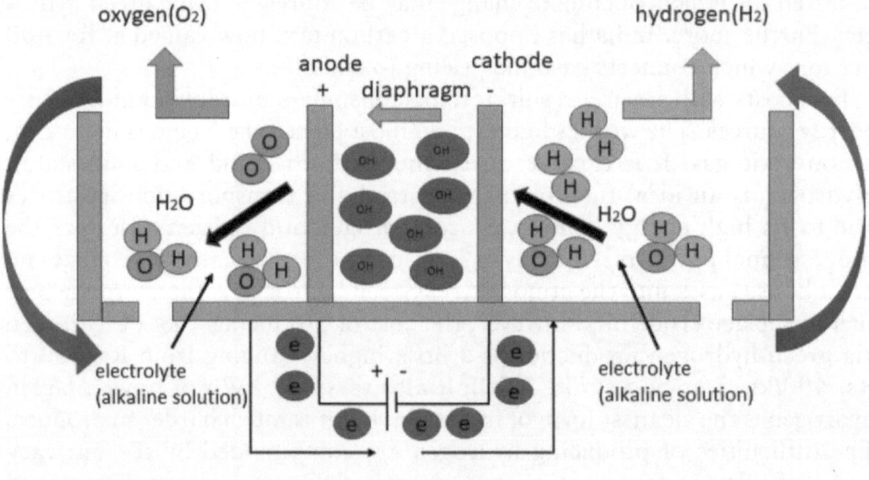

Figure 17.4 PEM method (Chowdhury et al., 2018).

Many industrialized and developing nations have already conceptualized and implemented green hydrogen in a variety of sectors and areas. It can be applied in several domains where it is impractical to use other technologies (Jacobson et al., 2005). For instance, because of their poor volumetric energy density, batteries in heavy-duty vehicles and aircraft are inefficient. Efforts are being made to address these problems because even light vehicle batteries include earth metals and other non-eco-friendly components, and disposing of these batteries pollutes the environment. Since transportation is one of India's top industries, here is where green hydrogen is most commonly used as a fuel. Trains, planes, and ships are examples of off-road vehicles that can use hydrogen powertrains in addition to automobiles, trucks, buses, motorbikes, and passenger cars. Hydrogen mixtures can already be used to power gas boilers and furnaces. Heating systems for homes and businesses also employ it. Therefore, hydrogen can be used in place of other fuels to power industries and reduce greenhouse gas emissions. India has enormous potential and capability in renewable energy resources of all kinds due to its vast geographic expanse and diversity. As a result, India stands to gain greatly from the generation of green hydrogen, which has a wide range of applications. Approximately 40% of India's installed power capacity currently comes from renewable sources. To boost the quantity of renewable energy that could be utilized to create green hydrogen, appropriate financial and operational commitments need to be made. From its current basic energy ecology, India should transform into a multisource, multi-medium energy ecosystem. It will be some time before there is a fully developed and functioning green hydrogen economy. According to the International Energy Agency (IEA), the number of projects related to electrolyzers—devices used for hydrogen production—has seen significant growth. In 2010, the total installed capacity of these electrolyzers was less <1 MW. However, by 2019, this capacity had surged to over 25 MW. The primary cause of this growth is the decline in the cost of electrolyzers and technology connected to renewable energy.

The gas industry proposes a gradual approach to developing the hydrogen economy. They suggest starting by creating technology that captures carbon from the atmosphere to produce blue hydrogen. Companies involved in resource exploration recognize valuable connections with hydrogen. However, these synergies have not been fully harnessed until we achieve a complete shift toward green hydrogen. These synergies include their gas reserves, evolving capture capabilities, carbon utilization and storage, and opportunities to repurpose existing technology and facilities to accommodate blue hydrogen. The government must take the lead in promoting business investments and creating a legislative framework that fosters the expansion of hydrogen-related projects due to the significant expenses associated with building hydrogen infrastructure. In order to satisfy their import requirements, they have openly advocated for the establishment of offshore hydrogen production facilities. A fully funded goal of producing 5 GW of electrolyzers by 2030 is part of Germany's latest €130 billion pandemic recovery program, which emphasizes environmental initiatives.

17.2 THE INFRASTRUCTURES REQUIRED FOR A SUSTAINABLE ECONOMY BASED ON GREEN HYDROGEN

The majority of the expenses involved in creating green hydrogen are energy-related, and these prices are mostly influenced by the availability of renewable resources (Edwardes-Evans, 2020). However, we also need to include the expenses associated with storage and shipping in addition to manufacturing expenditures. Green hydrogen produced locally using solar and wind energy can either be kept locally (in salt caves, for example) or transported via restricted pipes. Hydrogen imports by ship and pipeline, or perhaps simply long-distance pipelines, will be competitively priced. Large-scale green hydrogen production will occur in both cases where renewable energy is generated, not where this gas is required (hydrogen import or production). The main justification is that pipeline transportation of hydrogen is ten times more economical than cable transportation of electrical power. Additionally, pipeline capacities (15–20 GW) are more important than power cable capacities (1–4 GW). Delivering it via pipeline therefore avoids power grid capacity issues brought on by an increase in the supply of renewable energy. Therefore, hydrogen transfer would be more economical than producing power. Hydrogen also has a major advantage over electricity in that it can be kept year-round and used as a constant energy source. This is also true of natural gas. Transporting hydrogen from production sites—resources that are plentiful and renewable—to consuming areas requires a hydrogen pipeline infrastructure.

17.3 THE STATE OF THE HYDROGEN MARKET

The majority of hydrogen is utilized in the petrochemical sector to produce ammonia for fertilizers or to separate heavy oils into lighter petroleum products. BNEF reports that 90% of products are produced near to where they are used because of the large quantities required for consumption and the high expense of transportation. The majority of the hydrogen generated is "gray." Slightly more than half is produced in specialized facilities, according to the IEA. The remaining 41% is produced as a byproduct of other chemical reactions, typically in refineries and petrochemical factories. The production of "green hydrogen," which may form the cornerstone of a hydrogen economy devoid of carbon emissions, through the use of solar, wind, or other renewable energy sources, is still in its infancy. Figure 17.5 illustrates how the demand for hydrogen worldwide increased by about 5% by the end of 2021 over the prior year.

Additionally, it provides two distinct policies: the announced promises scenario (APS) and the stated policies scenario (STEP). The first shows how

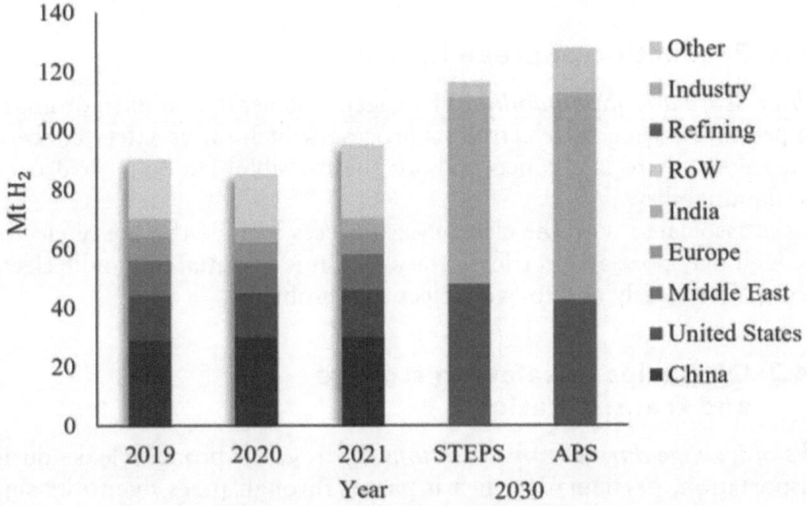

Figure 17.5 Demand of hydrogen by sector and region (Bairrão et al., 2023).

the policies are currently set up. The STEPS estimate indicates that by 2030, there may be a 115 million-ton demand for hydrogen. With low demand (less than 2 Mt) for new applications or for further replacing hydrogen derived from fossil fuels in conventional uses, the majority of this growth would come from conventional utilization. The second scenario assumes full and timely fulfillment of all international climate commitments, including longer-term net-zero targets and nationally determined contributions.

The IEA calculated that government investment on hydrogen research and development rose by 4.7% annually on average in the four years prior to 2018, despite the present supply crisis. With R&D investment jumping from $19 million in 2015 to $129 million in 2018, the Chinese government has been at the forefront of this trend. In 2019, fuel cells and hydrogen had the biggest gains (21% more than in 2018), according to an IEA assessment of government energy budgets. Japan is the greatest donor in the world to fuel cell and hydrogen research, contributing $303 million in 2019. The development of carbon capture and storage (CCS) technology to create "blue hydrogen" from natural gas, with the carbon dioxide byproduct trapped and stored rather than released, is the first step in the oil and gas industry's phased approach to the hydrogen economy. By the middle of this decade, natural gas will most certainly surpass coal as the world's most popular energy source. Resource companies think that green hydrogen can be reached through blue hydrogen as a "bridge" technology. Investing in blue hydrogen will facilitate the integration of hydrogen into natural gas networks and the construction of storage facilities to increase grid stability capacity.

17.4 SAFETY ISSUES

17.4.1 Production of green hydrogen

High pressure and flammability: Hydrogen is generated at extraordinarily high pressure during the electrolysis process, which raises safety concerns. Furthermore, there is a concern about the invisible flame and hydrogen's great flammability.

Risks associated with the electrolysis process include the use of electricity, which may pose an electrical safety risk. It is essential to handle electrical equipment safely and to avoid electrical problems.

17.4.2 Obstacles to safety in storage and transportation

Risks of leakage during transportation: Hydrogen is prone to leaks during transportation, particularly when it passes through pipes due to its small molecules and low viscosity. If a leak is not detected, it may be dangerous.

17.4.3 Challenges in the safety of use

Invisible flame and gas bubbles: The invisible flame of hydrogen and the formation of gas bubbles pose a challenge, especially in applications such as power generation, where leaks can go undetected.

Hydrogen sensors for vehicles: In the mobility sector, where hydrogen is used in fuel cell electric vehicles (FCEVs), it is important to ensure the safety of hydrogen sensors when detecting gas in confined spaces, such as wheel arches.

17.4.4 Technological innovations

Intelligent sensor networks: Develop interconnected smart sensor networks for real-time monitoring of hydrogen parameters. Integrate these networks into plants for comprehensive safety coverage and use data analytics for predictive maintenance.

Drones and robotics: Use drones and robots equipped with gas sensors for the remote inspection of hydrogen infrastructure. Use these technologies for routine inspections, reducing human exposure to potential safety risks in challenging environments.

17.4.5 Case studies

Initiatives for global cooperation: Showcase successful global collaborations to address hydrogen safety challenges. Promote knowledge sharing and collaborative efforts to establish global safety standards for the hydrogen industry.

Emergency response effectiveness: Analyze case studies that demonstrate effective emergency responses to hydrogen incidents. Incorporate lessons learned from successful responses into emergency response training programs and emphasize well-coordinated protocols.

17.4.6 Other challenges

17.4.6.1 Challenges of integrating hydrogen energy into daily human life

Hydrogen energy has the indisputable ability to revolutionize the global energy sector by providing a distinct, sustainable, and clean alternative. Investigating the possibilities of employing hydrogen energy for regular human needs is fraught with difficulties, though. This essay looks at the numerous obstacles people face while attempting to incorporate hydrogen energy into their daily lives across a range of energy-related topics, including production, storage, distribution, and consumption (Li et al., 2017).

The challenges summarized in Table 17.4 have been discussed in subsequent subheadings:

17.4.6.1.1 Health challenges

(1) *Hydrogen production technologies*: Particularly, methods of creating hydrogen through electrolysis could be detrimental to one's health. For electrolysis processes, specific materials or catalysts could be required; nevertheless, handling them incorrectly could pose health risks. Moreover, using natural gas for steam methane reforming (SMR) to produce hydrogen could pollute the air and have a negative impact on respiratory health.

(2) *Hydrogen storage and transportation*: The transportation and storage of hydrogen present some challenges. Due to its great flammability, hydrogen requires certain infrastructure as well as safety measures. Both public health and personal safety are at risk due to potential hydrogen leakage or combustion.

17.4.6.1.2 Psychological challenges

(1) *Anxiety during transition*: Psychological difficulties could arise from switching to hydrogen energy from conventional energy sources. People may feel uneasy or resistant because of concerns about the security, dependability, and social acceptance of hydrogen-powered technologies.

(2) *The overwhelm of technology*: One may get a feeling of technological overload due to the quick pace at which technology is developing in

Table 17.4 Challenges and subcategories in adopting emerging technologies

Sr. No.	*Challenges*	*Subcategories*	*Additional details*
1.	Health challenges	Hydrogen production technologies	Advances needed for efficient and sustainable hydrogen production methods.
		Hydrogen storage and transportation	Development of safe and cost-effective storage and transportation solutions for hydrogen.
2.	Psychological challenges	Transition anxiety	Addressing concerns and anxiety associated with transitioning to new technologies.
		Social adaptation	Navigating societal changes brought about by technological advancements.
3.	Societal challenges	Infrastructure accessibility	Ensuring equal access to necessary infrastructure for all communities.
		Economic disparities	Mitigating economic gaps resulting from technological shifts.
		Employment shifts	Managing workforce transitions due to automation and evolving industries.
4.	Environmental challenges	E-waste and resource use	Reducing electronic waste and optimizing resource utilization in technology.
		Energy consumption	Addressing the environmental impact of high energy consumption in various sectors.
5.	Technological challenges	Reliability and efficiency	Enhancing the reliability and efficiency of emerging technologies.
		Innovation pace	Coping with the rapid pace of technological innovation and its implications.
6.	Economic challenges	Cost affordability	Ensuring that new technologies are affordable for widespread adoption.
		Market viability	Assessing and ensuring the viability of new technologies in the market.

the hydrogen energy industry. Keeping up with the constantly changing appliances, safety procedures, and infrastructure can be mentally taxing for certain people, which can have an adverse effect on their mental health.
(3) *Social adaptation*: When hydrogen energy becomes a part of everyday life, social norms and behaviors might have to change. For example, people's daily commutes and appliance usage might change if they convert to cars or appliances fueled by hydrogen. Resistance to these changes can worsen social stress and adjustment issues.

17.4.6.1.3 Societal challenges

(1) *Infrastructure accessibility*: There are significant societal barriers pertaining to hydrogen infrastructure availability and accessibility. The absence of infrastructure for hydrogen production, storage, and delivery could potentially hinder its wider adoption.
(2) *Economic disparities*: It is possible that the transition to hydrogen energy could exacerbate existing issues. The costs associated with creating new infrastructure, updating antiquated systems, and developing technologies for producing hydrogen may be prohibitive for some individuals or groups, which could widen economic gaps currently present.
(3) *Modifications to job patterns*: The adoption of hydrogen energy may result in modifications to job patterns. Companies that mostly use traditional energy sources, such as fossil fuels, could lose their employees. The workforce will therefore need to be deliberately retrained and reskilled for new roles in the hydrogen economy.

17.4.6.1.4 Environmental challenges

(1) *E-trash and resource use*: The production and disposal of hydrogen-generating devices, such as electrolyzers, may result in issues with electronic trash (e-waste). When materials are extracted and processed for hydrogen technology, environmental degradation and resource depletion may also become problems.
(2) *Energy consumption*: The production of hydrogen may lead to increased energy usage, especially in the case of non-renewable electricity. Preventing the escalation of environmental issues requires preserving the balance between sustainable energy sources and efficient hydrogen generation (Ajanovic et al., 2022).

17.4.6.1.5 Technological challenges

(1) *Efficiency and reliability*: The efficiency and reliability of hydrogen technologies face numerous challenges. Ensuring the effective and dependable

functioning of hydrogen production, storage, and utilization technologies is vital for fostering daily trust in them (Carmo et al., 2013).
(2) *Innovation pace*: The hydrogen energy sector is experiencing a rapid rate of innovation, which may make it challenging to maintain compatibility and interoperability among various technologies. It could be difficult for users to update their gadgets or integrate new ones into their daily routines.

17.4.6.1.6 Economic challenges

(1) *Cost-effectiveness*: Currently, the production of hydrogen energy, particularly green hydrogen, may be associated with high costs. Cost becomes a significant barrier, especially for individuals with low incomes who would find it challenging to buy hydrogen-powered vehicles or appliances (Imasiku et al., 2021).
(2) *Market viability*: One important factor to take into account is how viable hydrogen energy systems are on the market. The current state of the economy and the market's reliance on natural gas and petroleum create doubts about the viability and competitiveness of hydrogen energy in the long run.

17.5 CONCLUSION

Hydrogen integration has the potential to accelerate the global energy transition toward a carbon-neutral society in the realm of energy systems. The study points out that there isn't much thorough research on the incorporation of hydrogen into energy systems. The current state of hydrogen technologies is thoroughly investigated in order to close this gap, and a forward-looking viewpoint is provided. Empirical evidence supports the features of fuel cells and electrolyzers, which are crucial to this paradigm shift. The study looks at actual projects from around the globe and offers examples of how hydrogen is used in cogeneration, energy storage, transportation, and power-to-gas. The study provides the groundwork for comprehending how hydrogen will influence the development of energy systems in the future by outlining current advancements and potential future prospects. It is crucial to thoroughly examine the techno-economic environment surrounding hydrogen technology for energy systems. Integrating hydrogen requires careful consideration of important factors like cost, efficiency, and sustainability. The research highlights the opportunities and challenges associated with the smooth integration of hydrogen technologies into contemporary energy systems based on a statistical analysis of the body of available literature. The document promotes continued efforts and breakthroughs in different dimensions of hydrogen integration, looking beyond the existing situation. The request goes out for consistent performance enhancement, scalability, and committed assistance with technical and policy development.

Continuous research and development is prioritized, with special attention paid to strengthening safety protocols, raising efficiency requirements, and tackling sustainability issues. A sustainable hydrogen economy is beginning to take shape in relation to renewable energy. Developments in upgrading and photocatalysis, two key technologies for renewable hydrogen production, mark a significant shift in the field. The emphasis is on the conversion of electricity to hydrogen, underscoring the critical role that solar and wind energy will play in determining the direction of hydrogen generation in the future.

17.6 FUTURE SCOPE

17.6.1 Using hydrogen energy everyday

17.6.1.1 Use in residential settings

The future of hydrogen in daily life includes a wide range of home applications. Among the most promising uses are those for cooking and heating. Residential buildings can be powered and heated by hydrogen-fueled fuel cells, which present a clean and efficient alternative to conventional natural gas systems. This modification might significantly reduce the amount of carbon emissions produced by households (F. Posso et al., 2022).

17.6.1.2 Power that can be transported

Hydrogen fuel cells may also change portable power options. A portable and environmentally friendly energy source for electronics and home backup power is offered by small-scale hydrogen fuel cells. By doing this, the need for traditional batteries may decline, and the harmful consequences of electronic trash may be mitigated.

17.6.1.3 Small-scale grid hydrogen

In microgrid systems, which allow for localized energy production and distribution, hydrogen can be utilized as a storage material and an energy carrier. When the production of renewable energy is at its lowest, hydrogen can be used as a clean and efficient energy storage solution to provide a consistent supply of power. By storing extra renewable energy generated during peak hours, this is made possible.

17.6.2 Using hydrogen as an automotive fuel

17.6.2.1 Vehicles powered by hydrogen

When it comes to using hydrogen as a sustainable energy source for transportation, the automobile sector is leading the way. Fuel cell vehicles (FCVs) that run on hydrogen are becoming more and more popular since they provide an emission-free substitute for conventional internal combustion engine vehicles (Steinhilber et al., 2013). Only water vapor is released as a byproduct of the electrochemical interaction between hydrogen and oxygen in FCVs, which generates electricity.

17.6.2.2 Development of infrastructure

Widespread adoption of hydrogen vehicles depends on a robust infrastructure for refueling. Public and corporate entities are investing in the development of a network of hydrogen recharging stations to enable the widespread use of fuel cell vehicles. As this infrastructure advances, consumers may find hydrogen-powered cars to be a more desirable alternative due to their ease of recharging and extended driving ranges.

17.6.2.3 Possibilities and challenges

Utilizing hydrogen-powered vehicles can help the transportation industry cut carbon emissions, but there are still challenges to be solved, such as the high cost of fuel cells, insufficient infrastructure, and the energy-intensive process of creating hydrogen. Overcoming these challenges and establishing hydrogen as a competitive and viable option for the automotive industry is the aim of current research and development.

17.6.3 Using hydrogen to fulfill industrial needs

17.6.3.1 Production processes

Hydrogen has uses in industry beyond the production of electricity. It is a crucial component of many industrial processes, including those that create chemicals, fertilizers, and metals. Because of its efficiency and cleanliness, hydrogen is a desirable substitute for traditional fossil fuels in various processes, aiding in the industry's decarbonization.

17.6.3.2 Green hydrogen's uses in industry

The concept of "green hydrogen," produced by the electrolysis of renewable energy sources, holds significant potential for the decarbonization of industrial processes. Green hydrogen, a sustainable replacement for conventional hydrogen production methods, is becoming more and more economically viable as the cost of renewable energy continues to decline (Kovač et al., 2021).

17.6.3.3 Convergence with current infrastructure

The integration of hydrogen with the current infrastructure must be done carefully when the industry adopts the technology. It costs money and takes careful planning to use hydrogen as the main energy source in retrofitting industrial buildings (Maggio et al., 2019). Governments and business leaders play a critical role in expediting this transition by providing incentives and encouraging collaboration between academic institutions and for-profit businesses.

17.7 IMPLICATIONS FOR THE ECONOMY AND ENVIRONMENT

17.7.1 Benefits for the environment

Hydrogen is sometimes referred to as a "clean" energy source because it doesn't release greenhouse gases when it burns or when it is utilized in fuel cells. Its ability to significantly reduce carbon emissions is consistent with global efforts to combat climate change. When hydrogen replaces conventional fossil fuels in many applications, it contributes to the development of a more sustainable and environmentally friendly energy landscape.

17.7.2 Financial prospects

When hydrogen technology is widely applied, there are a lot of business opportunities. The hydrogen economy has the potential to stimulate economic growth, foster innovation, and create employment. Investments in the hydrogen industry, which can be strongly promoted through infrastructure, research and development, and the production of hydrogen-related technology, can benefit both established and emerging economies (Abe et al., 2019).

In conclusion, hydrogen energy has a promising future due to its versatility and capacity to enhance numerous facets of human existence. The applications of hydrogen are rapidly growing, ranging from daily domestic use to vehicle fuel and meeting industrial demands (Liu et al., 2020). Infrastructure, technology, and policy support are all continuing to progress, indicating that hydrogen has a bright future as a ubiquitous and crucial component of sustainable energy. As we overcome its challenges and reap its rewards, incorporating hydrogen into our daily lives holds the promise of a more sustainable, eco-friendly, and clean civilization (Ji & Wang, 2021).

REFERENCES

Abe, J. O., Popoola, A. P. I., Ajenifuja, E., & Popoola, O. M. (2019). Hydrogen energy, economy and storage: Review and recommendation. *International Journal of Hydrogen Energy*, *44*(29), 15072–15086.

Ajanovic, A., Sayer, M., & Haas, R. (2022). The economics and the environmental benignity of different colors of hydrogen. *International Journal of Hydrogen Energy*, *47*(57), 24136–24154.

Appleby, A. J. (1994). Fuel cells and hydrogen fuel. *International Journal of Hydrogen Energy*, *19*(2), 175–180.

Bairrão, D., Soares, J., Almeida, J., Franco, J. F., & Vale, Z. (2023). Green hydrogen and energy transition: Current state and prospects in portugal. *Energies*, *16*(1), 551.

Carmo, M., Fritz, D. L., Mergel, J., & Stolten, D. (2013). A comprehensive review on PEM water electrolysis. *International Journal of Hydrogen Energy*, *38*(12), 4901–4934.

Chowdhury, T. H., Amin, R., & Khoshnaw, S. (2018). Technical analysis of hydrogen energy production. *Journal of Renewable and Sustainable Energy*, *4*(3), 40–46.

Dawood, F., Anda, M., & Shafiullah, G. M. (2020). Hydrogen production for energy: An overview. *International Journal of Hydrogen Energy*, *45*(7), 3847–3869.

Edwardes-Evans, H. (2020). *Green hydrogen costs need to fall over 50% to be viable: S&P global ratings. International energy agency global hydrogen review (2021)*. OECD.

Gondal, I. A., Masood, S. A., & Khan, R. (2018). Green hydrogen production potential for developing a hydrogen economy in Pakistan. *International Journal of Hydrogen Energy*, *43*(12), 6011–6039.

IEA—International Energy Agency. (2022, November 10). *IEA data and statistics*. https://www.iea.org/data-and-statistics.

Imasiku, K., Farirai, F., Olwoch, J., & Agbo, S. N. (2021). A policy review of green hydrogen economy in Southern Africa. *Sustainability*, *13*(23), 13240.

International Energy Agency. (2019). *The future of hydrogen*. International Energy Agency. https://www.iea.org/reports/the-future-of-hydrogen.

Jacobson, M. Z., Colella, W. G., & Golden, D. M. (2005). Cleaning the air and improving health with hydrogen fuel-cell vehicles. *Science*, *308*(5730), 1901–1905.

Ji, M., & Wang, J. (2021). Review and comparison of various hydrogen production methods based on costs and life cycle impact assessment indicators. *International Journal of Hydrogen Energy*, *46*(78), 38612–38635.

Kovač, A., Paranos, M., & Marciuš, D. (2021). Hydrogen in energy transition: A review. *International Journal of Hydrogen Energy*, *46*(16), 10016–10035.

Li, R., Wang, W., & Xia, M. (2017). Cooperative planning of active distribution system with renewable energy sources and energy storage systems. *IEEE Access*, *6*, 5916–5926.

Liu, W., Sun, L., Li, Z., Fujii, M., Geng, Y., Dong, L., & Fujita, T. (2020). Trends and future challenges in hydrogen production and storage research. *Environmental Science and Pollution Research*, *27*, 31092–31104.

Maggio, G., Nicita, A., & Squadrito, G. (2019). How the hydrogen production from RES could change energy and fuel markets: A review of recent literature. *International Journal of Hydrogen Energy*, *44*(23), 11371–11384.

Pielecha, I., Cieślik, W., & Szałek, A. (2018). The use of electric drive in urban driving conditions using a hydrogen powered vehicle-Toyota Mirai. *Combustion Engines*, *57*(3), 50–57.

Posso, F., Galeano, M., Baranda, C., Franco, D., Rincón, A., Zambrano, J., ... Lópes, D. (2022). Towards the hydrogen economy in paraguay: Green hydrogen production potential and end-uses. *International Journal of Hydrogen Energy*, *47*(70), 30027–30049.

Steinhilber, S., Wells, P., & Thankappan, S. (2013). Socio-technical inertia: Understanding the barriers to electric vehicles. *Energy Policy*, *60*, 531–539.

Yu, M., Wang, K., & Vredenburg, H. (2021). Insights into low-carbon hydrogen production methods: Green, blue and aqua hydrogen. *International Journal of Hydrogen Energy*, *46*(41), 21261–21273.

Index